Felix Z. Meerson

Adaptation, Stress, and Prophylaxis

Translated from Russian
by Joseph Shapiro

With 87 Figures and 61 Tables

Springer-Verlag
Berlin Heidelberg New York Tokyo 1984

Professor Dr. Felix Z. Meerson
Laboratory of Cardiac Pathophysiology
Institute of General Pathology and Pathological
Physiology of the Academy of Medical Sciences
Moscow, USSR

ISBN-13: 978-3-642-61730-0 e-ISBN-13: 978-3-642-61728-7
DOI: 10.1007/978-3-642-61728-7

Library of Congress Cataloging in Publication Data. Meerson, Felix Zalmanovich, 1926 –. Adaptation, stress, and prophylaxis. Includes bibliographical references and index. 1. Adaptation (Physiology) 2. Stress (Physiology) 3. Stress (Psychology) 4. Stress (Physiology) – Prevention. I. Title. QP82.M4313 1984 612 83-16787

Softcover reprint of the hardcover 1st edition 1984

2119/3020-543210

Foreword

This book deals with one of the fundamental problems in the fields of biology and medicine, i.e., the problem of adaptation of an organism to its environment. Essentially, adaptive responses enable animals and man to perform new tasks imposed by the environment and also to achieve timely avoidance of damage from environmental factors. Such are the roles of the defense response, adaptational reactions to high altitude and maximal loads, immunogenetic responses, etc. The author rightly views adaptive responses of the organism as reactions involving natural prophylaxis of disease, although processes of adaptation can also have detrimental consequences under certain conditions. It is, thus, understandable that problems of adaptation increasingly concern biologists, physiologists, and physicians.

A number of important points distinguish this treatise from others in the field. Most researchers concerned with individual adaptation of human and animal organisms pay major attention to the dynamics of the gradual development of adaptation, i.e., to gradual learning of new skills, training for endurance of new loads, or accomodation to high altitude. In the present monograph, abundant original and cited material is devoted to another problem: What is the concrete mechanism by which a nonadapted organism lacking resistance to a given factor is transformed into an adapted and resistant organism? In answering this question, the author concentrates mainly on the molecular basis of the adaptation mechanism. It is shown that enhanced function activates the genetic apparatus of cells in systems which take part in the adaptive response. Activation of nucleic acid synthesis, therefore, results in a selective increase in structures which govern physiologic cell function.

The reader encounters not only analyses of mechanisms of adaptation to such widely studied factors as high-altitude hypoxia, physical load, and complicated environmental situations requiring training, but also mechanisms of adaptation to a priori stressful, extreme situations. Central inhibitory and modulatory systems, e.g., the GABA-ergic inhibitory system of the brain, prostaglandins, natural antioxidants, adenine nucleotides, are shown to play an important role in such adaptation.

A further significant feature of the book is the detailed analysis of the practically important aspect of the adaptation process. For example, it is shown that adaptation to various environmental factors can be controlled with activators and inhibitors of nucleic acid and protein synthesis. The monograph presents original data showing that gradual adaptation to hypoxia inhibits the development of hereditary hypertension in animals and increases resistance to allergic disease as well as to stress-induced and ischemic heart damage. Special coverage is given to ample findings of experimental prevention of stress-induced damage using given chemical factors. The author and his co-workers show that nontoxic drugs, gamma-hydroxybutyric acid, antioxidants, and prostaglandins can effectively prevent stress-induced damage to the heart, brain, stomach, and retina. The findings undoubtedly open prospects for clinicians and again show that there is nothing more practical in science than fundamental theoretical research of an important problem.

Tübingen, August 1983 R. Jacob

Preface

The mechanism of phenotypic adaptation is at present a key issue in biology and medicine. Only a profound study of this mechanism can establish the grounds for controlling the adaptation of the human organism to main factors of the environment. The concept of phenotypic adaptation based on the investigations presented in this book can be applied, as will be shown, to the prevention of disease and to increasing the organism's resistance to stress-induced damage, physical loads, and other environmental factors. The book is thus aimed at a broad range of workers in medicine, biology, and related fields. At the same time, this concept naturally reflects only a certain stage in the study of this complex, and seemingly infinite problem.

The evidence provided in this work is based on comprehensive physiologic, biochemical, and cytologic studies performed at the Laboratory for Heart Pathophysiology, Institute for General Pathology and Pathological Physiology, USSR Academy of Medical Sciences, and by associated scientific groups. Relevant investigations by Arkhipenko, Belkina, Golubeva, Larionov, Malyshev, Manukhina, Markovskaya, Novikova, Pavlova, Pshennikova, Radzievsky, Rozhitskaya, Saltykova, and Yavich played an especially important role. Work involving peroxidation of lipids was performed with the participation of V. E. Kagan, Senior Research Fellow, Laboratory for Physico-Chemistry of Biomembranes, Moscow University.

I express my heartfelt thanks to all my colleagues for their fruitful cooperation.

Moscow, August 1983 F. Z. Meerson

Contents

List of Abbreviations

ADP	adenosine diphosphoric acid
ALT	alanine transaminase
AP	action potential
AST	aspartate transaminase
ATP	adenosine triphosphoric acid
cAMP	cyclic adenosine monophosphoric acid
CHH	compensatory heart hyperfunction
CP	creatine phosphate
CPK	creatine phosphokinase
EPS	emotional painful stress
GABA	gamma-aminobutyric acid
GABA-T	GABA-transaminase
GDC	glutamate decarboxylase
GOBA	gamma-oxybutyric acid
IPh	inorganic phosphate
LP	lipid peroxidation
MHG	malate dehydrogenase
NAD	nicotinamide adenine dinucleotide
NADH	reduced nicotinamide adenine dinucleotide
NAD-P	nicotinamide adenine dinucleotide phosphate
OAA	oxaloacetic acid
PR	phosphorylation regulator
SFI	structure functional intensity
SR	sarcoplasmic reticulum
TAC	tricarbonic acid cycle
TT	thyrosine transferase

Introduction

"One can conquer Nature only by obeying it."

Darwin

The understanding of adaptation as a process whereby the organism adapts to the environment or to changes occurring within the organism itself is widely relevant in biology.

To limit the bounds of my account, it would appear useful to remind the reader of the existence of genotypic adaptation, due to which recent animal and plant species have developed on the basis of inherited variability, mutation, and natural selection. I shall not examine this process, but will only underline that genotypic adaptation has become the basis of evolution because its effects are secured genetically and inherited. A set of specific hereditary features becomes the starting point of the next set of adaptations, namely those acquired in the course of the organism's individual life. These adaptations occur during the interaction of the individual with the environment and are quite often accompanied by deep structural changes within the organism. These variations acquired in the course of life are not inherited by subsequent generations; they are stratified on hereditary features and together with the latter form the organism's individual appearance, the phenotype.

Phenotypic adaptation may be defined as a process which develops in the course of life and causes the individual organism to acquire resistance to a given environmental factor, allowing the organism to live in conditions theretofore incompatible with life and solve previously insoluble problems.

In the above definition, the ability to live in conditions theretofore incompatible with life may correspond to complete adaptation, which in conditions of cold or oxygen shortage makes it possible to retain a broad range of behavioral responses and to procreate, or conversely, it may signify a situation far from complete adaptation which for a long or short period of time simply permits preservation of life itself. Similarly, the ability to previously insoluble problems includes the most primitive and the most complex tasks, from avoiding a predator by the passive defensive stock-still reflex to space travel and conscious control of the organism's vital activity. To my mind, this intentionally wide definition covers the real meaning of the process of adaptation, which is an inseparable attribute of all living matter and is characterized by the same variety as life itself.

The definition focuses attention on the results of adaptation, viz enhanced resistance and resolution of the problem, and in a way puts to one side the essence of the process that develops under the influence of environmental factors in the organism and leads to adaptive achievements. In my view, this reflects the actual state of affairs in adaptology, the science of adaptation, in which a remarkable diversity of external manifestations is often not helpful in elucidating the basic, general mechanisms of highly variable cases of adaptation. Hence, at present the question, "At the expense of specifically what mechanism and chain of events does an unadapted organism turn into an adapted one?" appears to be the major issue in phenotypic adaptation, an issue which remains largely unresolved.

Lack of clarity in this field inhibits the solution of a number of applied issues, e.g., control of adaptation among large groups of people encountering new conditions, adaptation to the simultaneous effects of several factors, provision of complex forms of intellectual activity in intentionally changed environmental conditions, adaptation to extreme situations which over a long period are either unavoidable or should not be avoided, and the use of preliminary adaptation and chemical factors to enhance resistance and prevent injuries caused by extreme, essentially stress, situations. In view of the existing state of the problem, this book deals principally with the general, basic mechanism of phenotypic adaptation, the concept that developed during the study of this mechanism being used as a foundation for applying adaptational and chemical factors with a view to enhancing the organism's resistance, primarily to prevent stress injuries.

In examining gradual, long-term adaptation, one should keep in mind that until the factor to which adaptation takes place starts to act the organism has no ready fully formed mechanism for complete and consummate adaptation, only genetically determined prerequisites therefore. If the factor has no effect, the mechanism remains unformed. For example, an animal removed from its natural habitat during early development and reared by humans may complete its life cycle without acquiring adaptation to physical loads, the elementary skills for avoiding danger, or the ability to pursue prey. Similarly, a human removed from his natural social medium in the early stage of development and raised among animals fails to form most of the adaptive responses constituting the behavioral basis of a normal person. All animals and humans avoid collisions with injurious environmental factors by means of defensive responses; hence, in many cases, they can do without actuation of long-term adaptive responses inherent in an injured organism, e.g., not developing specific immunity acquired as a result of illness.

In other words, the organism's genetic program envisages not a preformed adaptation, but only the possibility of its formation under environmental action. This affords only vital adaptive responses, resulting in efficient, environment-oriented expenditure of the organism's energy and structural resources, and also gives rise to the specifically oriented formation of the whole phenotype. Hence, the fact that the results of phenotypic adaptation are not inherited should be regarded as advantageous for preserving a given species. In a rapidly changing environment, the next generation of each species risks encountering absolutely new conditions, in which it will require its temporarily latent potential of adapting to a broad spectrum of factors, not the specialized responses of its ancestors.

In effect, the question of the mechanism of phenotypic adaptation is how the organism's genetically determined potentials are realized in response to environmental needs. Precisely this realization of potentials represents the mechanism of phenotypic adaptation discussed in Chaps. 1 and 2, which show that new environmental factors or situations result relatively quickly in the forming of functional systems that could seemingly ensure the organism's adaptive responses to these environmental requirements. However, for complete adaptation, the development of a functional system is in itself insufficient; it is necessary that structural changes securing the system and enhancing its physiologic potency occur in the component cells and organs. The key link in the mechanism ensuring this process, and consequently in all the forms of phenotypic adaptation, is the relationship between function and the genetic apparatus existing in the cells. Through this relationship, the functional load induced by environmental factors results in increased synthesis of nucleic acids and proteins and hence in so-called structural traces in systems specifically responsible for the organism's adaptation to a given environmental factor. In this case, the mass of membrane structures responsible for the perception by the cell of monitor signals, ion transport, and energy supply, i.e., precisely those structures that restrict the cell function as a whole, are characterized by maximum growth. The resultant systemic structural trace represents a set of structural changes affording expansion of the link that restricts the cell function to enhance the physiologic potency of the functional system responsible for adaptation; this trace constitutes the basis of reliable, long-term phenotypic adaptation. After the said environmental factor ceases to act on the organism, the activity of the genetic apparatus in the cells of the system responsible for adaptation declines quite abruptly to result in disappearance of the systemic structural trace constituting the basis of deadaptation.

Chapters 1 and 2 show how the synthesis of nucleic acids and proteins activates and how the systemic structural trace actually forms in the cells of the functional system responsible for adaptation. These chapters also compare the architecture of systemic structural traces at relatively simple and higher adaptive responses of the organism and particularly examine the role of the stress reaction in the formation of the systemic structural trace. This reaction has been shown to provide not simply mobilization of the organism's energy and structural resources, but a directed transfer of all these resources to a dominant functional system responsible for adaptation, the system wherein the systemic structural trace forms. Thus the systemic structural trace, which plays the main role in specific adaptation to a given concrete environmental factor, forms with the essential involvement of a nonspecific stress reaction occurring on any substantial change in the environment. In this case, on the one hand the stress reaction potentiates the forming of a new systemic structural trace and adaptation, and on the other, at the expense of its catabolic effect, it promotes eradication of old structural traces which have lost their biologic significance. This reaction is thus a necessary link in the integral adaptation-deadaptation mechanism of the organism in a changing habitant; it plays an important part in reprograming the organism's adaptive possibilities for solving new problems posed by the environment.

As the systemic structural trace and reliable adaptation form, the stress reaction, having played its part, disappears, to reappear when a new situation requiring new adaptation arises. Understanding of this lifelong dynamic phenotypic adaptation was a basis for distinguishing the principal stages of the process and the adaptational diseases most probably related with each of the stages.

Chapters 3–5 show how the above mechanism and the stages of adaptation are realized with known long-term adaptive reactions as diverse as adaptation to altitude hypoxia, compensatory adaptation to injury, and higher adaptive responses developing in the form of conditioned reflexes and behavioral responses. In assessing the development of these concrete adaptive responses, it is clear that the realization of potential, genetically determined possibilities of the organism, resulting in a systemic structural trace, makes the organism acquire new qualities, namely adaptability in the form of resistance to hypoxia; a high degree of training for coping with physical loads, new skills, and so on. These new qualities show primarily in that the organism cannot be injured by the factor against which adaptation was acquired; hence adaptive responses are in effect responses that prevent injury. It may be said without exaggeration that adaptive responses constitute the foundation of the natural prevention of diseases, the basis of natural prophylaxis. The role of adaptation as a preventive factor essentially grows because long-term, structurally conditioned adaptive responses possess only relative specificity, i.e., they enhance the organism's resistance not only to the factor that caused adaptation, but also simultaneously to certain other factors. For instance, adaptation to physical loads enhances resistance to hypoxia, adaptation to toxic chemical substances improves the organism's ability to oxidize cholesterol, and adaptation to painful stress increases resistance to ionizing radiation. Numerous phenomena of this type, usually designated as crossed adaptations or crossed resistance, are the results of relative specificity of phenotypic adaptation.

The relative specificity of phenotypic adaptation is based essentially on the fact that the ramified systemic structural trace, upon which adaptation to a definite factor is based, quite often contains components which can enhance the organism's resistance to the effects of other factors. For example, the increase in the liver cell population in adaptation to hypoxia probably causes increased efficiency of the detoxicating system of microsomal oxidation in the liver as well as higher resistance of adapted animals to various poisons (see Chap. 3). Partial atrophy of the supraoptic nucleus of the hypothalamus and the zona glomerulosa of the adrenals, observed upon adaptation to hypoxia, facilitates the loss of sodium and water by the organism and is the basis for the increased resistance of adapted animals to hypertension-inducing factors (see Chap. 3). Such phenomena of relatively specific adaptation play an important role in the natural prevention of diseases and may play an even greater part in man's active prophylaxis of noninfectious diseases such as hypertension, aftherosclerosis, and cardiac ischemia. In other words, there is a possibility that adaptation could play its role in the prevention of the so-called noninfectious or endogenous diseases.

The reality of this outlook may be assessed most successfully in adaptation bases on a ramified systemic structural trace encompassing both higher regulatory regions and executive organs, for precisely this type of adaptation would be characterized to the largest extent by relative specificity and could with high probability lead to crossed resistance. Based on this, the present author and his colleagues have obtained evidence (Chaps. 3 and 5) on the use of adaptation to the periodic effect of hypoxia in preventing experimental circulatory and brain diseases. Preliminary adaptation to hypoxia was found to activate fixation of temporary correlations, change animal behavior in conflicting situations to the advantage of the organism, and increase resistance to extreme irritators, hallucinogens, factors causing epileptiform convulsions, and alcohol. It was further shown that this adaptation prevents acute cardiac insufficiency in experimental heart failure and myocardial infarction, largely prevents heart injury in emotional stress, and inhibits hereditary hypertension in animals.

This growth of resistance to a broad spectrum of known injurious factors as a result of adaptation to some specific factor is apparently only part of what may be obtained by adaptation to a set of prescribed and individually selected environmental factors. Hence adaptive prophylaxis and enhancement of resistance by means of adaptation should become subjects of purposeful research in human physiology and in clinical practice.

Another aspect of the problem stems from the accepted tenet that all adaptive responses are only relatively functional. In specific conditions, i. e., when the environment poses excessive demands, the adaptive responses that have developed during evolution become dangerous for the organism and begin to play a role in the development of various injuries in the organs and tissues. A major example of this transformation of adaptive responses into pathologic reactions is the exceedingly intensive and prolonged stress reaction, which occurs in so-called hopeless situations, when the system responsible for adaptation cannot perform; as a result, the systemic structural trace fails to form and adaptation does not successfully develop. These conditions are characterized by disorders of homeostasis, which develop under environmental effects and constitute the stimuli of the stress reaction. Correspondingly, the stress reaction itself proves to be unusually intensive and lengthy. Long-term high catecholamine and glucocorticoid concentrations may induce multiform stress injuries, from ulcerous lesions of the gastric mucous membrane and serious focal damages of the cardiac muscle to diabetes and blastomatous growth. This transformation of the stress reaction from a general, nonspecific link in the adaptation to various factors into a general, nonspecific link in the pathogenesis of various diseases is the main subject of Chap. 6. What appears important to note in analyzing this transformation is that even in serious stress, death from stress diseases, though possible, is not inevitable. The majority of animals and humans subjected to serious stress do not die, but adapt in one way or another to the resultant situations. In full conformity with this, it was shown experimentally that the stress reaction was less pronounced with repeated stress situations that the animals could not avoid.

A study of adaptation to stress effects and the organism's responses thereto have led me to assume the existence in the organism of modulatory systems which restrict the stress reaction and prevent stress injuries. Chapters 7 and 8 show that these systems may function at the brain level, restricting excitation of stress-realizing systems and preventing an excessive and lengthy rise in catecholamine and glucocorticoid concentrations; they may also function at the tissue level, confining the hormone effect to the cell. Prostaglandin and antioxidant systems and a gamma-aminobutyric inhibitory system of the brain are discussed as examples of such modulatory systems of natural prevention. In addition to theoretical results, a study of these systems has shown them to have practical results.

Administration to animals of active metabolites of modulatory systems and their synthetic analogs effectively prevents stress injuries of the heart and other internal organs. Chemical prevention of stress injuries deserves special attention in human pathology.

Chapter 1
Basic Regularities of Phenotypic Adaptation

"Everything truly great is accomplished by slow, inconspicuous growth. . . ."

Seneca

Despite all the diversity of phenotypic adaptation, its development in higher animals is characterized by common features, which I shall examine in the account that follows.

Short-term and Long-term Stages of Adaptation

Two stages may be definitely traced in the development of most adaptive responses, namely initial, short-term (or urgent), but incomplete adaptation, and the subsequent complete, long-term adaptation. The urgent stage of adaptive response arises directly after the irritant starts to act, and consequently may occur on the basis of ready, preformed physiologic mechanisms. Obvious manifestations of urgent adaptation are an animal's flight in response to pain, increase in heat production in response to cold, greater heat release in response to warmth, and greater lung ventilation and minute volume in response to oxygen shortage. A major feature in this stage of adaptation is that the organism functions at the limit of its physiologic possibilities – at almost complete mobilization of its functional reserve – to provide, far from sufficiently, the needed adaptive effect. For example, an unadapted animal or human runs at almost maximum cardiac output and pulmonary ventilation minute volume, involving maximal mobilization of the glycogen reserve in the liver; due to insufficiently rapid pyruvate oxidation in the muscle mitochondria, the lactate level in the blood grows. This lacticemia restricts load intensity; as a result, the motor response can be neither sufficiently quick nor of sufficient duration. Thus, adaptation is achieved 'momentarily,'

Quite similarly, when adaptation to new complex environmental situations is achieved at the level of the brain, the stage of urgent adaptation occurs at the expense of cephalic preexistent mechanisms and is manifested as the "period of generalized motor responses" or the "period of emotional behavior," which is well known in the physiology of higher nervous activity. In this case, the necessary adaptive effect, dictated by the organism's need for food or self-preservation may stay unrealized or be ensured by some fortuitous action, i. e., be inconstant.

The long-term stage of adaptation arises gradually as a result of lengthy or repeated environmental actions on the organism. In fact, it develops on the basis of repeated realization of urgent adaptation. As a result of gradual quantitative accumulation of certain changes, the organism acquires a new quality: from a nonadapted organism it turns into an adapted one. Such adaptation can permit the organism to perform physical tasks previously too difficult, to develop resistance/to considerable altitude hypoxia previously incompatible with life, and to develop resistance to cold, heat, and large doses of poisons that were formerly fatal. The same applies to the qualitatively more complex adaptation to environment, which develops in the course of training based on brain memory and is evident in the occurrence of new stable temporal relationships and their realization in the form of corresponding behavioral responses.

By comparing urgent and long-term adaptation, one can readily arrive at the conclusion that the transition from the urgent and largely incomplete stage to the long-term stage represents the key element in adaptation, for precisely this transition makes possible the organism's permanent existence in new conditions, allowing it to enlarge its habitant and giving it behavioral freedom in changing biologic and social media.

The mechanism of this transition should preferably be examined in terms of the concept, accepted in physiology, that the organism's responses to environmental factors are provided not by individual organs but by systemically organized and coordinated systems. This concept has been universally developed in the works of Descartes, Harvey, Sechenov, Pavlov, Ukhtomsky, Winer, Bertolamfi, Anokhin, and Selye and is therefore not given special attention in this book, but it makes it possible to state that response to any new and sufficiently strong environmental effect—to any disruption of homeostasis—is ensured first, by a system that reacts specifically to a given irritant and second, by the stress-realizing adrenergic and hypophyseal-adrenal systems, which react nonspecifically to very different changes in the habitat.

In using the concept "system" in the study of phenotypic adaptation, it should essentially be emphasized that formerly Ukhtomsky, one of the greatest physiologists of the century and founder of the theory of dominants, was closest to revealing the essence of systems that are active in solving the organism's principal tasks at a given stage of individual life. He studied in detail the role of the body's intrinsic needs, which are realized via hormones and intero- and exteroceptive afferent signaling in the formation of dominants; at the same time, he regarded the dominant as a system involving a constellation of nerve centers subordinating the executive organs and determining the direction of behavioral responses, the organism's vector.

Ukhtomsky [553] wrote: *Outwardly, the dominant expresses itself through definite work or working posture of the organism, supported in a given moment by diverse irritants and excluding other work and postures in the said moment. One has to assume that, behind this work or posture, there is not simply a single local focus, but a whole group of centres, possibly widely scattered within the nervous system. The sex dominant conceals excitation of centres in both the cortex and the subcortical apparatuses of vision, hearing, smell, and sense of touch, and also in the medulla oblongata, the lumbar sections of the spinal cord, and in the secretory and vascular systems. Hence, every natural dominant presumably conceals excitation of a whole constellation of centres. In an integral dominant, one has to first of all distinguish cortical and somatic components* (p. 34).

In developing the idea that the dominant combines working centers and executive organs situated at various levels, Ukhtomsky sought to underscore the unity of this newly emerged system and, not infrequently, called the dominant the "organ of behaviour." He noted: *Every time a dominant symptom complex is present, its specific behavioural vector is also there. And it is natural to call it the "organ of behaviour", albeit it is mobile like Descartes' eddy motion. A definition of the notion "organ" as something, which I would call a dynamic mobile worker or a working combination of forces, is, I think, exceedingly valuable for the physiologist* (p. 80).

Ukhtomsky subsequently made the next step, to designation of the dominant as a system [553].

These tenets and the entire creative power of Ukhtomsky's school unequivocally provide evidence that in his works the dominant system is presented as one that differs basically from what we understand as the anatomophysiologic systems of circulation, digestion, motion, etc. This system was presented by Ukhtomsky as something that forms in the organism in response to the action of the environment and combines into a single unit the nerve centers and executive organs belonging to different anatomophysiological systems in order to adapt them to a perfectly definite environmental factor, i.e., resolve a problem posed by the environment. It was precisely such systems that Anokhin [6] subsequently designated as functional emphasizing that the information on response results (adaptive effect supplied to the nerve centers through feedback) is the principal system-forming factor.

In examing the transition of urgent adaptation into long-term adaptation in terms of the functional system, one can readily note the important but not always duly considered fact that the availability of either a ready functional system or its formation does not in itself signify stable and effective adaptation. Indeed, the initial effect of any unconditioned irritant causing considerable and lengthy motor response is to excite corresponding afferent and motor centers and mobilize skeletal muscles, circulation, and respiration, which together form a single functional system specifically responsible for realizing a given motor response. However, the efficiency of that system is quite low. For example, untrained people cannot run quickly or for long distances; this needs repeated practice, leading to development of long-term adaptation.

Under the effects of oxygen shortage, the influence of hypoxemia on chemoceptors, directly on nerve centers and executive organs, entails a response in which the functional system specifically responsible for eliminating oxygen shortage in the organism is the circulation and external respiration organs, which form a single control system and achieve an advanced function. The initial result of mobilizing this functional system after an unadapted person has climbed to a height of 5,000 m is that heart hyperfunction and lung hyperventilation are quite strongly pronounced, but nevertheless prove insufficient to eliminate hypoxemia and combine with more or less pronounced adynamia, apathy, or euphoria and, consequently with declined physical and intellectual efficiency. In order that this urgent, but incomplete adaptation be replaced by complete, long-term adaptation, it is essential to stay for a long time or repeatedly at elevated sites, i.e., there is need for lengthy or repeated mobilization of the functional system responsible for adaptation.

Quite similarly, when poison, say Nembutal, is administered, the factor specifically responsible for its destruction is mobilization of the microsomal oxidation system localized in the liver cells. Activation of the microsomal oxidation system undoubtedly limits the injurious effect of the poison, but does not eliminate it completely. As a result, the clinical intoxication picture is quite pronounced and adaptation is correspondingly incomplete. Subsequently, after repeated administration of Nembutal, the initial dose no longer causes intoxication. Thus, the presence of a ready functional system responsible for adaptation to a given factor and instantaneous activation of that system do not in themselves imply immediate adaptation.

When more complex environmental situations act on the organism (for instance, previously unencountered irritants, danger signals, or situations arising in the course of teaching new skills), the latter has no ready functional system capable of providing a response that would correspond to environmental requirements. In fact, the organism's response is already ensured by the above-mentioned genealized reaction against a background of sufficiently intense stress. In this situation, some of the numerous motor responses prove adequate and are reinforced. This initiates the forming in the brain of a new functional system, namely the system of temporal relations that becomes the basis of new skills and behavioral responses. However, when it directly appears, the system usually lacks strength; it may be eliminated by inhibition caused by the occurrence of other behavioral dominants, periodically realized in the organism's activity, or extinguished by repeated non-reinforcement.

In order that adaptation be stable and guaranteed for the future, time is needed as well as a certain number of repetitions, i.e., consolidation of a new stereotype. All the above essentially boils down to the following: *The availability of a ready functional system with relatively simple adaptive responses and the development of that system with more complex responses realized at the level of the cortex do not in themselves lead to the instantaneous occurrence of stable adaptation but are the basis of the incipient, so-called urgent and incomplete stage of adaptation. For urgent adaptation to turn into guaranteed long-term adaptation within a newly developed functional system, a certain important process must take place to secure the established adaptive systems and to enlarge their power to a level demanded by the environment.* Investigations performed over the past 20 years at our laboratory [315, 317, 320] and by other researchers have shown this process to be the activation of nucleic acid and protein syntheses in cells responsible for the adaptation of the respective systems that provide systemic structural traces therein.

Systemic Structural Trace: The Basis of Adaptation

In recent decades, researchers working with very different subjects studied using the same of methods in contemporary biochemistry have unambiguously shown that increased functions of organs and systems naturally entail activated synthesis of nucleic acids and proteins in the cells that form these organs and systems. Inasmuch as in response to environmental requirements the function of the system responsible for adaptation grows, activation of the synthesis of nucleic acids and proteins develops precisely in those systems. It results in structural changes, which basically increase the power of the systems responsible for adaptation. Precisely this constitutes the foundation of the transition from urgent emergency adaptation to long-term adaptation, the decisive factor in the development of the structural basis of long-term adaptation.

The sequence of events involved in the forming of long-term adaptation involves the increased physiologic function of the cells of systems responsible for adaptation, which causes, as a first shift, a higher RNA transcription rate over DNA structural genes in the nuclei of those cells. Increase in the amount of messenger RNA leads to increase in the number of RNA-programed ribosomes and polysomes, the sites of intensive synthesis of cell proteins. As a result, the structure mass grows to increase the functional possibilities of the cell, which is the shift constituting the basis of long-term adaptation.

It is significant that the activating effect of the increased function, mediated by the mechanism of intracellular regulation, is addressed precisely to the cell's genetic apparatus. Administration to animals of actinomycin, an antibiotic which adheres to the DNA guanyl nucleotide to make transcription unfeasible, deprives the cell's genetic apparatus of the possibility to respond to the increase in function. As a result, urgent adaptation can no longer turn into long-term adaptation; in other words, adaptation to physical loads [331], hypoxia [353], formation of new temporary interrelations [349], and other adaptive responses become unfeasible under the action of nontoxic doses of actinomycin, which do not impair ready, previously formed adaptive responses.

From these and other facts, we designated the mechanism via which a given function regulates the quantitative parameter of the genetic apparatus activity, viz the transcription rate, as the 'interrelation between the function and the cell genetic apparatus' [315]. This relationship is bilateral. The direct relationship is in that the gentic apparatus – the genes situated in the nuclear chromosomes – indirectly, via the RNA system, provide protein synthesis – they 'generate structures,' and the structures 'generate' the function. The feedback is in that the 'intensity of structure functioning' – the amount of function per organ unit mass – somehow governs the activity of the genetic apparatus.

It turned out that a major property of hyperfunction (hypertrophy of the heart at the narrowing of the aorta, hypertrophy of a sole kidney after the other one had been removed, hypertrophy of part of the liver after resection of its other portions, and hypertrophy of a sole lung after removal of the other one) is that nucleic acid and protein synthesis activation, which occur in the hours and days immediately after the beginning of hyperfunction, gradually stop after hypertrophy had developed and the organ mass had become enlarged (see Chap. 3). This dynamic situation is determined by the fact that in the beginning of the process hyperfunction is achieved by a still nonhypertrophied organ and an increase in the amount of function per unit mass of cell structures causes the genetic apparatus of differentiated cells to activate. After the organ has developed complete hypertrophy, its function is distributed over the enlarged mass of cell structures; as a result, the amount of function performed by the structure unit mass returns to normal [316].

If one were to eliminate the hyperfunction of an organ that had already undergone hypertrophy, the amount of function performed by 1 g of tissue would become abnormally low. As a result, protein synthesis in differentiated cells would drop and the organ mass would start to decrease. This would cause the amount of function per unit mass to gradually increase and, after it had returned to normal, inhibition of protein synthesis in the organ cells would cease since the mass would no longer decrease.

This evidence afforded grounds for surmising that in differentiated cells and in the mammalian organs they form the amount of function performed by the unit mass of a given organ [structure functional intensity (SFI)] plays a major role in controlling the activity of the cell's genetic apparatus. Increased SFI corresponds to a situation whereby the 'function lacks room within the structure.' It causes protein synthesis to activate and the mass of cell structures to grow. A decline in this parameter would correspond to a situation in which the 'function has too much room within the structure' to cause synthesis intensity to drop with subsequent elimination of excess structure. In both cases, SFI returns to a certain optimal value inherent in a healthy organism. Thus, the intracellular mechanism of bilateral interconnection between the physiologic functional and the genetic apparatus of a differentiated cell provides a situation in which the SFI is simultaneously the genetic apparatus activity determinant and the physiologic constant maintained at constant level due to timely changes in the activity of that apparatus [316].

In relation to a helathy organism, this regularity is corroborated in the works of a whole series of researchers. For example, Zak [595] showed the dependency of the myocytic genetic apparatus of a healthy organism on the level of its physiologic function. He compared the functions of three different muscles with the intensity of protein synthesis and the RNA content in the muscle tissue. The myocardium, which continuously contracts at high speed was shown to possess intense protein synthesis and the highest RNA content; respiratory muscles contracting with a slower rhythm were found to have a lower RNA concentration and a lower protein synthesis intensity. Finally, skeletal muscles contracting either periodically or episodically have the lowest intensity of protein synthesis and the lowest RNA content, though they develop a tension considerably higher than that in the myocardium.

Essentially similar evidence was obtained by Margreth and Novello [298], who showed that the RNA concentration, protein/RNA ratio, and intensity of protein synthesis in various muscles of the same animal are directly dependent on the functions of those muscles; in the rabbit's chewing muscles and the rat's diaphragm, all these indices are approximately twice as high as in the musculus gastrocnemius of both species. This apparently depends on the fact that the duration of the average daily activity in the chewing and phrenic muscles considerably exceeds that in the musculus gastrocnemius.

On the whole, the studies by Zak, and also those by Margreth and Novello, make it possible to emphasize that as a factor determining the activity of the genetic apparatus SFI should be measured not by the maximum attainable functional level (for instance, not by maximal muscle tension) but by the average of function achieved by cell unit mass per day. In other words, the mean daily SFI inherent in a given organ and forming its differential cells is apparently the factor that controls the power and activity of the cell's genetic apparatus not the maximum episodic SFI, which is very conveniently determined with functional tests stipulating maximal load on the organ. Naturally, with equal duration of average daily activity, i.e., with the same period during which the organ functions, the mean daily SFI would be higher in the organ that functions at a higher level. For instance, in a healthy organism the tension developed by the myocardium of the right ventricle is known to be slightly lower than that developed by the myocardium of the left ventricle, the daily duration of ventricle function being equal; correspondingly, the content of nucleic acids and the intensity of protein synthesis in the right ventricular myocardium are also lower than in the left [348].

Based on our concept of SFI, Matsumoto and Krasnow [307] performed an interesting study, which, to my mind, shows the different functional structure intensities in different tissues during ontogenesis to influence not only RNA synthesis intensity over DNA structural genes but also, via RNA, protein synthesis intensity. Structure functional intensity was found to act deeper, specifically to determine the number of DNA matrices in tissue unit mass, i.e., the total power of the genetic apparatus of the tissue-forming cells or the number of genes per tissue unit mass. This influence was evident in that for the left ventricular muscle DNA concentration constitutes 0.99 mg/g, for the right ventricular muscle 0.93 mg/g, for the diaphragm 0.75 mg/g, and for the skeletal muscle 0.42 mg/g, i.e., the number of genes per unit mass changes in different types of muscle tissue proportionally to the SFI. The number of genes is one of the factors determining the intensity of RNA synthesis. In conformity with this, in subsequent experiments the above researchers established that the RNA synthesis intensity determined by inclusion of labeled ^{14}C in glucose constitutes 3.175 counts/min for the left ventricle, 3.087 counts/min for the right ventricle, 2.287 counts/min for the diaphragm, and 1.154 counts/min for the skeletal muscle of the extremities per RNA contained in 1 g of muscle tissue. Thus, the SFI developing in the course of ontogenesis in juvenile animals, whose cells have retained the ability to synthesize DNA and divide, may determine the number of genes per tissue unit mass and, indirectly, the intensity of RNA and protein synthesis, i.e., perfect structural provision of cell function.

The above unambiguously shows that the correlation between the cell function and genetic apparatus, which we shall henceforth term the $G \rightleftarrows F$ interrelation, is a continuous mechanism of intracellular regulation realized in the cells of various organs. At the stage of transition of urgent to long-term adaptation, in hyperfunction of the system specifically responsible for adaptation, $G \rightleftarrows F$ naturally ensures activation of nucleic acid and protein synthesis in all the cells and organs of a given functional system. As a result, it starts to accumulate definite structures to realize a systemic structural trace. For instance, in the adaptation to physical loads, pronounced activation of nucleic acid and protein synthesis and pronounced structural changes naturally occur in motor center neurons, adrenals, skeletal muscle cells, and the heart [49, 54]. The essence of these changes is that they provide selective increase in the mass and power of structures responsible for control, ion transport, and energy supply.

In adaptation to physical loads, moderate heart hypertrophy was found to be combined with higher activity of the adenyl cyclase system [589] and an increased number of adrenergic fibers per myocardium unit mass [279]. As a result, the heart's adrenoreactivity increases as does the possibility to mobilize the former urgently. At the same time, myosin heads are observed to increase the number of their H chains, the carriers of ATPase activity; ATPase activity grows to cause increased velocity and amplitude of cardiac muscle contraction [33, 585, 586]. This is followed by increased power of the calcium pump of the sarcoplasmatic reticulum and, consequently, the velocity and degree of the heart's diastolic relaxation [321]. In parallel with these changes in the myocardium, the number of coronary capillaries was noted to increase [447], the myoglobin concentration to grow [407], and the activity of enzymes responsible for substrate transport to the mitochondria to increase [573]; at the same time, the mass of mitochondria also increased. This growth in the power of the energy supply system naturally entails higher heart resistance to fatigue and hypoxemia [321].

This selective increase in the power of the structures responsible for regulation, ion transport, and energy supply is not characteristic solely of the heart but is naturally realized in all the organs responsible for adaptation. In the course of the adaptive reaction, they form an integral functional system, while the structural changes developing therein represent a systemic structural trace that constitutes the basis of adaptation.

In relation to the discussed adaptation to physical loads, this systemic structural trace at the level of nerve regulation shows in hypertrophy of motor center neurons and in the higher activity of respiratory enzymes therein [49, 191]. At the level of endocrine regulation, it shows in hypertrophy of the cortical substance and the medulla of the adrenals [32, 510]; and in executive organs, in hypertrophy of skeletal muscles and a 1.5–2-fold increase in the number of mitochondria therein [139, 179]. The last change is exceedingly important, for, in combination with increase in the power of the circulation and external respiratory systems, it provides an increase in the organism's aerobic power (growth of its possibility to utilize oxygen to achieve aerobic resynthesis of ATP) essential for intensive functioning of the locomotive apparatus. An increase in the number of mitochondria causes the organism's aerobic power to grow, together with increased muscle ability, and pyruvate is utilized, which forms in increased amounts under strain on account of glycolysis activation. This prevents an increase in lactate concentration in the blood of adapted humans [567] and animals [111]. Growth in lactate concentration is known to be a factor that limits physical work; at the same time, lactate inhibits lipases and, correspondingly, lacticemia inhibits the utilization of fats. With developed adaptation, increased utilization of pyruvate in the mitochondria prevents lactate concentration in the blood from growing, ensures mobilization and utilization of fatty acids in the mitochondria, and ultimately enhances maximum intensity and duration of work. *Hence, a ramified structural trace expands the link that limits the organism's efficiency and thus precisely consitutes the basis underlying the transition of urgent, but unreliable, adaptation to long-term adaptation.*

The forming of a systemic structural trace and the transition from urgent to long-term adaptation under the long-term effects of altitude hypoxia compatible with life take place in a quite similar manner. The adaptation to this factor is discussed in detail below and is characterized by the initial hyperfunction and subsequent activation of nucleic acid and protein synthesis; it covers simultaneously many systems of the organism and, correspondingly, the systemic structural trace formed proves more ramified than is the case with adaptation to other factors. Indeed, hyperventilation is followed by activation of nucleic acid and protein synthesis and by subsequent hypertrophy of the respiratory center neurons, the respiratory musculature, and the lungs themselves, the number of alveoli increasing in the latter. This results in more power to the external respiratory apparatus as well as a greater lung respiratory surface area and oxygen utilization coefficient to make respiration more economical. In the hematopoietic system, activation of nucleic acid and protein synthesis in the bone marrow causes increased formation of erythrocytes and polycytemia to provide higher oxygen capacity. Finally, activation of nucleic acid and protein synthesis in the right and, to a lesser extent, left regions of the heart ensures a set of changes largely similar to those described for adaptation to physical loads. As a result, the heart's functional possibilities grow, particularly in its resistance to hypoxemia.

Synthesis is also activated in systems whose functions are not elevated, but, on the contrary, are impaired by oxygen shortage, primarily in the cortex and the underlying regions of the brain. Like the activation caused by increased function, this activation is apparently caused by ATP shortage, since it is precisely through change in the balance of ATP and its decomposition products that the $G \rightleftarrows F$ interrelation is realized; the detailed structure of this interrelation is discussed below. It should be noted that the activation of nucleic acid and protein synthesis developing under brain hypoxia becomes the basis or vessel growth and stationary increase in glycolysis and thus contributes to the forming of a systemic structural trace constituting the basis or adaptation to hypoxia. This systemic structural trace and adaptation to hypoxia causes adapted people to acquire the ability to achieve under oxygen shortage physical and intellectual work which unadapted individuals can never perform. In a well-known example, Hurtado [187] showed that adapted natives of the Andes could play chess in an altitude chamber at 7,000 m, while nonadapted inhabitants of the plains lost consciousness.

In adaptation to certain factors, the systemic structural trace proves spatially very limited, being localized in definite organs. For example, in the adaptation to increasing doses of poison, nucleic acid and protein synthesis in the liver naturally grows. As a result, the microsomal oxidation system, in which the cytochrome ^{450}P plays the major role, becomes stronger. Outwardly, the systemic structural trace may show in an increase of liver mass; this is the basis of adaptation that is evident in essentially higher resistance to poisons like barbiturates, morphia, alcohol, and nicotine [10, 396]. The effect of the power of the microsomal oxidation system upon resistance to chemical factors appears to be rather high. For example, after having smoked a regular-size cigarette, nonsmokers were found to have a nicotine concentration in the blood 10–12 times higher than in blood of smokers, whose microsomal oxidation system is enhanced and provides adaptation to nicotine.

By means of chemical factors inhibiting the microsomal oxidation system, one may lower resistance to any chemical substances, specifically to narcotics; conversely, by means of factors that induce an increase in microsomal oxidation, one may enhance resistance to highly different chemical substances. In principle, the possibility of this type of cross-adaptation at the level of the microsomal oxidation system in the liver was demonstrated by Salganik and collaborators. Manankova and Salganik [293] showed phenobarbital-16-dehydroprednalone and 3-acetate-16α-isothiocyanopregnenolone (ATCP) to boost the activity of cholesterol 7α-hydroxylase by 50%–200%. From this observation, in their subsequent work Salganik et al. [492] used ATCP to stimulate cholesterol oxidation in whole organism and thus reduce alimentary hypercholesterolemia. It turned out that, after having been kept on an atherogenic diet, control animals retained an increased cholesterol level for over 15 days after returning to a normal diet; on the other hand, animals given ATCP for 5 days showed a normal cholesterol level by that time. This evidence shows that the power of the microsomal oxidation system in the liver is a factor that influences the cholesterol level in the blood and, hence, the probability of atherosclerosis.

Thus, *we have an interesting possibility of stimulating the power of the microsomal oxidation system and preventing diseases connected with excess accumulation of a concrete endogenous metabolite in the organism. This task is resolved on the basis of formation of a systemic structural trace localized in the liver.*

The structural trace developing during adaptation to injury of one of a pair of organs also has limited localization, i.e., it is evident in compensatory hypertrophy of the preserved organ, e.g., kidney, lung, adrenal. As was indicated above, hyperfunction of the remaining organ via the G $\rightleftarrows$ F mechanism leads to activation of nucleic acid and protein synthesis in its cells. Further, as a result of hypertrophy and hyperplasia of those cells, the organ develops pronounced hypertrophy and by increasing its mass the organ acquires the ability to cope with the same strain as that previously endured by two organs. Below, we shall examine compensatory adaptation in greater detail (see Chap. 4).

Hence, *a systemic structural trace constitutes the common basis of various long-term responses; yet, adaptation to various environmental factors is simultaneously based on systemic structural traces with varying localization and architecture.*

Interrelation between Function and Genetic Apparatus: The Basis for Forming the Systemic Structural Trace

In examining the G⇄F interrelation, it would be advisable to evaluate first the basic features characterizing the realization of this phenomenon and then the mechanism at the expense of which function affects the activity of the genetic apparatus of a differentiated cell. We shall analyze these general regularities, citing the heart, a vitally important organ, as an example.

1) *The response of the genetic apparatus of a differentiated cell to lengthy continuous function increase is a stepwise process.* The evidence characterizing this process was presented in detail in my previous monographs [317, 320, 322] and now permits the distinction of four basic stages, which are most clearly apparent in the continuous compensatory hyperfunction of internal organs, e.g., the heart upon stenosis of the aorta, a single kidney after removal of the other kidney; however, these stages can also be traced when some environmentally caused function is responsible.

In the initial, urgent stage, increased load on the organ, i.e., increased SFI leads to mobilization of the functional reserve, for instance, bringing into operation all the actomyosin power-generating bridges in the cardiomyocytes, all the nephrons, or all the alveoli. In this case, the ATP expenditure per function exceeds its resynthesis and a more or less pronounced ATP shortage develops, not infrequently accompanied by labilization of lysosomes, injury of cell structures, and functional insufficiency.

In the second, transition stage, activation of the genetic apparatus leads to increase in the mass of cell structures and whole organs. Even in highly differentiated cells and organs, this process occurs very rapidly. For instance, 5 days after stenosis of the aorta, a rabbit's heart may increase its mass by 80% [314], and a human heart increases its mass by over twofold 3 days after abruption of the aortal valve [563]. The increase in size of a given organ signifies the distribution of augmented function in the grown mass, i.e., decline in SFI. At the same time, the functional reserve is restored and the ATP content begins to approach normal. Due to a decrease in SFI and restoration of ATP concentration, the transcription rate of all types of RNA also starts to decline. Thus, the protein synthesis and organ growth rates are retarded.

The third stage of stable adaptation is characterized by the organ mass increasing to a certain stable level, while the SFI, the functional reserve, and the ATP concentration are close to normal. The activity of the genetic apparatus (RNA-transcription and protein-synthesis rates) is near to normal, i.e., at a level required to cope with the increased mass of cell structures.

The fourth stage of wear and 'local aging' is realized only with a highly intensive and lengthy load, especially under repeated loads, when the organ or system is confronted with the necessity to go through repeatedly the above-mentioned step-wise process. In the conditions of protracted, exceedingly tense adaptation and readaptations, the ability of the genetic apparatus to generate increasingly new RNA portions may prove exhausted. As a result, the hypertrophied organ or system cells are characterized by decreased RNA- and protein-synthesis rates. The outcome of such disturbed structural renovation is the destruction of certain cells and their substitution by connective tissue, i.e., the development of sclerosis in the organ or system and events of more or less pronounced functional insufficiency.

The possibility of this transition from adaptive hyperfunction to functional insufficiency has now been proven for compensatory hypertrophy of the heart [316], kidney [108, 346], and liver [491]; for hyperfunction of nerve centers [163] and the hypophysial-adrenal complex as a result of the long-term effects of strong irritants; and for hyperfunction of stomach secretory glands under the long-term effects of a stimulating hormone (gastrin).

There is need to examine the question whether or not this 'wear from hyperfunction' in genetically inferior systems is an important link in the pathogenesis of such diseases as hypertension and diabetes. At present, it is known that the administration to animals and consumption by humans of large amounts of sugar may cause hyperfunction and hypertrophy of Langerhans' islet cells in the pancreas to be replaced by wear and the development of diabetes. Similarly, salt hypertension in animals and humans develops as the concluding stage of long-term adaptation to excess salt, the process being characterized by hyperfunction, hypertrophy, and subsequent functional exhaustion of certain kidney medullary layer structures responsible for the removal of sodium and playing a highly important role in controlling vascular tension [548].

Thus, in this stage, the question is one of *transforming adaptive response into pathologic response, of transforming adaptation into disease. We have designated this common pathogenic mechanism observed in highly diverse situations as 'local wear of systems dominant in adaptation;' not infrequently, this type of local wear has widely generalized consequences for the organism* [320].

The fact that at increased levels of a cell's function its genetic apparatus is characterized by a stepwise response is essentially an important regularity for realizing the $G \rightleftarrows F$ interrelation constituting the basis of the stepwise nature of adaptation as a whole (see below).

2) The $G \rightleftarrows F$ interrelation is an exceedingly autonomous, phylogenically ancient mechanism of intracellular self-control. With a whole organism, our experiments have shown it to be corrected by neuroendocrine factors; however, it may also be realized without their involvement. This was corroborated by Schreiber et al. [502], who observed nucleic acid and protein synthesis to be activated with increased contractile function of an isolated heart. In creating an elevated load on the isolated heart of a rat, in the first stage they reproduced our result obtained under the influence of load-activated protein and RNA synthesis and they prevented activation by introducing actinomycin into the perfusion fluid.

Subsequently, it became clear that the extent to which ribosomes are programed with messenger RNA and are capable of synthesize protein grows within an hour of increasing the load on an isolated heart. In other words, both in isolated and whole conditions, the increased contractile functions of myocardial cells quite rapidly entail speedier transcription and transport to the ribosomes of the messenger RNA formed in the process and increased synthesis of the protein constituting the structural supply to the grown function.

3) *Activation of nucleic acid and protein synthesis with increased cell function does not depend on greater supply to the cell of amino acids, nucleotides, and other starting synthesis products.* In experiments on an isolated heart, Hjalmarson et al. [178] showed that if the amino acid and glucose concentrations in a perfused solution are increased fivefold, the load on the heart against a background of such excess oxidation substrates continued to cause activation of nucleic acid and protein synthesis. With a whole organism examined in the initial stage of compensatory hyperfunction of the heart caused by stenosis of the aorta and the naturally concomitant tremendous activation of RNA and protein synthesis, the amino acid concentration in myocardial cells did not differ from the control [445]. Hence, in no way does a grown function activate the genetic apparatus via increased supply of amino acids and oxidation substrates to the cells.

4) *The function index on which genetic apparatus activity depends is usually the same parameter upon which the ATP consumption in the cell depends.* A whole organism and an isolated heart were used to show that increase in the amplitude and velocity of isotonic myocardial contractions acompanied by slightly increased oxygen and ATP consumption moderately activate nucleic acid and protein synthesis. Conversely, greater isometric tension in the myocardium due to higher resistance to blood expulsion is accompanied by an abrupt increase in ATP and oxygen consumption, which naturally entails pronounced activation of the cell's genetic apparatus.

5) The $G \rightleftarrows F$ interrelation is realized so that in response to functional increase accumulation of various cell structures takes place asynchronously. This asynchronism is apparent as rapidly renewed, short-lived proteins of the sarcolemma membranes, sarcoplasmic reticulum, and mitochondria accumulate quicker, and slowly renewed, long-lived contractile myofibril proteins accumulate more slowly. As a result, the initial stage of heart hyperfunction reveals an increase in the number of mitochondria [343], in the activity of the principal respiratory enzymes [458], and also in the membrane structures isolated in the microsomal fractions per myocardium unit mass. A similar phenomenon was proven in the neurons and cells of the kidneys, the liver, and other organs with considerably increased function [511].

If the load on a given organ and its function is within the physiologically optimal limits, this selective increase in the mass and power of membrane structures responsible for ion transport may become secure; under exceedingly high loads, myofibril growth normalizes or even decreases the specific weight of these structures in the cell (see below). In any event, the priority increase in the mass of the structures responsible for ion transport and energy supply plays an important role in the development of long-term adaptation, this role being determined by the fact that under high loads an increase in myocyte function is limited: first, by insufficiently powerful membrane mechanisms responsible for the timely removal from the sarcoplasma of Ca^{2+}, supplied thereto at every excitation cycle and, second, by insufficiently powerful resynthesis mechanisms of ATP, consumed in increased amounts with each contraction. The priority, selective increase in the mass of membranes responsible for ion transport and ATP-resynthesizing mitochondria expands the link that limits the function to become the basis of stable long-term adaptation.

6) In humans and certain animal species, the G⇄F interrelation in highly differentiated myocardial cells is realized so that functional increase results not only in higher RNA reading speed at the available genes but also in DNA replication and increase in the number of chromosome sets and the genes they contain.

The figures in Table 1 [596] show that with physiologic growth DNA biosynthesis in the heart of higher apes and humans causes increased ploidy of hypertrophied myocyte nuclei. For instance, in a child with a heart mass equal to 150 g, 45 % of the myocytic nuclei contain diploid and 47 % triploid DNA. In a grown man with heart mass of 250–500 g, only 20 % of the nuclei are diploid, and 40 % contain octaploid and 16-ploid DNA. With very high compensatory hypertrophy, when the heart mass is 500–700 g, the number of octaploid and 16-ploid nuclei reaches 60 %–90 %. Hence, the myocytes of the human heart retain the ability to replicate DNA and increase the number of genomes localized in the nuclei throughout their lifetime. This renovates an increased area of the hypertrophied cell and, possibly, is also a prerequisite for division of certain polyploid nuclei and even the cells themselves.

The physiologic significance of polyploidization is that it provides an increase in the number of structural genes over which the messenger RNA, which is the matrix for membrane, mitochondrial, contractile and other individual proteins, is transcribed. In differentiated animal cells, structural genes are unique; the genetic set contains several genes that code for a given

Table 1. Ploidy of myocytes in left ventricle of various mammal species [596]

Subject	Number of chromosome sets in nuclei				
	2	4	8	16	32
Rats					
Age 6½ weeks	96		4		
Age 17–18 weeks	98		2		
Rhesus monkey					
Age 3–4 years	88	8–14			
Age 8–10 years	29	55	16		
Humans					
Heart weight 150 g	45	47	8		
Heart weight 250–500 g	20	50	35	5	
Heart weight 500–700 g	0–10	10–45	45–65	15–30	0–5

protein, e.g., genes that code hemoglobin synthesis in the erythroblast genetic set. In polyploid cells, the number of unique genes is increased to the same measure as that of genetic sets.

With increased function, the increasing demands made of the synthesis of specific proteins and corresponding messenger RNA may be satisfied by numerous polyploid cell genomes not only by increasing the reading intensity from each structural gene but also by increasing the number of those genes. As a result, opportunities appear for greater activation of transcription and, correspondingly, for greater cell growth at less intense exploitation of each genetic matrix.

The examined features of the G ⇄ F interrelation do not describe it exhaustively but make it possible to pose the basic question relating to the very essence of this control mechanism, namely how does the SFI control the activity of the cell's genetic apparatus? At present, this process may be examined most effectively by using heart activity as an example, for long-term cardiac adaptation to changing loads has been the subject of constant attention of theoretical cardiology for the past decade.

With reference to the cariomyocyte, the above question may be concretized as follows: How does an increase in myofibril tension activate the genetic apparatus located in the nucleus? In answering, one should keep in mind that when diverse irritants requiring motor response, hypoxia, cold and emotional stresses act on the organism, neurohormonal control and automatic heart control provide an increase in the contractile function virtually instantaneously. As a result, ATP utilization in the myocardial cells instantly grows to overtake for a certain short period of time the ATP resynthesis in the mitochondria. This results in a lower concentration of the energy-rich phosphorus compounds in the myocardial cells and in a higher concentration of their decomposition products. The [ADP] : [AMP] : [IPh]/[ATP] ratio increases. Since ATP suppresses oxidative phosphorylation and its decomposition products activate that process, the above correlation may conventionally be designated as the phosphorylation regulator (PR) to signify that it controls the ATP resynthesis rate in the mitochondria.

The diagram of the cell link of long-term adaptation in Fig. 1 shows that the load and increased myocardial cell function signifies a decline in the CP and ATP concentrations and that the PR increase entails an increase in ATP resynthesis in the mitochondria of the myocardial cells. As a result, the ATP concentration ceases to decline and stabilizes at a definite level to restore the cell's energy balance. Thus, energy is provided for urgent adaptation. This mechanism of energy supply for urgent adaptation is quite well known.

The principal element in the diagram, which makes it possible to understand both urgent and long-term adaptation, is that the same PR actuates another more complex control contour, which indirectly via a certain intermediary link designated the 'control factor,' controls the activity of the cell's genetic apparatus and determines the nucleic acid and protein synthesis rate. In other words, function increased under load reduces ATP concentration; the PR value grows, and this shift via certain intermediary control links

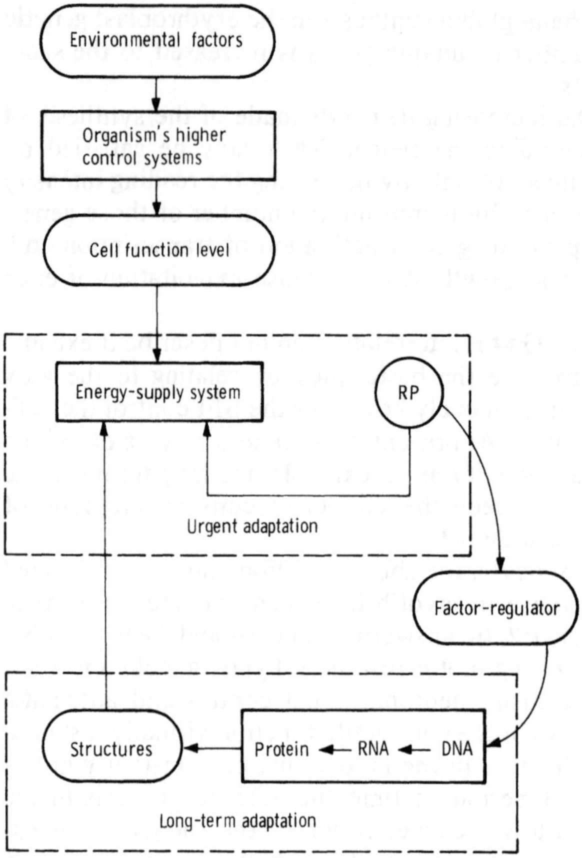

Fig. 1. Scheme of cellular links in long-term adaptation. Explanation in text

activates nucleic acid and protein synthesis, i.e., leads to growth of the myocardial structure. Lower function leads to a conserve result.

The reality of this control contour was grounded relatively recently and is based on the following facts.

1) Considerably increased heart function is naturally accompanied by a decline in ATP concentration and, even more, in CP concentration. This shift is followed by an increase in the nucleic acid and protein synthesis rate in the myocardium and by growth of the heart mass, i.e., by its hypertrophy [112, 318].

2) Considerable cardiac hyperfunction caused by stenosis of the aorta usually leads to lower ATP and CP concentrations and, further, to higher activation of nucleic acid and protein synthesis. However, if the aorta is narrowed in animals adapted to hypoxia or physical loads, the concentrations of energy-rich phosphorous compounds do not decline since these animals have a stronger ATP resynthesis system in the myocardial cells. As a result, on the 1st day after the beginning of hyperfunction, adapted animals do not develop activated nucleic acid and protein synthesis (Fig. 2); this means that when there is no signal in the form of energy shortage to activate the genetic apparatus, there is no genetic apparatus activation itself [319].

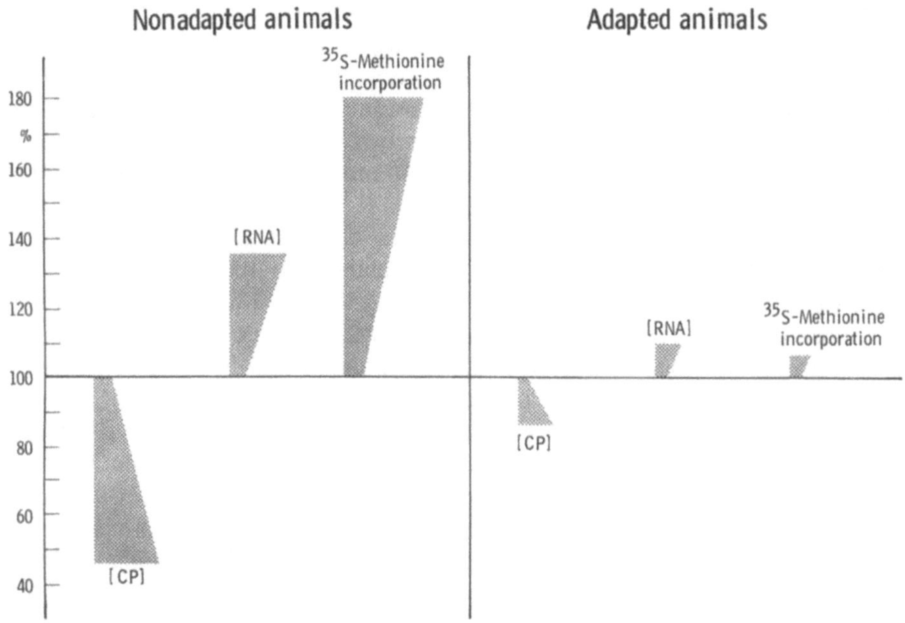

Fig. 2. Effect of preliminary adaptation to hypoxia on creatine phosphate (CP) concentration and activation of RNA and protein synthesis in CHH emergency stage. *Ordinate axis*, changes in CP and RNA concentrations and in protein synthesis activation as percentage of amount prior to CHH

3) Genetic apparatus activation, which is evident in higher nucleic acid and protein synthesis and in considerable cardiac hypertrophy, may be caused without any load increase, i.e., by any action that reduces the concentration of energy-rich phosphorous compounds in the myocardium. This result was obtained specifically by moderately narrowing coronary arteries [150], by isoproterenol, a synthetic analog of norepinephrine, which uncouples oxidation and phosphorylation [558], and by cold, also acting via the sympathoadrenal system [574]. This result also develops as a result of an inferior sarcolemmal membrane and increased calcium influx to the cell, this being ultimately also due to decline in the CP and ATP concentrations [277, 503].

4) In a myoblast culture, decline in oxygen tension, known to be accompanied by a drop in the ATP and CP content, naturally entails increased histone acetylation and higher nucleic acid and protein synthesis rates.

5) Increased ATP and CP content naturally entails lower nucleic acid and protein synthesis rates in the myocardial cells. This effect can be reproduced by hyperoxia in myoblast culture [64] and also naturally develops in a whole organism after parasympathetic innervation has been shut down. In this case, disturbed ATP utilization and increased ATP concentration in the myocardium are naturally accompanied by a decrease in the RNA and protein synthesis rate and a diminishing of the heart mass [59].

These facts unambiguously show that the contents of the energy-rich phosphorous compounds control not only their synthesis but also the activity of the cell's genetic apparatus, i.e., the forming of cell structures.

It is important that this type of $G \rightleftarrows F$ interrelation, the structure of the key link of long-term adaptation, is not a unique property of the heart. The role of energy shortage in activating the genetic apparatus has been shown in the cells of highly diverse organs: skeletal muscles [207], neurons [181], kidney cells [184], etc.

These facts unambiguously show that, via concentration of energy-rich phosphorous compounds, the cell energy balance controls not only ATP synthesis but also the activity of the cell's genetic apparatus, i.e., the forming of cell structures. In conformity with the general principle of rigid structural organization of control mechanisms in the organism and in each of its cells from early research, it appeared probable that the ratio of the ATP and its decomposition products controls the genetic apparatus activity not in itself but via a certain control metabolite. Hence, in 1973 we introduced the 'metabolite-regulator' concept to suggest that this molecular signal, which reflects the function level, removes physiologic repression of structural genomes in the cell nuclear chromosomes, activating the transcription of messenger RNA, then ribosomal RNA, and consequently protein translation [320, 359]. It had already been noted that the biosynthesis and accumulation of short-lived membrane proteins takes place before everything else and to the highest degree in response to function growth. This fact led us to consider that transcriptions coding the synthesis of precisely these key cell proteins, due to their highest affinity with the control metabolite or some other structural features, prove accessible to RNA polymerase at lower metabolite-regulator concentrations, i.e., at lower loads on organs and systems. As a result, repeated moderate loads selectively increase the mass and potency of structures responsible for control, ion transport, and energy supply and, consequently, for the increase in functional power of the organs and systems constituting the basis of adaptation. This hypothesis serves as a foundation for a mathematical adaptation model analyzed in a special monograph [322]; in response to various 'loads' it satisfactorily reproduces the dynamics and final structural relationship at adaptation and deadaptation.

The physical essence of the metabolite-regulator and the question of whether this hypothetical metabolite can exist at all have become subjects of many-sided investigations. One possibility was that cAMP could play the role of such a metabolite-regulator. The assumption was based on the fact that in microbes a state of energy hunger caused by insufficient glucose in the environment is naturally accompanied by an increase in the content of cAMP, which induces adaptive synthesis of enzymes essential for utilizing other substrates [359], thus performing the role of a signal that actuates adaptation to hunger. In higher animals, mammals for example, cAMP is also a powerful inducer capable of activating transcription [170] in cells and thereby increasing nucleic acid and protein synthesis. Norepinephrine and especially isoprotorenin, its analog, both of which specifically activate adenyl cyclase and thereby cAMP synthesis in the whole organism, naturally cause transcription to activate and increase RNA concentration in the myocardium with subsequent heart hypertrophy. All the other factors causing cardiac hypertrophy, e.g., cold, physical loads, and hypoxia, activate adrenergic control of the heart and, consequently, can also increase cAMP accumulation and, via this metabolite-regulator, activate transcription.

In recent years evidence has been produced concerning the role of cAMP in the origin of activated synthesis of nucleic acids and proteins during hypertrophy. For example, Limas et al. [276] established that directly after the beginning of cardiac hyperfunction caused by stenosis of the aorta, prostaglandin synthesis is stimulated in the myocardium; in turn, prostaglandins activate adenyl cyclase, the result being increased cAMP concentration in the myocardial cells. Further, it was shown that as with hyperfunction, the ATP deficit arising with cardiac hypoxia causes the accumulation of cAMP. Yet another important fact was established: cAMP proved to activate RNA polymerase and RNA synthesis in the cell nuclei of the cardiac muscle.

These important data did not exclude the possibility that the ATP and CP contents control the activity of the genetic apparatus not only via cAMP but via other metabolites as well. For example, the results of studies on cell cultures permitted the assumption that a magnesium ion could play a substantial role in controlling the activity of the genetic apparatus. This cation represents the essential cofactor of transcription and translation and is contained in the cells in a complex with ATP. Release of magnesium ions with ATP decomposition and lower concentration was shown to lead to activation of the cell's genetic apparatus, growth of cellular structures, and increased intensity of fibroblast proliferation in the culture; bonding of magnesium ions with excess ATP led to the opposite result. Hence, it may be that the relationship ADP · P_{inorg}/ATP controls the genetic apparatus activity in the cell via a magnesium ion [482].

Another observation made in recent years was that the ATP deficit in the myocardium naturally causes greater activity of ornithine decarboxylase, the key enzyme in the system of aliphatic amine synthesis [7], viz spermin and spermidin. These substances activate RNA and protein synthesis in the myocardial cells [54].

The most interesting work to confirm directly our original concept that a specific intracellular metabolite-regulator plays a decisive role in realizing the interrelationship between function and the genetic apparatus was published in 1979 by Hammond et al. [164], who reproduced in dogs a compensatory hyperfunction of the heart, by stenosing the aorta, or a compensatory hyperfunction of one kidney, by removing the other. After 1 or 2 days, in the hyperfunction trouble stage when the ATP deficit and the metabolite concentration we had postulated were to be highest, they prepared aqueous extracts liberated from the cellular structure. In the next experimental step, these extracts were introduced into the perfusive blood stream of another dog heart, which functioned in an isotonic mode, i.e., with constant minimal load. Prior to introducing the extracts and after different periods following introduction, the RNA was extracted from the isolated heart myocardium to examine its ability to activate protein synthesis in an extracellular system containing rabbit reticulocyte lysate. The system includes all the components essential for protein biosynthesis with the exception of messenger RNA; consequently, the biosynthesis activation occurring in response to the addition of myocardial RNA samples was the quantitative criterion for estimating the messenger RNA content in the myocardium. The heart and kidney extracts that performed the compensatory hyperfunction were found to increase the ability of the RNA from an isolated heart to activate protein synthesis to a considerably higher degree than the extracts from control organs. In other words, with compensatory hyperfunction of these organs, the amount of the organ-nonspecific metabolite activating messenger RNA synthesis, i.e., transcription of structural genes, naturally increased.

Further it became clear that inclusion in the perfusion system of isolated hearts of donor dogs with narrowed aortas or single kidneys does not reproduce the effects of extracts, i.e., does not increase the ability of isolated heart RNA to activate protein biosynthesis. Thus, the metabolite-regulator activating transcription in cells of intensively functioning organs normally does not enter the blood and, in conformity with the initial hypothesis, functions as a link in intracellular control. Finally, investigators established that kidney and heart extracts lose their ability to activate transcription after treatment for 1 h at 60 °C. This means that the extract-activating effect does not depend on the presence therein of RNA, nucleotides, and amino acids, the most probable 'candidates' to qualify as metabolite-regulators, being thermolabile proteins or polypeptides.

Concepts on the structure of the control mechanism through which the cell function affects the activity of the genetic apparatus are now being formed. At present, there is no doubt that this influence is realized via the cell energy balance, i.e., ultimately through the contents of ATP and its decomposition products. The next link, viz the metabolite-regulator, directly affecting the activity of the genetic apparatus is so far a subject of investigations and suppositions, which are gradually becoming increasingly concrete. The effect of this metabolite is undoubtedly achieved via a complex system of regulator proteins in the cell nucleus. In terms of our exposition, it is significant that *via the above-mentioned G⇄F interrelation the cell function determines the formation of essential structures; hence, this interrelation is an essential link for providing physiologic functions with structures in general and for forming the structural basis of adaptation in particular.*

Interrelationship of Cell Structures: A Parameter Determining the Functional Possibilities of the System Responsible for Adaptation

The concept that the function level controls the activity of the genetic apparatus via the cell energy balance and concentration of energy-rich phosphorous compounds in itself explains solely the hypertrophy of organs under long-term load and atrophy resulting from idleness. Meanwhile, in the course of adaptation, considerable changes in the power of a given functional system are not infrequently coupled with slight variations in mass. Hence, there are no grounds for thinking that one can expand the function-limiting link and increase the potency of the systems responsible for adaptation simply by increasing the organ mass.

To understand the real mechanism providing the expansion of the limiting link, one should keep in mind that the actual consequences of different loads on a given organ and of changes in the PR value in its cells are not exhausted by simple activation of the genetic apparatus and by increase in the organ mass. Depending on the size of additional load, the synthesis rate of specific structural proteins and the correlation between cellular structures proved to vary in different degrees. For instance, in studying the heart we established that, depending on the load, it develops three versions of long-term adaptation differing in the cellular structure ratio.

1) Under periodic loads of increasing intensity, i.e., with natural or athletic training, the heart develops moderate hypertrophy accompanied, as was already indicated above, by an increase in the power of adrenergic innervation, higher coronary capillary to muscle fiber ratio, greater myoglobin concentration and activity of enzymes responsible for substrate transport to the mitochondria, and higher heavy H chain to light L chain ratio in the myofibril myosin heads and increase in the myosin ATPase activity. At the same time, the cells are characterized by an increased content of the sarcoplasmic reticulum membrane structures, physiologic changes indicative of higher potencies of the mechanisms responsible for calcium ion transport, and relaxation of the myocardium. As a result of this preferential increase in the power of the systems responsible for regulation, ion transport, energy supply, and energy utilization, the maximum rate and amplitude of myocardial contractions in adapted animals increases, and the relaxation rate grows even more [344]. The effectivity of oxygen utilization also improves. As a result, the maximum amount of external work that the myocardium unit mass can generate and the maximal work of the whole heart with formed adaptation considerably increase [172, 321].

2) With heart failure, hypertension, and other circulatory diseases, the load on the heart proves to be continuous and correspondingly causes continuous compensatory heart hyperfunction (CHH). One version of this process caused by increased resistance to blood expulsion to the aorta entails a major increase in the activity of the genetic apparatus of the myocardial cells, pronounced cardiac hyperfunction, and a 1.5–3-fold increase in the heart mass [321].

This hypertrophy is an unbalanced form of growth, as a result of which the mass and functional possibilities of structures responsible for neural control, ion transport, and energy supply increase to a lesser extent than the organ mass. This results in a set of changes opposite to those just described for heart adaptation and is examined in detail in Chap. 4. The concomitant decline in the functional possibilities of the myocardial tissue is for a long time compensated by an increase in its mass; however, it may subsequently cause heart insufficiency. Such exceedingly tense adaptation characteristic of CHH has been designated as overadaptation.

3) With protracted hypokinesia and lower loads on the heart, the protein synthesis rate in the myocardium and the mass of the heart ventricles decline [109, 452]. This atrophic process is characterized by a preferential decrease of the mass and power of the structures responsible for neural regulation [246], energy supply [236, 288], ion transport, etc. As a result, the structural correlation in the myocardium and the functional possibilities of that correlation in the myocardial tissue prove to be changed as great as in CHH. Since the mass of this tissue is decreased, the functional possibilities of the heart are diminished; this state is defined as heart deadaptation.

A comparison of these states, which are apparently inherent not only in the heart but also in other organs and systems, leads to the concept that the same intracellular control mechanism, viz a G⇄F interrelation dependent on the size of the load determined by the requirements of whole organism, governs the development of the following three states of the system: (a) adaptation itself, (b) deadaptation, and (c) overadaptation. The difference between these states is determined by the structural correlation in the cells. It would be desirable to assess the truth of this concept by directly analyzing the correlation of ultrastructures in the myocardial cell with the principal parameters of the contractile function of the heart in animal adaptation caused by training.

Empirical experience gained from practice and experimental evidence unambiguously show that a relatively slight increase in heart mass in the adaptation to physical loads entails a big increase in the maximal cardiac output and external work that the heart can perform. Similarly, a relatively small hardly determinable decrease in heart mass under hypokinesia is accompanied by a pronounced decline in its functional possibilities. In other words, the enormous advantages enjoyed by an adapted heart and the functional failure of a deadapted heart cannot be explained by simple changes in the myocardial mass. Equally, this adaptation result cannot be attributed to the effect of extracardial control factors since it is vividly evident in an isolated heart and on papillary muscles when the myocardium does not depend on the control factors of a whole organism. Thus, the mechanism responsible for the increase in the functional possibilities of a trained heart and for the failure of an untrained heart, the main issue of long-term heart adaptation, has remained open until recently.

My hypothesis implies that with protracted increased loads on the heart, a developed G⇄F interrelation leads to selective increase in the biosynthesis and mass of the key structures that limit the function of the myocardial cell, i.e., membrane structures responsible for ion transport and providing ATP ultilization in myofibrils, and ATP resynthesis in the mitochondria. As a result, the functional possibilities of the heart considerably increase with a slight increase in mass. The protracted decline of loads on the heart in conditions of hypokinesia entails a selective decline of biosynthesis and atrophy of the same key structures; the heart's functional possibilities decrease, again with a slight change in mass.

This condition appears sufficiently important to warrant illustration by means of our evidence on the correlation of ultrastructure with the heart contractile function in adaptation to physical loads [361]. The experiments were performed on male Wistar rats, the function of the papillary muscle being studied by Sonnenblick's [529] method. The volume of myocyte structures was measured by an electron-microscopic stereological study. This procedure permits the quantitative estimation of not only the mitochondria and myofibril volumes but also the volume of the sarcolemma and the sarcoplasmic reticulum membrane system responsible for Ca^{2+} transport.

Table 2. Effect of adaptation to physical loads on contraction of thin papillary muscle strips under low (0.2 g/mm²) and high loads

Index	Control (n = 11)	Adaptation (n = 8)	p
Contraction amplitude under low loads, % of initial length	6.9 ± 1.4	13.8 ± 2.3	< 0.05
Contraction rate under low load, unit muscle length/s	1.1 ± 0.17	2.1 ± 0.32	< 0.02
Maximal load, g/mm²	3.8 ± 0.27	3.2 ± 0.36	> 0.1

To obtain adaptation, the animals were compelled to swim 1 h daily at a water temperature of 32 °C for 2 months. From Table 2 it is apparent that the maximum rate and amplitude of isotonic myocardial contraction in adapted animals was twice as high as in controls. With these high-amplitude rapid contractions, adaptation was obtained quite convincingly, this result being in good agreement with the fact that myosin ATPase activity was proven to increase considerably in the course of adaptation to physical loads due to a rise in the content of myosin heads of short-lived H-chains, the carriers of the ATPase activity.

In terms of our presentation, it is important that this increase in the basic contraction parameters apparently presumes an increase in the amount of calcium that had entered the sarcoplasma to interact with the myofibrils. This signifies that long-term stable functioning under rapid high-amplitude contractions may be achieved only with intensive work of the mechanisms responsible for calcium removal and relaxation.

In subsequent experiments, we obtained evidence showing relaxation to be activated in a trained heart. First of all, the relaxation rate of the myocardium of trained animals is more increased than the contraction amplitude. This resulted in a higher relaxation index, i.e., greater ratio of the relaxation rate to the contraction amplitude. The curves in Fig. 3 show that with a highly variable contraction rate this ratio in the myocardium of trained animals was increased by 20%–25%. Second, the elasticity of the myocardial strips of trained animals was increased compared with controls. Figure 4 shows that to stretch the myocardial strips of control animals by 10% one would have to apply a load 1.5 times greater than that needed to expand the myocardium of trained animals. In my view, this increased elasticity was due to a more

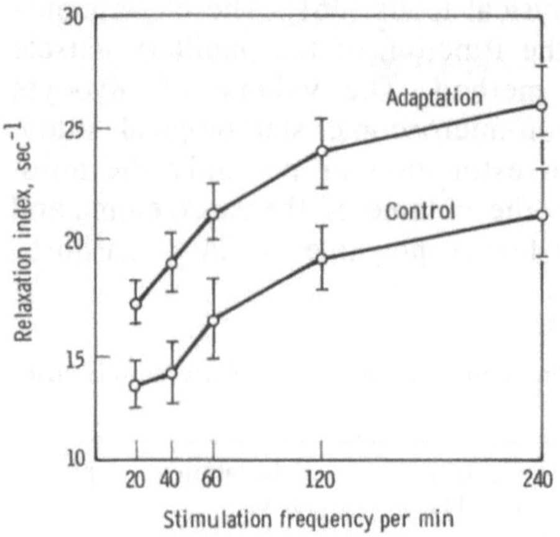

Fig. 3. Effect of adaptation to physical load on the ratio of the relaxation velocity to concentration amplitude in strips of rat papillary muscle

complete calcium removal from trained heart sarcomeres, resulting in a lower quantity therein of residual calcium-troponine complexes and actomyosin bridges, which are retained after completion of relaxation.

Logically, increased power of the calcium pump in the sarcoplasmic reticulum (SR) should promote accelerated accumulation of calcium in the SR cisterns; in turn, this may increase positive inotropic effects on the heart. In full conformity with the above, the curves in Fig. 5 show that a high-frequency positive inotropic effect in the myocardium of trained animals proved considerably higher than in controls.

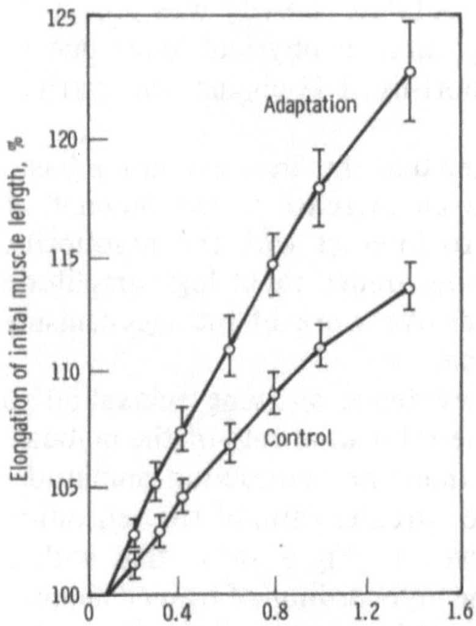

Fig. 4

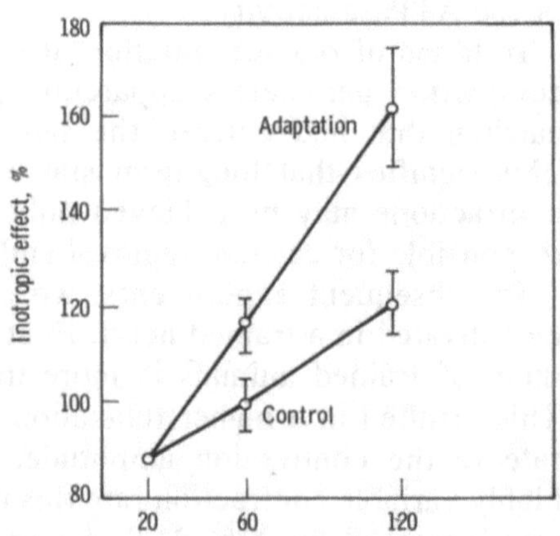

Fig. 5

Fig. 4. Effect of adaptation to physical load on elasticity in strips of rat papillary muscle
Fig. 5. Effect of adaptation to physical load on inotropic effect of increased contraction frequency in strips of rat papillary muscle

These facts seem to be indicative of the increased power of the membrane mechanism of calcium transport in the myocardium of animals adapted to physical loads. At the same time, the sum of the physiologic evidence presented leads to the consideration that selective increase in the potency of these mechanisms may play an important role in raising the functional possibilities of a trained heart.

It is clear that pure reasoning should be verified by directly measuring the structural volumes of the sarcolemma, the SR, and also of the mitochondria and myofibrils in adapted and deadapted animals. This was performed by G. Gussky (unpublished data) during studies performed jointly by our laboratory and the Laboratory of Morphometry, W. Humboldt University, Berlin.

The principal results of this research boil down to two features. First, with the adaptation regimes used, the heart mass as a whole and the mitochondria and myofibril volumes per myocardium unit mass remained essentially unchanged. Only the size of each mitochondrion proved diminished and, naturally, the number of such smaller mitochondria per section unit area increased by 30% and the mitochondrial membrane surface per tissue unit volume increased by 28%. Second, the volumes and surfaces of the membrane structures of the sarcolemma and SR changed quite substantially.

Table 3. Structural ratios in myocardial cells in heart adaptation

Index per tissue unit volume	Control	Swimming	p
Mitochondrion volume	0.314 ±0.011	0.361 ±0.038	
Number of mitochondria[b]	0.670 ±0.041	0.872 ±0.088	<0.05
Surface of mitochondrial external membranes	2.043 ±0.125	2.616 ±0.055	<0.01
Surface of longitudinal channels	1.787 ±0.037	2.636 ±0.362	<0.05
Golgi apparatus volume	0.00105 ±0.00016	0.00296 ±0.0009	<0.005

[a] Each group contained three animals; analysis was performed using 72 electron graphs.
[b] Per section unit area.

From Table 3 it follows that in animals adapted to physical loads the surface of the longitudinal channels of the SR was increased by 50%. At the same time during adaptation to physical loads, an almost threefold increase was observed in the volume of the Golgi apparatus, where according to modern concepts cell-membrane forming processes take place, specifically those involving sarcoplasmic reticulum membranes. This may be indicative of the considerably activated formation of myocyte membrane structures during of the heart to physical loads.

According to current concepts, the membranes of the longitudinal channels of the SR and the sarcolemma are the sites where the Ca-activated Mg-dependent ATPases are localized; these membranes play a decisive role in relaxation, by removing the calcium from the sarcoplasma and myofibrils, and also in contraction by transporting calcium to sites from which it escapes during excitation. Thus, the results of electron-stereological and physiologic research coincide in affording grounds for the conclusion that the preferential increase in the mass and power of the membrane system of calcium transport plays an important part in enhancing the functional possibilities of the heart during adaptation to physical loads.

When assessing this statement, one should keep in mind that the preferential accumulation of membrane structures responsible for calcium transport is but one of the manifestations of the principle of selective increase of biosynthesis and accumulation of key structures, the principle upon which heart adaptation is based. At present, it has been proven that moderate hypertrophy during adaptation is combined with greater activity of the adenyl cyclase system and with increase in the number of adrenergic fibers per myocardium unit mass. As a result, the adrenoreactivity of the heart, i.e., the possibility to mobilize urgently its function, is increased.

At the same time the number of H chains, the carriers of ATPase activity in the myosin heads, is found to be increased; ATPase activity grows to result in an increased rate and amplitude of myocardial contraction. As was shown above, the power of the SR is augmented, increasing the rate and depth of diastolic relaxation. In parallel with the above shifts in the myocardium, the number of coronary capillaries, the concentration of myoglobin, and the activity of the enzymes responsible for substrate transport to the mitochondria were found to increase. This rise in the potency of the energy supply system naturally entails higher heart resistance to fatigue and hypoxemia.

Thus, during adaptation of the heart, in no way does the organism respond by global growth of the myocardium, but rather by the economical way of preferentially or selectively increasing the mass or potency of the key structures that limit cardiac function. *This selective growth of the power of structures responsible for control, ion transport, and energy supply is not a unique feature of the heart. This process is naturally present in all organs of the functional systems responsible for protracted adaptation when highly diverse environmental factors act on the organism. Precisely due to this fundamental and genetically realized mechanism, the organism, when adapting to the environment, increases the functional strength of its systems without significantly changing their mass. In effect, this signifies that structural correlation is the principal intrinsic characteristic of an adapting system,*

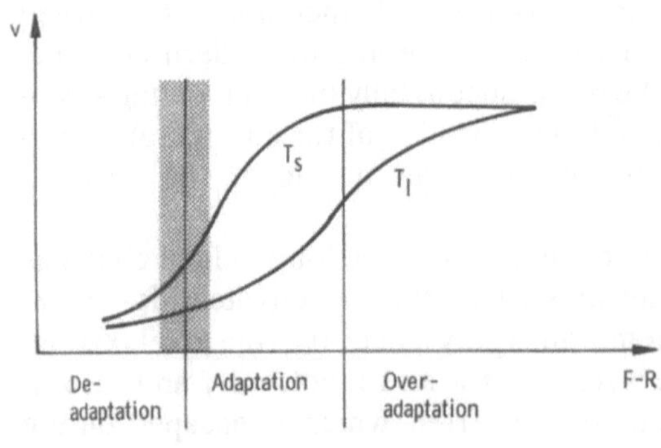

Fig. 6. Incorporation into the cell of genetic apparatus transcription depending on load and factor-regulator. T_s, short-lived proteins; T_l, long-lived proteins. *Abscissa axis*, CF-dependent control factor; *ordinate axis, V*, degree of incorporation, relative units

the factor that predetermines its 'target parameter,' maximum achievable work.

This concept and the one developed by Georgiev [132] on the structure of higher animal transcriptons have led us to assume that changes in cell-structure correlations in adaptation, overadaptation, and deadaptation are determined, which code for short-lived and long-lived proteins quantitatively differing in the structures of their regulatory regions. This, in turn, leads to different involvement of transcriptons that code for short- and long-lived proteins with equal PR and PR-dependent control factor.

The difference is seen when with equal PR and factor-regulator the involvement of transcriptons that code for short-lived proteins is higher than that of transcriptons that code for long-lived proteins. In accordance with the theory developed by Jacob and Moneau [194a] for the bacterial cell genome, the dependence of operon involvement upon control metabolite concentration has the form of a constantly growing function with an inflection and subsequent saturation plateau. Given that a different structure in the repressor and derepressor systems in mammals does not basically change this dependence, our principal assumption may be diagrammatically presented by Fig. 6, which compares the dependence of the involvement of transcriptons that code short- and long-lived proteins upon factor-regulator concentration.

The curves show that in normal conditions when the load the PR value, and the resultant control factor value are low, the involvement of the transcripton that codes for short-lived proteins is higher than that which codes for long-lived proteins. With an increased load within physiologic limits, i.e., at adaptation, the amount of control factor and the difference in the involvement of both transcriptons grow. With an excessive increase in the load, i.e., with CHH, the amount of the PR-dependent control factor grows still more and ceases to limit the involvement of these transcriptons. In addition, the involvement of the transcriptons approaches maximum and the difference in the involvement of transcriptons coding for short- and long-lived proteins proves to be minimum. With a considerable decline in the load and PR, i.e., at hypokinesia and deadaptation of the heart, the amount of the control factor is substantially decreased; the involvement of all transcriptons and the difference in the involvement of transcriptons that code for short- and long-lived proteins is also decreased to a minimum.

In the light of the tenet sufficiently substantiated by molecular biology that the involvement of each transcripton predetermines via the transcription and translation mechanism the content of the transcripton-coded protein in the cell, the curves satisfactorily explain the change in the ratios between short- and long-lived structures in the course of heart adaptation-deadaptation. Indeed, in accordance with this hypothesis, the increased involvement of all the transcriptons observed during adaptation (with increased difference in the involvement of transcriptons that code for short- and long-lived proteins) should lead to a certain increase in the whole cardiac mass and to an increased ratio of the mass of short-lived structures to the mass of long-lived structures compared with the norm; precisely this situation actually exists in an adapted heart.

The maximal involvement of all transcriptons and the lesser differences between the involvement of transcriptons that code for short- and long-lived proteins should at overadaptation (CHH) lead to a large increase in the heart mass and to a considerable decrease in the ratio of the mass of short-lived structures to long-lived ones; this situation exists in reality in CHH. Finally, decreased involvement of all transcriptons and lesser difference between the involvement of transcriptons that code for short- and long-lived proteins may with deadaptation lead to reduced heart mass and smaller ratio between the short- and long-lived structures in the myocardium; this is actually the case during hypokinesia-caused deadaptation.

Thus, the developed hypothesis on the key link of long-term adaptation is that the load via PR and the intracellular control mechanism determines first the general activity of the cell's genetic apparatus and thence the protein synthesis rate and cell mass, and second the ratio of involvement of transcriptons that code short- and long-lived proteins and thence the ratio of structures in the cells of organs and systems. This hypothesis, which needs verification, implies that the mechanism connecting the level of function with the activity of the genetic apparatus operates so that in response to an increase in function maximum increase occurs in the transcription rate of the genes that code short-lived enzymatic proteins and membrane proteins, which, in effect, are the function-limiting cell proteins.

One of the possible ways of verifying the hypothesis is to compare the dynamics of adaptational accumulation of various isozymes of a definite enyzme, one of these isozymes being less stable and at the same time functionally more effective under load. Inasmuch as subunits that form various enzyme isozymes are normally coded by various genes, preferential accumulation of such a short-lived fuctionally active isozyme would be a substantial argument in favor of the given idea.

Myocyte CPK, which represents a vitally important enzymatic system responsible for the transport of energy-rich phosphate groups from mitochondria to myofibrils, is an enzyme that may be utilized in this type of research. Creatine phosphokinase activity sharply grows with hyperfunction of the heart [570] and skeletal muscles [436]. This increase in activity is naturally prevented by introducing puromycin and is, therefore, the result of the enzyme population increase, of selective enzyme accumulation during adaptation to load.

In animal tissues, CPK is represented by three isozymes: MM, which consists of two M-type (muscle type) polypeptide chains; BB, which consists of two B-type chains characteristic of embryonal tissues and the brain; and MB, which consists of two different subunits inherent solely in the myocardium [416]. In addition, the isozyme BB synthesizes more rapidly [436] and is apparently less long-lived than MM [472], also, having higher affinity to CP and ADP [76, 540], may function more effectively under heart overload, when the content of energy-rich compounds in the myocardium is reduced [571]. From these data, we studied in collaboration with Yavich [342] the activity dynamics and the isozyme spectrum of myocardial creatine kinase in the course of heart adaptation to continuous load, i.e., at CHH caused by stable narrowing of the aorta.

With irreversibly increasing loads, these types of compensatory hyperfunctions are interesting in that in the course of development they are characterized by the rapid mutual replacement of the three states that can steadily exist in various organs: the initial stage, characterized by preferential accumulation of key structures that limit cell function; the stage in which structural correlation returns to the starting norm; and the final overadaptation stage, in which stable structures prevail over enzymic and membranous structures that limit cell function. This signifies that in measuring the CPK activity and its isozyme ratio in the course of compensatory hyperfunction we can really evaluate the truth of the concept on the role of the ratio between the activity of short- and long-lived proteins in the adaptation mechanism [342].

Experiments were performed on white male rats. Two groups of animals were used: juvenile rats aged 3–4 months, weight 180–200 g; and rats aged 9–10 months, weight about 400 g. Compensatory heart hyperfunction was generated by coarctation of the abdominal aorta by a previously described technique [224]. Animals with CHH were represented by two groups. Group 1 included those affected by early hyperfunction occurring 1, 2, 3, 5, 7, and 9 days following coarctation of the aorta, and Group 2 included those with late hyperfunction (6–7 months following coarctation of the aorta). The control group included rats of corresponding age and weight. To separate CPK isozymes, we used a slightly modified method suggested by Ogunro and Hearse [416].

The results of total CPK activity determined in the heart homogenates are shown in Table 4 and Fig. 7. Creatine phosphokinase activity in the myocardium of control animals amounted to 1 100 IU per 1 g wet weight and had already increased on day 2 after the beginning of CHH. Maximal increase in CPK activity develops on days 3–5 after the beginning of CHH to attain 40%–50% per unit mass. Total CPK activity calculated for the whole heart increased by 42%–63% during that period. Subsequently, the CPK activity per unit mass decreased to the norm, and for the whole heart remained

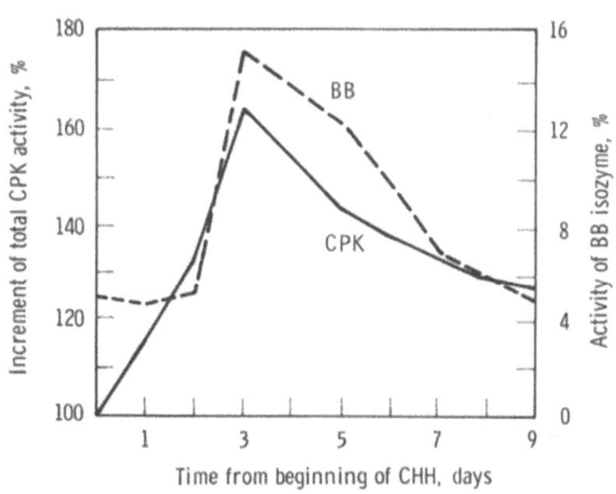

Fig. 7. Effect of CHH on total CPK activity and relative activity of BB isozyme in myocardium

slightly increased as late as on day 9 of CHH; 6 months after the beginning of CHH, CPK activity per myocardium unit mass decreased by 15% and for the whole hypertrophied heart increased by 25% compared with control.

Table 4. Creatine phosphokinase activity and its isozyme pattern in rat heart

Group	Heart mass, g	Total CPK activity		Isozyme ratio, %		
		IU/g wet mass	IU/heart	MM	MB	BB
Group 1 (age 3–4 months)						
Control	650 ± 15	1,100 ± 80	715 ± 50	65	30	5
CHH, day 1	670 ± 20	1,170 ± 80	784 ± 45	65	31	4
CHH, day 2	710 ± 30	1,320 ± 90[a]	937 ± 50[a]	64	31	5
CHH, day 3	750 ± 20	1,560 ± 90[a]	1,170 ± 60[a]	45	40	15
CHH, day 5	800 ± 30[a]	1,276 ± 80	1,020 ± 61[a]	50	38	12
CHH, day 7	840 ± 32[a]	1,155 ± 65[a]	970 ± 49[a]	62	31	7
CHH, day 9	830 ± 25[a]	1,100 ± 70	913 ± 52	65	31	4
Group 2 (age 9–10 months)						
Control	1,300 ± 30	1,020 ± 80	1,326 ± 90	65	31	4
CHH, 6 months:	1,900 ± 50[a]	870 ± 60	1,672 ± 100	62	31	7

[a] $p < 0.01$; < 0.05.

The electrophoretogram of CPK isozymes on agarose is shown in Fig. 8, from which it is apparent that the myocardium of control rats contained three CPK isozymes, viz MM, MB, and BB. In accordance with the fluorimetric analysis result, the MM, MB, and BB activities were 65%–66%, 30%–31%, and 4%–5%, respectively. It is important that fluorescence in enzymograms in the enzyme localization sites was caused by NAD-P reduction since no fluorescence was observed in samples lacking CP. The amount of enzyme is estimated by its activity; hence, one may regard our evidence as reflecting the quantitative relationship of isozymes in the total CPK protein population.

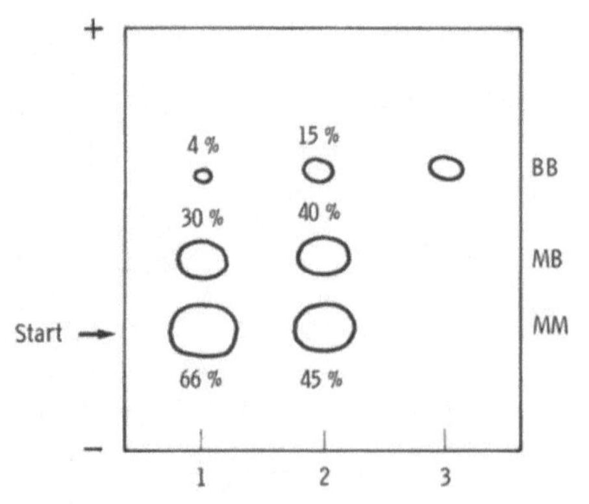

Fig. 8. Distribution of CPK isozymes in agarose (1.2% solution), pH 8.6. *1*, control heart; *2*, heart 3 days after beginning of CHH; *3*, brain. Tissue extracts were prepared on 0.25 *M* sucrose with mercaptoethanol (2,000 × *g*, 20 min, 3°C) and subjected to electrophoretic analysis

Electrophoretic separation of hyperfunctioning myocardium extracts was performed 1, 2, 3, 5, 7, and 9 days after the start of CHH to show that 1 and 2 days after the beginning of hyperfunction, the creatine kinase proteins do not undergo perceptible changes. However, on days 3–5, the hyperfunctioning heart was observed to show considerable transformation of the isozyme spectrum. The activity of the BB isozyme abruptly increased, and that of the hybrid MB increased; at the same time, MM activity declined. The BB activity variation curve is shown in Fig. 9, from which it is apparent that unlike normal BB activity ranges of 4%–5%, on day 3 after the beginning of heart hyperfunction, the BB activity in the myocardium reached 15%. On day 5 after coarctation of the aorta, BB activity was 12%. On the day 7 after the start of hyperfunction, it declined to 7%, and after 9 days it did not differ from the control. Six months after the beginning of hyperfunction, the total CPK activity per heart unit mass had declined in the same way as is the case with other key enzymes of the heart at overadaptation and major, protracted heart hypertrophy.

Thus, the BB activity peak develops on days 3–5 of CHH to coincide with the CPK total activity peak. From Table 4 and the curves in Fig. 9 it follows that during this time the activity of the heterodimer MB also increases to reach 39%–40%, while MM activity conversely declines to 45% to form less than half on the CPK total activity. With further development of CHH, changes in the CPK isozyme spectrum gradually level out to approach control level by day 9.

The principal result of the experiment was that in the CHH emergency stage an increase in the total CPK activity is simultaneously accompanied by substantial changes in its isozyme spectrum toward an increase in BB and decrease in MM. To understand this fact, it would seem expedient to examine the following three questions: (a) In what heart cells do the changes established in the CPK system take place? (b) what is the probable mechanism of the dynamics of the activity and variation in the CPK isozyme spectrum? and (c) what is the biologic significance of that phenomenon?

Basically, the changes in CPK activity and in its isozymic spectrum may

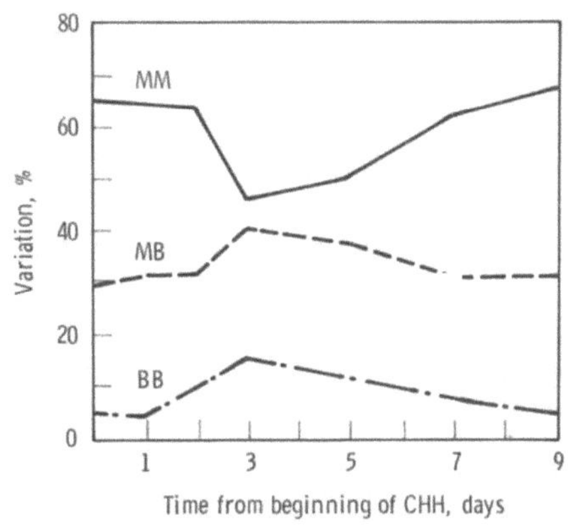

Fig. 9. Effect of CHH on activity of CPK isozymes, percentage of total enzymic activity

occur in cardiomyocytes, as in the fibroblasts of its interstitial connective tissue, which also contains CPK BB isozymes [434]. In our view, there are at least three factors which permit the inference that just as in the total increase in CPK activity this shift in the isozyme spectrum chiefly takes place in the cardiomyocytes.

First, according to evidence obtained in tissue culture, the CPK activity in myocytes is over 16 times higher than in fibroblasts [434], and the myocyte mass constitutes about 80% of the myocardial mass [405]. This means that myocytes were the source of the basic CPK molecule population which we separated into isozymes.

Second, CHH was observed to involve activation of DNA synthesis and mitotic activity in the cells of the interstitial connective tissue in the myocardium; this process attains its peak on day 7 of CHH [324]. The peaks of creatine kinase activity, of relative increase of B subunits, and of relative decline of M subunits occur much earlier – on day 3 of CHH – since it is connected with events in myocytes, not in connective tissue cells.

Third, transformation of the CPK isozyme spectrum during CHH is evident in the increased activity not only of BB, but also of MB. This permits the assumption that both isozymes are synthesized in the same cells and since MB is present only in myocytes [434] the latter are the most probable loci for the biosynthesis of both isozymes.

Thus, the established increased CPK activity and higher share of the BB homodimer in the CPK isozyme spectrum is most likely localized in the myocardial myocytes.

As has already been noted, the increase in CPK activity naturally observed with higher myocyte function is caused by activation of CPK molecule biosynthesis and growth of the CPK molecule population [436]. To understand the changes in the CPK isozyme spectra established by ourselves in the course of this process, it would be expedient, from the obtained data, to quantitate the effect of CHH on the content of M and B subunits in the myocardium. In a normal heart, the MM, MB, and BB content amounts to 65%, 30%, and 5%, respectively; correspondingly, the B:M ratio is 20:80. On day 3 of CHH, the MM, MB, and BB isozymes constitute 45%, 40%, and 15%, respectively, in the myocardium; correspondingly, the ratio of subunits B:M is 35:65. Total CPK activity has increased by that time to 153% of the normal value; hence, the total amount of subunits B and M increases 2.5 and 1.5 times, respectively, compared with controls. Inasmuch as DNA synthesis in rat myocytes does not occur and the number of genes does not change, one may assume that the significantly augmented contractile function of the myocardium (this increase occurring in the form of CHH), caused a multiple increase in the transcription rate of the gene that codes for the relatively short-lived and functionally effective isozyme BB and that this contractile function had a lesser effect on the transcription rate of the gene that codes for isozyme MM, which is less effective for the given situation. This explanation is in agreement with the fact that an increase in CPK and its BB-isozyme activity is not achieved in the CHH emergency stage in an isolated form since this increase is but a component in a set of adaptational changes which become apparent in the accumulation of structures limiting myocyte function.

As an example, the important role of genetically determined changes in the isozyme spectrum of myosin in adaptation to various typed of loads was most specifically demonstrated by Jacob et al. [194], Medugorac et al. [312], Rupp [487]. They compared the ATPase activity of myofibril myosin, the maximal velocity of isotonic shortening, and the maximal isometric tension of the myocardium with the isozyme spectrum of myosin, taking into account that there are three myosin isozymes: V_1, which possesses the highest ATPase activity; V_2, which has intermediate ATPase activity; and V_3, which possesses the lowest ATPase activity.

In the initial stage of research, suppression of the thyroid function in juvenile rats with propyl thiouracyl was found to result in a considerable decline of V_1 in the myocardium and in sharply predominant V_3. As a result, the ATPase activity of myosin and the velocity of shortening of glycerinized muscle fibers dropped by 30%–40%; conversely, the isometric tension developed by these fibers remained essentially unchanged. Thus, the V_1 content predetermines the myosin ATPase activity and the myocardial contraction velocity; however, it does not affect the developing tension.

During the second stage of research, it was shown that upon adaptation of the heart to filling load, namely when training animals by swimming, the V_1 content, the myosin ATPase activity, and the myocardial contraction velocity naturally increase; logically speaking, this should cause an increased amplitude of myocardial contraction and stroke volume, the increase being characteristic of the given type of adaptation. However, during adaptation to loads with resistance, namely during aortic coarctation and renal hypertension, the V_1 content was noted to decline and the V_3 to be prevalent. This resulted in a decrease of myosin ATPase activity and in a lower velocity of isotonic contraction of the cardiac fibers. The developing pressure of glycerinized fibers changes little and may even rise in an intact myocardium. The authors arrive at the conclusion that in this case the adaptive significance of a lower V_1 content, weaker myosin ATPase activity, and a lower contraction velocity is that the tension in the heart ventricle generally lasts longer so as to retain the stroke volume notwithstanding the increased resistance to expulsion.

The significance of such studies, which directly reveal the molecular mechanisms of adaptation, can hardly be overestimated. It is essentially important that adaptation primarily involves the accumulation of short-lived proteins and their isozymes. Indeed, the average half-life for the principal proteins of the heart myofibril myocytes is known to equal about 12 days [457] and for proteins that are components of the mitochondria respiratory chain 5–6 days [12]; for heavy chains of the myosin heads, which are carriers of ATPase activity, the half-life is shorter than for the light chains that inhibit the activity [598].

Correspondingly, in the CPK initial stage and during adaptation to physical loads, one naturally observes a per-heart-unit mass increase in the membrane proteins and Ca-activated SR ATPase [433, 525], an increase in the mitochondria mass and mitochondrial enzyme activity [343, 597], prevalence of H chains over L chains in the myosin heads and a concomitant increase in myofibril ATPase activity [585], and finally, increased CPK and its BB isozyme activity. This accumulation of short-lived structures provides the heart with an opportunity to adapt to a high and continuous load and thus to prevent death.

The concrete adaptive significance of increased CPK activity and the transformation of its isozyme spectrum are determined in the given situation by the fact that the CHH emergency stage initially gives rise to a delay in ATP resynthesis in the mitochondria after ATP consumption, and ATP resynthesis is restored at the expense of considerably increased respiration and oxidative phosphorylation in the mitochondria. In these conditions, the increased activity of CPK and its isozyme BB, which has increased affinity with CP and ADP [76, 540], provides a solution for the vitally important task of increasing the transport rate of energy-rich phosphate groups from the mitochondria to the myofibrils.

Thus, *a change in the ratio of transcription rates for definite genes and the resultant change in the ratio of cell structures become the basis for adaptation to protracted increased loads. The fact that this change in the ratio develops toward the accumulation of short-lived structures is apparently not a specific property of the heart, but a more general feature of the molecular mechanism of adaptation.* For example, 46 enzyme proteins from the liver tissue of 12 patients were found to possess the shortest half-lives and to form precisely either the initial or rate-limiting links of the major metabolic pathways. The enzymes possessing the longest half-lives never fulfilled this role [135].

In full accordance with this, the adaptive activation of the biosynthesis of enzymes of neoglucogenesis, which develops in the liver under the influence of a stress hormone, corticosterone, is achieved so as to result in the accumulation of short-lived isozymes, which can rapidly degrade following the termination of the stress situation [60, 392]. *As a result, we arrive at the fact that precisely these short-lived proteins are functionally the most important ones, limiting the function along with cell proteins.* The significance of this feature of the biologic machine is that the high biosynthesis and decomposition rates of these proteins make possible rapid changes in their content and, hence, changes in the functional possibilities of organs and systems in the course of adaptation to the environment.

Economical Functioning: Principal Characteristic of Adapted Systems

The development of behavioral adaptation responses is conspicuous for its first energy-wasting phase of generalization, which manifests itself in abundance of superfluous motions, mistakes, as well as there being a distinct vegetative component. Subsequently, as a result of strengthening of the system of temporary relationships constituting the basis of a developing skill, superfluous motions and mistakes disappear and the vegetative component of the response diminishes considerably and the desired result with maximum economy of the organism's resources is attained.

A simple analysis shows that in the successful adaptation to greatly differing environmental factors, highly economical functioning is a characteristic feature of the systems responsible for adaptation. This important feature is revealed equally clearly at the level of the cells and organs in which it is determined by correlating the cellular structures at the level of the whole system, where it is determined by a correlation of the organs. It can also be correlated at the level of neurohormonal control, where economy proves to be the outcome of the higher reactivity of adapted organs to mediators and hormones.

On the whole it has been shown that at the level of the heart as a whole organ, the hearts of humans adapted to manual work [172] or altitude hypoxia [404] consume about one-third less oxygen and oxidation substrates for an equal amount of work. The most probable cause of this increased oxygen utilization efficiency is the increased potency of membrane mechanisms responsible for Ca^{2+} transport, which was proven for the cells of an adapted heart. Accelerated removal from the sarcoplasma of Ca^{2+}, which enters it with every excitation cycle, means lesser Ca^{2+} absorption by the mitochondria, i.e., a decline of the alternative process to ATP resynthesis. As a result, the degree of respiration and phosphorylation grows to increase the ATP output per unit of utilized oxygen. As was indicated above, upon over- and deadaptation, the power of the membrane mechanisms of Ca^{2+} transport was reduced to create the converse result. The efficiency of heart oxygen utilization is substantially diminished at high compensatory hypertrophy and deadaptation due to hypokinesia. Thus, the correlation of cell structures largely determines the effectivity of energy conversion at the cellular level. During adaptation this leads to an economical functioning of organs; during over- or deadaptation the reverse is true.

An adapted system also becomes economical through decreased decomposition of the organism's structures under high load. Injury and decomposition of the structures naturally develop due to a lower ATP concentration, irrespective of what had caused the given shift. The injurious effect of a shortage of energy-rich phosphorous compounds is most distinctly evident in myocardial infarctions, with which an ischemic ATP-deficient myocardial necrosis is naturally accompanied by the entry into the blood of large numbers of enzymes, viz transaminases, creatine kinases, lactate dehydrogenases. The degree of this fermentemia correlates with the size of the necrosis to be used for assessing the size of the infarction and the efficiency of treatment. It was proved in full conformity with this that in the hypoxic injury of an isolated heart the degree of decline in ATP concentration predetermines the enzyme yield from injured cardiac cells; there is a strong correlation between the decrease in ATP concentration in the myocardium and the yield of enzymes to the perfusate [531].

When the organism is whole, ATP concentration in skeletal muscles naturally decreases under the effect of loads. In full conformity with this, Highman and Altland [174] established that the urea nitrogen concentration in the blood plasma of unadapted rats increased twice after 12 h of running and the activity of aldolase and transaminase cell enzymes increased two- to fivefold. Evidently, these enzymes could enter the plasma from dead or injured cells. Animals 'trained' for 1 month by running sessions showed no increase in nitrogen concentration and enzyme activity in the blood after 12 h loads. In their electron-microscopic study, Sarkisov and Vtyurina [498] showed that single 6 h swimming sessions involving untrained animals caused injury to the myocardial cells, namely swelling of the mitochondria, decomposition of crysts, sarcoplasmic edema, occasional decomposition of the sarcolemmal membrane, and swelling of SR elements. In animals 'trained' by swimming for 3–4 months, swimming loads of equal intensity caused no such injuries.

Evidently, the absence of fermentemia and structural injuries in the above situations was caused by the fact that preceeding adaptation based on the $G \rightleftarrows F$ interrelation led to a selective increase in the mass of mitochondria, concentration of myoglobin, and number of capillaries, i.e., to higher potency of the system of ATP resynthesis.

Thus, *adaptation enhances the potency of the ATP synthesis system to prevent ATP shortage and its harmful effect. This renders unnecessary* subsequent activation of nucleic acid and protein synthesis, which provides *intracellular regeneration* and *restoration of decomposed structures.* As an outcome, adaptation permits the organism to avoid a situation involving injury and regeneration and ensures economical utilization of its structural resources preventing its organs and systems from wearing. With de- and overadaptation, in which the potency of the ATP resynthesis system is decreased to create the opposite situation, in which ATP shortage, structural injury, and expenditure of the organism's structural resources may occur at relatively low loads. Thus, changed correlations of structures that develop during adaptation at the cell and tissue levels enhance the efficiency of ATP resynthesis, decrease oxygen consumption, and prevent ATP shortage under loads; as a result, the organism's energy and structural resources are utilized economically.

As was indicated above, at the systemic level the economizing effect of adaptation is determined by changes in the correlation between the organs that form a given system. For instance, it was shown that the increased potency of the mitochondrial system in trained humans would permit the tissues to extract an increased amount of oxygen from every liter of circulating blood. Sufficient oxygen supply to tissues under physical load is provided not only by increasing pulmonary ventilation and minute volume but also by a more complete liberation of circulating blood from oxygen. Some investigators long ago noted arterial hypoxemia in trained athletes subjected to loads [129, 241].

Rowel et al. [480] took blood samples directly from the aorta at the end of intensive 3 min on-the-spot running to show that distinct hypoxemia, expressed in 12% lesser saturation with oxygen of the arterial blood, was observed only in trained people, i.e., in those who responded to such moderate loads by less increased minute volume. Taken together, all this evidence leads to the idea that enhanced potency of the mitochondrial system and the correspondingly increased ability of the tissues to extract oxygen from the blood may provide a situation in which the organism's oxygen need could be satisfied at a lower minute volume and more economical mobilization of the contractile function of the heart.

When hypoxia acts on the organism, the latter first displays pronounced mobilization of external respiration and heart function. However, as adaptation develops, an increase takes place in the lung respiratory surface, the amount of blood in the lung capillaries, and the oxygen capacity; at the same time, arterial resistance declines (see Chap. 3). As a result of these changes, the necessary amount of oxygen is supplied to the blood and thence to the tissues, external respiration and the heart contractile function being increased insignificantly. *Thus, the changes developing in certain links of a system responsible for adaptation provide economical functioning of the other links and the system as a whole.*

At the level of neurohormonal control, economical functioning of the system responsible for adaptation is apparent in higher reactivity of the organs that form the given system so as to control signals, hormones, and mediators. This shift ensures a situation in which the system can be mobilized under environmental factors so that a lesser release of regulatory metabolites *occurs* with lower excitation of control mechanisms.

This phenomenon was studied in its most distinct form in the adaptation to hypoxia and physical loads. The hearts of hypoxia-adapted animals, both in whole organisms and in an isolated form, possess enhanced adrenoreactivity and respond to a standard dose of norepinephrine by more increased strength and contraction rate than the hearts of control animals. This increase in adrenoreactivity is accompanied by another important event, namely by a 2.5–3-fold increase in the disappearance rate of the positive inotropic effect of norepinephrine, i.e., in adapted animals the response to the control signal is more intense and shorter [330].

Biochemical studies established that at adaptation of activity that phenomenon is also possibly based on the presence of two enzymes of the sarcolemmic membrane of the myocardial cell, namely adenyl cyclase, which under the effect of norepinephrine increases cAMP synthesis, and phosphodiesterase, which degrades cAMP [371].

Figure 10 shows that after injection of norepinephrine, the cAMP concentration in the myocardium of hypoxia-adapted animals becomes steeper to reach a higher level due to increased adenyl cyclase activity and then drops more steeply owing to an increase in phosphodiesterase activity. From this type of evidence, we came to think that adaptation caused by increased potency of the adrenoceptor – adenyl cyclase system enhances the inotropic effect of catecholamines to make it shorter at the expense of higher phosphodiesterase activity. *In other words, selective increase of the mass and power of the membrane cell systems responsible for reception and transformation of the control signal led to greater signal efficiency and created a prerequisite for more economical functioning of the control system, specifically the adrenergic system.*

Further investigations showed that this economizing effect of adaptation on the functioning of control systems is not limited to enhancement of the adrenoreactivity of the executive organs. After complete heart desympathiza-

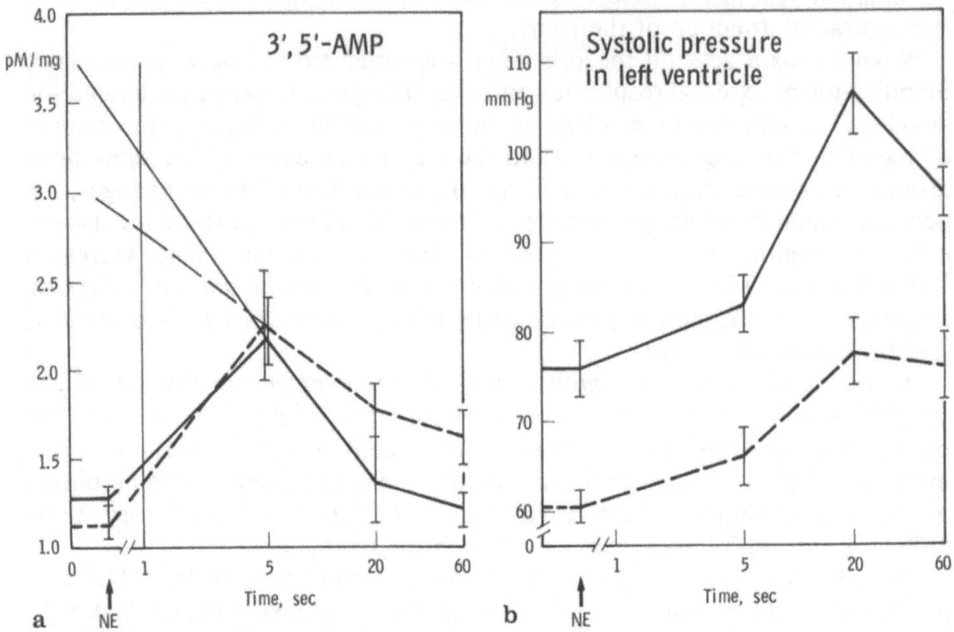

Fig. 10. Dynamics of cAMP concentration in myocardium **a** and of systolic pressure in left ventricle **b** under the influence of single noradrenalin injection (0.5 μg/kg) in control rats *(dashed bold lines)* and hypoxia-adapted rats *(continuous bold lines)*. **a** *Thin lines* show extrapolation of dynamics of cAMP concentration during first 5 s after injection of noradrenalin; **a, b** cAMP concentration in myocardium 1 s after injection of noradrenalin in adapted and control animals, respectively. *Abscissa axis,* time elapsed since noradrenalin *(NA)* injection, in logarithmic scale

tion in animals preadapted to hypoxia or physical loads, depression of its contractile function and its ability to endure maximal loads were found to have decreased much less than in unadapted animals [360]. Hence, it is quite probable that during adaptation to physical loads and altitude hypoxia the self-control mechanisms of myocardial cells are less dependent on the activating effect of adrenergic control, which is known to come about via cyclic 3′- and 5′-AMP.

When evaluating this phenomenon, one should keep in mind that in normal conditions cyclic 3′- and 5′-AMP potentiate the principal responses underlying the contractile function of the myocardial cell and are controlled by Ca^{2+} concentration in the sarcoplasm. Indeed, the increase in Ca^{2+} concentration occurring during the passage of the action potential activates protein kinases, which are structurally localized in the key points of the cell. This results in troponin phosphorylation, which predetermines systolic contraction [240], phosphorylation of phosphorylase B to phosphorylase A, which predetermines activation of glycolysis [403], and phosphorylation of a definite component of the SR membrane, which predetermines diastolic relaxation [493]. Cyclic AMP potentiates the Ca^{2+} effect via these protein kinases; hence, the formation of lesser amounts of cATP after pharmacologic desympathization reduces the rates of the above responses causing depression of the heart's contractile function.

In this case, adapted animals regain their heart function to a higher degree than control animals. Possibly, this was caused by lesser dependence of the calcium-controlled responses, namely protein kinases and protein kinase-phosphorylated components of the troponin complex, the phosphorylase system, SR, and cAMP. A hypothesis needing experimental verification implies that increasingly potent regulatory protein complexes, localized in the cell membrane and in other key points of the cell, constitute the foundation of enhanced efficiency of the self-regulatory mechanisms, this increase being actually revealed at denervation of the heart in animals adapted to physical loads and altitude hypoxia.

Now, if the postulated adaptive increase in the power of self-regulation mechanisms really does exist, then combined with the increased response of the executive organs to mediators it should provide diminished involvement of higher levels in the control hierarchy of an adapted organism, i.e., a lower load on its neuroendocrine mechanisms during intensive responses. The results obtained by many researchers show that this situation is observed in long-term stable adaptation to hypoxia, physical loads, and other factors.

For instance, Myles and Ducker [408] showed that in animals preadapted for 5 weeks to altitude hypoxia at 5,400 m an ascent to 8,000 m caused no increased release of norepinephrine and epinephrine into the blood, something normally observed in intact animals. According to Le Blanc [259], rat adaptation to cold prevents the sympathoadrenal system from acutely activating under intense hypoxia. Chin et al. [62] established that in rats adapted to moderate physical loads, a sharp effect of yet another factor, namely cold (for 3 h at 20 °C), caused lower release into the blood of catecholamines and corticosterone than in unadapted animals. This evidence agrees with the investigation results, which established that in animals adapted to physical loads the norepinephrine metabolism rate is decreased and its release in the myocardium is diminished under extreme effects against a background of its normal or increased content therein. Thus, in adapted animals the normal content of mediators of the sympathoadrenal system in the myocardium is combined with reduced metabolism and lesser centrogenic mobilization of these substances under extreme situations.

The fact that control mechanisms are less involved when adaptation already exists is not only evident in the catecholamine system; it is also revealed with equal certainity in relation to adrenal cortex hormones. Such results were obtained by Stolyarova [537], who revealed that after short-term swimming sessions the corticosterone content in the blood plasma and adrenals increased only in nontrained animals and remained unchanged in trained ones. Frenkl et al. [121] also noted a reduced response of the adrenal cortex to acute loads in humans and rats trained for physical loads. In their experiments, Ottovicz et al. [427] established that preliminary 5-week training on a treadmill stabilized adrenal response during maximum effort; in trained animals, increase of corticosterone and decrease of ascorbic acid in the adrenals was less than in controls.

Numerous experiments on rats involving the regular use of diverse acute effects, including muscular loads, showed that the functional activity of the hypophysial-adrenal system develops in two phases. The first phase continues for about 7–12 days and is characterized by more intens activity of the hypophysial-adrenal system, while the second phase, which forms after 1 month, is characterized by reduced response both to moderate and strong irritants [488].

In our view, the above evidence leads to the following important conclusion: Long-term adaptation is on the one hand characterized by increased power of self-regulatory mechanisms in individual systems and on the other by their increased reactivity to monitoring singals, namely mediators and hormones. As a result of this, equilibration of an adapted organism with the environment is achieved with lesser involvement of the higher levels in the control 'hierarchy' with more economical functioning of the neuroendocrine control of the systems responsible for adaptation.

It is significant that the above concerns an increase in the functioning efficiency of an adapted system responsible for adaptation, not of the organism as a whole. Indeed, so far there is no reason to think that the heart and its neural regulation would function more efficiently in humans or animals that had developed a system of habits unconnected with the mobilization of locomotive functions. Neither has it been proven that the most perfect adaptation to physical loads increases the efficiency and economical functioning of the mechanisms of higher nervous activity. Thus, adaptive change in structural correlation, increased efficiency of functioning, and economical utilization of the organism's structural resources characterize precisely that system responsible for adaptation, the system in which the systemic structural trace was realized.

System Responsible for Adaptation as the Organism's Dominant System

Thus far, the subject of scrutiny has involved the changes that develop inside the system responsible for adaptation, principal attention having been focused on structural changes, which while gradually accumulating therein increase its potency to provide the whole organism with new qualities, namely resistance to specific environmental factor and the possibility to solve definite problems. Now it should be emphasized that adaptation is not a simple outcome of the accumulation of structural changes within specific functional systems. This more complex phenomenon represents also the result of a change in the correlations between various systems.

In the course of adaptation, intersystemic relations are known to change usually in accordance with the dominant principle discovered by Ukhtomsky [553] who regarded the dominant as a constellation of excited centers, which intensifies its excitation at the expense of any environmental irritants, inhibits the excitation of all other centers, and, by subjugating definite executive organs, comes out as an integral system that determines the organism's vector, its behavior in conditions of existing realities. Development of the dominant represents a process with a pronounced struggle and unity of opposites. Indeed, in an intact organism, excitation and intense functioning of the dominant system inhibit the activity of other systems. These two opposite states are in unity since, first of all, they are interconnected by cause-and-effect relationships and, second, can only jointly concentrate all the resources of the organism for fulfilling a specific task that had arisen during adaptation.

The evidence obtained over the past decade permits us to augment our ideas about the role of the dominant in the mechanism of long-term adaptation and to assess its significance in forming the animal and human phenotypes. In this case, the above concept on the $F \rightleftarrows G$ interrelationship in the organism's differentiated cells acquires substantial significance. In accordance with this concept, an increase in the function of the dominant system naturally entails activation of nucleic acid and protein synthesis in the cells that form a given system. Inhibition of the functions of other systems entails a decline in nucleic acid and protein synthesis in the cells. In other words, it appears possible to provide structurally the dominant systems preferentially at the expense of other systems.

This structural equivalent of the dominant should develop most rapidly at the level of the higher regions of the nervous systems, for the neuron is the cell wherein the correlation between function and the genetic apparatus occurs most urgently also on a wide scale (see Chap. 4). It has been proven that as early as in the immediate seconds of excitation, the neuron may develop a certain shortage of energy-rich phosphorous compounds, mobilize glycolysis, and accumulate lactic acid; at the same time, proteolytic enzymes become activated and proteins begin increasingly to decompose. This is followed by activation of nucleic acid and protein synthesis. Due to this, repeated and protracted increase in the functions of highly different neurons may lead to growth in their cell mass, to pronounced cell hypertrophy.

The biosynthesis activation product (the proteins forming in the nucleus and body of the nerve cell, specifically protein enzymes of mediator metabolism) migrate by means of an acsoplasmatic current to the neuron processes; biosynthesis is activated in the processes themselves as well, and also in the synaptic structures that provide interneuron contacts. As a result, activation of biosynthesis becomes the factor that forms and expands interneuronal synaptic relations, the basis of structural consolidation of multineuron systems forming in the brain during adaptation to the environment. These data signify that the prevalence in the brain of definite multineuron systems, a factor apparently realized at the expense of other systems, corresponds to stationary or periodically recurring dominant responses in the organism's behavior.

The possibility of specific dominants forming in the animal and human brain is genetically determined. Yet, the question as to whether this possibility is utilized or not is decided in the course of individual development during contact with the environment and adaptation with it. The development during adaptation of structurally secure and, quite often, highly stable dominant systems is associated with inhibited development of other systems and, in our view, is an important mechanism of phenotypic formation.

This regularity is most distinctly evident in man's individual development. The uniqueness of the process is that the human being develops in society. Correspondingly, beginning from the very first days of development, man's adaptation is not only to environment but also to the existing social medium, this adaptation being specifically inherent in humans.

Facts show that if a child in his 1st or 2nd year is removed from the surrounding social medium and reared among animals, *Homo sapiens* in the full meaning of the term does not develop. Children who had lived among wolves for several years possessed a whole system of skills inherent in the latter: They jumped on all fours, hunted, preyed upon, and devoured live animals. On the other hand, these creatures could not walk erectly; neither could they speak nor do other things inherent in man. Even several years of special education among people did not help to develop fully human traits and eliminate wolfish habits. Hence, the dominance of wolfish habits and inhibition of human habits were highly stable and apparently structurally secured in the brain. In essence, these facts signify that only life among people, i.e., social adaptation makes it possible to realize genetically determined human features.

Social adaptation is a decisive factor in the development of the human phenotype, the decisive factor in the forming of modern man. Hence, in our view, Dubinin was right in pointing out that "A nascent human is a specific form of animal capable of becoming a man; the former's peculiarity is that natural selection had in its time 'recorded' in his genes the features of physical and mental organization without which an intelligent human cannot form" [90, pp. 103–104]. From this, one may add that the transformation of a 'specific form of animal' into man takes place in the course of social adaptation and, apparently, through preferentially structural systems that dominante in the course of adaptation with the simultaneous inhibition of other systems.

The dominant is a constellation of excited nerve centers, which subjugates executive organs taking part in a given behavioral response. This means that the contradictory relations developing in the course of adaptation between various systems, viz preferentially structural provision of dominant systems and inhibition of other systems, should be evident during individual development not only at the brain level but also at the level of executive organs. Indeed, recent years have witnessed the appearance of data that show how this preferentially structural provision of dominant systems and inhibition of other systems are concretely developed at the level of executive organs. It turned out that when young growing animals trained for physical loads the hypertrophy of myocardial cells they usually experience was replaced by cell division (hyperplasia); as a result, the total number of myocytes proves increased by over one-third. The structural reserve of the heart, which plays a major role in adaptation to physical loads, essentially increases. At the same time, the opposite changes are observed in the kidneys, adrenals, and liver. The number of nephrons, the principal functional units of the kidney, decreases by 25%, and the number of cells in the adrenals and liver by 20%–30%. The structural reserve of these vitally important organs is apparently reduced [36].

This fact, the biologic significance of which is not discussed by the above authors, led us to the idea that in early ontogenesis the influence of adaptation may result in a redistribution of the cell pool between the systems of the organism. Since the DNA content in one genome (in one diploid set of the chromosomes of a given species) is strictly constant, it appeared advisable to verify this assumption by measuring the DNA content in the principal animal organs in the course of quite tense long-term adaptation to a definite environmental factor. The changes in DNA content that occur in this case in various organs would probably reflect changes in the number of diploid chromosomal sets or in the number of conventional diploid cells.

Basing our work on this approach, we studied the effect of adaptation to considerable altitude hypoxia the DNA content in rat internal organs. The work was performed jointly with Durova, a researcher from the Tadzhik Medical Institute. The rats together with feeding females were lifted to an altitude of 3,400 m 1 day after birth (series 1); adult rats were lifted to high altitude 45 days after birth (series 2). Both series were kept there for 45 days before being killed. The DNA concentration in the organs was determined by the method of Tsanev and Markov [550] modified by Galkin and Berdyshev [127], and the DNA content was calculated as the product of the concentration by the organ weight.

The results of these experiments are published for the first time below and evidence that 45-day adaptation of newborn animals to high-mountain conditions, i.e., primarily to altitude hypoxia, did not essentially affect growth: The weight of control rats, like the weight of adapted rats, equalled about 80 g. However, the weight of internal organs and the DNA concentration therein substantially changed. As a result, the total DNA content in adapted animals increased by almost 25%; in some organs, the DNA content and, hence, the number of genomes, substantially increased, while in others it decreased. These data are shown in Table 5.

Table 5. Effect of adaptation to hypoxia in newborn animals upon DNA content in internal organs

Organ	DNA content, μg		p	DNA content, % of total	
	Control (n = 20)	Adaptation (n = 20)		Control	Adaptation
Large hemispheres	41.7 ± 1.2	33.8 ± 0.8	< 0.001	1.83	1.17
Cerebellum	81.9 ± 4.8	62.2 ± 2.2	< 0.001	3.60	2.20
Thymus	382.9 ± 13.2	553.0 ± 20.4	< 0.001	16.90	19.60
Thyroid gland	3.9 ± 0.8	3.9 ± 0.5	–	0.17	0.14
Heart	63.7 ± 3.2	90.5 ± 3.5	< 0.001	2.80	3.2
Lungs	383.6 ± 12.3	467.0 ± 20.5	< 0.001	16.90	16.50
Liver	973.9 ± 38.6	1,293.0 ± 43.3	< 0.001	42.80	45.80
Kidneys	183.4 ± 8.4	139.4 ± 11.9	< 0.01	8.04	4.90
Adrenals	8.7 ± 0.8	12.0 ± 2.1	> 0.2	0.38	0.40
Testes	152.4 ± 2.0	170.6 ± 3.1	< 0.01	6.70	6.00
Total	2,275.7	2,826.1		100	100

The data in Table 5 indicate that the DNA content increased most pronouncedly in the organs that constitute the system that dominates during adaptation to hypoxia, i.e., the system in which the structural trace forms. For example, the DNA content in the heart is increased by 42%, in the adrenals by 38%, and in the lungs by 16%. It is also sharply increased in the liver, which synthesizes the blood albumin and can play a role in hemopoiesis, as well as in the thymus, one of the principal organs of immunogenesis, whose role in adaptation to hypoxia is still obscure. Conversely, in organs whose hyperfunction is not a vitally important link in the adaptive response to oxygen shortage (namely, in the large hemispheres of the brain, the cerebellum, and the kidneys), the DNA content and correspondingly the cell genome content is decreased by 23%–26%. With regard to the kidneys, it should be specially noted that hypoxia and adaptation thereto were characterized by diuretic and natriuretic effects; these effects are caused by a decrease in the water and sodium readsorption in the renal convoluted tubules and are examined below. The decline in the function of very important nephron regions probably causes the relative atrophy of the kidney parenchyma and the decreased number of cells there.

The effect of adaptation on distribution of cell genomes between various organs is definitely confirmed by the figures in the last two columns in Table 5, where the total DNA content in all the internal organs is taken to be 100% to show what percentage of DNA is contained in each organ. From Table 5, it is apparent that adaptation to hypoxia led to a certain increase in the DNA located in the heart and liver and abruptly reduced the DNA portion contained in the large hemispheres of the brain and in the kidneys.

Further experiments showed the effect of hypoxia on genome distribution between organs and systems to depend on what ontogenetic stage the organism was compelled to adapt to altitude hypoxia. Table 6 shows data on the effect of adaptation to altitude on the DNA content in the organs of animals that were lifted to 3,400 m when 45 days old and, like the rats in the above series, were kept there for 45 days. Their total body weight by the end of adaptation proved to be 10% less than the original one and the DNA concentration in many internal organs was considerably increased; as a result, the total DNA content in internal organs grew by 25%. At the same time, the DNA content in the heart and liver increased much more significantly, namely by 35% and 48%, respectively, while the DNA content in the large cerebral hemispheres, thyroid gland, and kidneys conversely decreased by 35%, 33%, and 14%, respectively. The data in the last two columns in Table 6 confirm the effect of adaptation to hypoxia upon DNA distribution between organs and show that the DNA portion contained usually in the liver and heart had somewhat increased under the influence of adaptation, conversely, in the cerebral hemispheres and kidneys it had decreased from 2 to 1.5-fold.

The proven adaptive redistribution of cell genomes between the organism's systems is obviously a quite important phenomenon. A possible explanation may be that the total number of cell genomes forming in the organism and, correspondingly, the total cell population are not infinite, but in a way

Table 6. Effect of adaptation to hypoxia in adult animals on DNA content in internal organs

Organ	DNA content, μg		p	DNA content, % of total	
	Control (n = 20)	Adaptation (n = 20)		Control	Adaptation
Large hemispheres	68.1 ± 2.3	43.9 ± 1.4	< 0.001	1.78	0.91
Cerebellum	70.3 ± 2.8	87.8 ± 5.2	< 0.01	1.84	1.83
Thymus	485.1 ± 22.1	529.0 ± 12.8	> 0.05	12.70	11.00
Thyroid gland	7.7 ± 1.2	5.1 ± 0.6	< 0.05	0.20	0.11
Heart	84.8 ± 2.3	115.0 ± 1.9	< 0.001	2.12	2.40
Lungs	632.5 ± 18.5	657.7 ± 34.2	> 0.1	16.50	13.70
Liver	1,920.0 ± 52.5	2,852.4 ± 10.0	< 0.001	50.20	59.50
Kidneys	268.9 ± 12.2	230.7 ± 9.8	< 0.01	7.03	4.80
Adrenals	11.2 ± 0.9	13.9 ± 1.2	> 0.05	0.29	0.29
Testes	276.0 ± 14.4	258.6 ± 5.9	< 0.1	7.20	5.40
Total	3,832.0	4,794.7		100	100

limited. Upon adaptation, this limited cell pool is distributed under the dominant systems; the latter obtain more genomes at the expense of other systems. For example, as a result of adaptation to physical loads in early ontogenesis, the cell population of the heart, lungs, and skeletal muscles proves to be increased, but the liver and kidney cell population conversely decreased. Upon adaptation in early ontogenesis to considerable hypoxia, the cell populations of the heart, lungs, liver, and thymus increased, but those of the brain hemispheres and kidneys decreased. One can imagine that in the early assimilation of complex skills, the cell populations in the brain systems responsible for those skills would be increased, while those in other systems would decrease. In other words, the structural reserves and functional potencies of dominant systems irreversibly increase, while the structural reserves and functional potencies of inhibited systems irrevocably decrease.

The idea of the role of preferential structural supply to the dominant systems upon adaptation in early individual development is in good agreement with the familiar assumption that this period is very suitable for obtaining high results in teaching and training animals. At the same time, it permits one to emphasize the relative advisability of adaptation that may be revealed in a later stage of life, when, due to changes in environmental conditions, the organism is compelled to adapt to absolutely different factors so that the systems whose cell populations are decreased may prove dominant, i.e., responsible for adaptation. In this case, the organism's resistance to acting environmental factors is diminished, and this may become the cause of vital failures or diseases.

Thus, exceedingly tense adaptation to a given factor, when taking place successfully for a long time, nonetheless has a high 'structural cost' and involves at least two potential dangers. First, the possibility of functional exhaustion of the system's dominant in the adaptive response, something already mentioned above and, second, decrease in the structural and, correspondingly, functional reserves of other systems that do not directly participate in the adaptive response and prove to be detrained.

A natural measure that could prevent both these events would be for the organism to use behavioral responses to limit the intensity and duration of each environmental factor and simultaneously to increase the diversity of those factors. This transition from unilateral adaptation and unilateral development of the phenotype to combined adaptation and relatively harmonious development essentially means to account for environmental requirements not only at a given time interval but also in the future; this transition implies a more promising version of adaptation 'from the viewpoint of the organism.'

In real conditions, change of dominants determining the behavior vector is normally predetermined by changes in the environment; for protracted factors and long-term adaptation it means the substitution of one system responsible for adaptation for another. This makes it possible for the organism to adapt itself successively to numerous changes in the habitant and to many external factor arising in the course of life. Naturally, the organism can adapt simultaneously to many external factors with corresponding specialized responses in time. However, its adaptive possibilities are not infinite. Hence, an important prerequisite for new long-term adaptive responses in ontogenesis would be to eliminate, either partially or fully, previously formed adaptation, the deadaptation examined below.

Reversibility of Adaptation: Physiologic and Pathologic Deadaptation

The possibility for long-term adaptation to develop is evident in a gradual disappearance of adaptation to physical loads, hypoxia, and chemical factors after these factors have ceased to act. This has been shown to involve a rather rapid disappearance of the systemic structural trace that constitutes the basis of each of these adaptation versions, viz a decrease in the skeletal muscle mass and number of mitochondria therein, a decrease in the lung and myocardium mass after the termination of physical loads, a decrease in the right ventricular mass and a gradual disappearance of polycytemia after the organism has emerged from hypoxia, and a decreased activity of the microsomal oxidation system and a decrease in the liver mass after poisons have ceased to be present.

The mechanism of this phenomenon involves an abrupt decline in RNA synthesis and a decrease in the number of polysomes, this taking place directly after an abrupt decrease of loads on any of the organs forming the functional system responsible for adaptation. As a result, protein synthesis in the organ declines and begins to lag behind its decomposition rate, causing the organ mass to decrease.

After our experiments involving 45-day adaptation in an altitude chamber adjusted to an altitude of 7,000 m, the rats' ventricular weight was found to be 42% higher and protein synthesis in the heart by the inclusions of labeled amino acids was 20% higher. Three days after hypoxic effects were terminated, the ventricular weight remained increased by 40% and the intensity of protein synthesis decreased to control level. On days 7–15 of deadaptation, the ventricular weight was increased by 35% and 10%, respectively, while protein synthesis in the heart was reduced compared with controls by 40% and 30%, respectively. On day 30 of deadaptation, the ventricular weight and protein synthesis in the heart were at control level [317].

In another study performed in our laboratory, the right ventricular mass and the RNA content in the myocardium of that ventricle were determined in rats at different periods following the termination of a 20-day altitude chamber adaptation at 6,000 m. After completing a course of effects, the right ventricle was found to be hypertrophied and its RNA content increased. By day 10 of deadaptation, hypertrophy decreased by 54% and RNA adaptive growth by 57%. After 20 days, 81% of the hypertrophy was lost, and the RNA content returned to normal. Forty days after hypoxic effects were stopped, the right ventricular mass and its RNA content did not differ from controls [230]. The investigations of other laboratories have revealed approximately the same rate for the inverse development of the systemic structural trace in adaptation.

For example, in their experiments, Cutiletta et al. [72] caused hypertrophy of the left ventricle of rats by narrowing the ascending aorta. Twenty-eight days after the operation, the left ventricular mass and its RNA content had increased by about 50%. After that, the heart hyperfunction was abruptly terminated by removing the aorta-narrowing ring. As a result, the ventricular mass and its RNA content sharply decreased in the subsequent 3 days; they then decreased more slowly, and in 20 days were the same as in the controls. This kind of reversibility has now been proven to exist also after protracted compensatory adaptations in humans. For example, strong hypertrophy of the left ventricle of the human heart caused by failure of the aortal valves completely disappears after eliminating the defect by transplantation of artificial valves. This was seen in a decrease in the ventricular mass by about 90 g and in the elimination of the contractile function [239].

These facts indicate that the decline in nucleic acid and protein synthesis plays an important role in the inverse development of the adaptive structural trace. The possible involvement in this process of mechanisms responsible for degradation of cell structures, e.g., of lysosomal enzymes or lipid peroxidation products is presently being investigated. With regard to the heart, it has been shown that inverse development of hypertrophy caused by narrowing of the aorta is not accompanied by the activation of lysosomal enzymes, whereas inverse development of hypertrophy caused by introduction of thyroxine is.

For another adaptive structural trace, namely the increased potency of the microsomal oxidation system arising upon the introduction of poison into the organism, it was shown that the peroxidation of membrane lipids plays a role in its inverse development. For example, the increased power of the system of microsomal oxidation and the higher content of cytochrome P-450 in the liver induced by 20-methylcholonthrene undergoes inverse development after the administration of this carcinogenic substance is stopped. In the course of inverse development, i.e., the decomposition of excess cytochrome

P-450 and dismantling of the SR membranes that had become superfluous, and important part is played by the activation of membrane lipid peroxidation and the effect of lipid hydroperoxides on these structures [29, 198]. Thus, reduced intensity of protein synthesis and, possibly, activation of the structure degradation mechanisms lead rather quickly to the inverse development of the adaptive trace.

Salganic and his collaborators [392, 60] obtained evidence that indicated that apart from the above-mentioned factors a genetically determined instability of the proteins forming during adaptation against proteolytic enzymes may play an important role in eliminating the adaptive structural trace. For example, during stress or under the influence of glucocorticoids administered from outside, the liver develops activated synthesis of enzymes of gluconeogenesis which provide for the conversion of the organism's structural resources into energy resources, something expedient in extreme conditions. After the extreme situation has elapsed, this conversion becomes superfluous to say the least and, moreover, hampers the restoration of lost structures and the development of reparative processes. In confirmity with this, a study of tyrosine transferase, an enzyme of neoglucogenesis, revealed that during the first 5 h after administration of hydrocortisone its activity grew tenfold and succeeded in declining again to normal in the subsequent 10–15 h. Tyrosine transferase is represented in the liver by the so-called A isozymes, which in electrophoresis shift to the anode, and by the K isozymes, which under the same conditions shift to the cathode. These isozymes are coded for in various cistrons of the liver cell genome. The cistrons that code exclusively A isozymes are activated under the effect of glucocorticoids; correspondingly, adaptive increase of A isozyme biosynthesis becomes the basis for generally increased tyrosine transferase activity during stress. This isozyme was found to be incommensurably less stable to the effects of proteolytic enzymes than isozyme K; it rapidly decomposes, and precisely for this reason tyrosine transferase activity rapidly decreases to return in time to its initial level. A similar situation was proven for adaptive insulin-induced activation of the biosynthesis of hexokinase and other enzymes.

The question, therefore, concerns general regularity in which the organism genetically determines not only a possibility for structural change, which constitutes the basis of adaptation, but also for a timely inverse development of those changes.

In terms of our presentation, it is significant that *there are at least three types of molecular mechanisms responsible for inverse development of the systemic structural trace, namely decline of RNA and protein synthesis, activation of specialized mechanisms of structural degradation, and a genetically determined instability to natural degradation mechanisms of structures that had formed during adaptation.* The existence of various more or less mutually duplicating mechanisms, which 'erase' the systemic structural trace and provide inverse development of adaptation, is, in our view, indicative of the major biologic significance of this process. Indeed, inverse development of adaptation, or deadaptation, expresses the organism's remarkable ability to eliminate nonutilized structures. This is an essential prerequisite for utilizing released structural resources in other systems and, therefore, a prerequisite for transitions from one adaptation to another under environmental effects.

Thus, in regarding deadaptation as a physiologic process and its high rate as a prerequisite for new long-term adaptive responses, one should nonetheless keep in mind that the organism of higher animals is evolutionally determined to live in relatively constant media, to which it gradually adapts. Probably for that reason, the deadaptation rate is only relatively high compared with the status quo; yet, at the same time it is less than the adaptation rate. This important factor is the basis for accumulation of adaptive changes under periodic and, occasionally, not too frequent environmental effects, the traces of each of those effects disappearing incompletely and, when taken together, constituting the foundation of gradual adaptation. This aspect was the subject of a special study [322].

It should be emphasized here that disappearance of the adaptive structural trace is slower in its final stage than in its initial stage; adaptive traces persist in a concealed form for quite a long time, particularly with higher long-term adaptive responses based on brain memory, a region well known for 'revival' of long nonutilised skills, 'forgotten' foreign languages, and so on. Compared with traces of simpler adaptive responses, the higher stability of adaptive structural traces in the brain is specifically evident when after physical loads have been terminated the organism's aerobic potency and corresponding endurance gradually disappear in the course of deadaptation and special skills, viz the ability to fulfil a specific exercise, are retained for quite a long time and, not infrequently, may be demonstrated by people no longer trained.

The question as to how long the systemic structural trace can be retained with relatively simple long-term adaptations based on preneural memory (see Chap. 4) is being debated in the literature. According to Motorin [406], as a result of 14-day adaptation to mountains at 2,500 m, the changes in the respiratory system of athletes underwent completely inverse development for 45 days at low altitudes; after staying for 30 days in the mountains, adaptive changes also continued to be apparent on day 45 of deadaptation. Turkmenov and Imankulov [551] reported that people subjected to 40-day adaptation at 4,000 m retained increased physical efficiency and useful hemodynamic changes for 60 days. Improved execution of physical loads on a cycling ergometer following 30-day adaptation in mountains at 2,100–4,700 m was recorded on day 8 of deadaptation and was absent during the next test on day 90 of deadaptation [235].

According to Mirrakhimov et al. [401], the maximum endurable degree of acute hypoxia, the so-called 'altitude ceiling,' continued to be considerably elevated in rats after 15- and 30-day adaptation in mountains at 3,200 m for 30 days of subsequent deadaptation, and for 60 days after 45- and 60-day adaptation at 4,000 m. The elevated 'altitude ceiling' in rats resulting from 30-day adaptation in an altitude chamber developed inversely in less than 30 days [233]. These authors [233] developed adaptation to altitude hypoxia at various heights and during different periods; however, all the evidence appears to indicate quite lengthy (30–60 days) retention of useful changes acquired by an adapted organism. Other studies cite shorter retention times for adaptive changes during deadaptation.

After 14-day adaptation in an altitude chamber at 7,00 m, the adaptive increase in mouse, rat, and guinea pig survival following considerable overloads in a centrifuge was retained for 10 days after the end of adaptation, and was absent during tests on days 11 and 18 [560]. An increase in the 'altitude ceiling' of mice, achieved after 30-day adaptation in mountains at 2,000–3,800 m, disappeared on days 12–14 of deadaptation; after mice were trained for 14 days in an altitude chamber at an 'altitude' of 3,000–7,000 m, the same effect disappeared on day 18 [561]. In this case, increased erythrocyte and hemoglobin contents were retained for only 7–9 days.

Similar evidence is cited by Loginova [281], who found that 30-day adaptation in mountains at 2,000–3,800 m resulted in a considerably elevated 'altitude ceiling' being retained in mice and rats for 10–12 and 7–12 days, respectively. Adaptive increase in the erythrocyte number and hemoglobin content returned to normal in 7 days of deadaptation.

Korolkov et al. [234] showed that 1 day after 30-day adaptation of rats to an altitude chamber at 3,000–7,000 m, the animals were found to have a higher resistance to acute hypoxia. In this case, acute hypoxia led to the accumulation of less lactate and the retention of a higher CP content in the brain of adapted animals than in controls. Ten days after the end of adaptation, lactate accumulation and decreased CP content in response to acute hypoxic effects were the same in preadapted animals as in controls; adaptive changes were no longer observed in the blood.

Naturally, the different results obtained in these works largely depended on the different adaptation levels and criteria for assessing both adaptation and deadaptation. Yet, despite this fact, one may apparently note at least three points: *first, inverse development of adaptation based on elementary, preneural memory is more rapid than the inverse development of higher adaptive responses based on brain memory; second, after lengthy and stable adaptation, deadaptation develops more slowly than after short-term adaptation; and, third, the systemic structural trace disappears at different times in various links of the system responsible for adaptation.*

Disappearance of the adaptive structural trace and of adaptation itself with the return to a conventional norm corresponding to a given species and ontogenetic period cannot essentially be designated other than as *physiologic* deadaptation. This deadaptation should be distinguished from a situation caused by specific conditions of hypokinesia and zero gravity, with which the process in a normal organism begins to lead to unusually reduced mass and potency of the key cell structure, reduced against the efficiency norms of whole functional systems. This deadaptation is an obvious prerequisite of numerous diseases and may apparently be identified as *pathologic* deadaptation. Pathologic deadaptation develops on the basis of the same molecular mechanisms as physiologic deadaptation; however, quantitatively its result substantially reduces the functional possibilities of the organs and systems. This has been shown especially convincingly for deadaptation to physical loads owing to lengthy hypokinesia.

In accordance with our concept on the decisive role of the F⇄G interrelation in the mechanisms of adaptation and deadaptation, the essence of the events that take place during hypokinesia is that decreased loads on the functional system responsible for motor responses, specifically on the heart and skeletal muscles, reduce the functions of the cells forming that system. In response to that, the decrease in the activity of different transcriptons of the cell's genetic apparatus is far from being the same. Most reduced is the activity of the transcriptons that code for short-lived proteins; this results in selective atrophy of short-lived membrane cell structures responsible for the reception of mediators and hormones, and for ion transport and energy supply. Since these structures constitute only a small portion of the cell mass, but are essential links in the vital activity of the cell, hypokinesia at a certain stage of development may be characterized by a relatively slight decrease in mass and by an abrupt decline in the functional possibilities of the heart and skeletal muscles. In other words, if the above theory is correct deadaptation should involve: Relatively minor atrophy of the myocardium and the skeletal muscles; abrupt decrease in the masses of the sarcolemmic, sarcoplasmic and, to a lesser extent, mitochondrial membranes in the myocytes; and considerably reduced functional possibilities of the muscles. This corresponds to the basic concept of this book that *adaptive increase in the potency of organs and systems may really be provided by a genetically determined increase in the mass of preferentially short-lived cellular structures, the decline in the potency of organs and systems with the above-mentioned pathologic deadaptation being based on the opposite situation involving selective atrophy of short-lived membrane cellular structures.*

One should keep in mind that not only deep a fortiori pathologic adaptation can be a prerequisite of a disease. Situations may arise when physiologic deadaptation assumes a definite role in developing pathologic states. One such situation is observed when the organism, having been adapted throughout all its life to a specific factor or set of factors, at the late stage of ontogenesis changes its living conditions and loses the systemic structural trace constituting the basis of adaptation. In an effort to return to its previous living conditions, i.e., to realize adaptation, it proves essential to restore the previous systemic structural trace by activation of nucleic acid and protein synthesis. However, in conditions of aging, the activity of the genetic apparatus of differentiated cells and, correspondingly, the renovation of structures is diminished; the organism cannot build all the necessary structures, for the 'structural cost' of adaptation proves too high.

The same type of situation apparently underlies many cases of chronic high altitude sickness, which develops in elderly residents of the Peruvian Andes when they return to an altitude of 4,000 m after living in the lowlands. This sickness, which Hurtado [186, 187] identified as the loss of natural adaptation by people who were born and had always lived in mountains, is revealed in reduced external respiration (hypoventilation) and in compensatory mobilization of hemopoiesis and circulation. When studying the mechanism of hypoventilation, Hurtado found that patients affected by chronic mountain disease had reduced hyperventilation, arising in response to injections of carbonic acid, i.e., they had diminished sensitivity of the respiratory center to carbon dioxide. From this, he assumed that wear of respiratory center neurons constitutes the principal mechanism of the chronic mountain disease. Upon adaptation to hypoxia, the respiratory center performs moderate, but continuous, hyperfunction and becomes the object of hypoxemia. Both these factors, via energy shortage, may activate the neuronal genetic apparatus and cause nucleic acid and protein synthesis to become more active. This increased synthesis is apparently not infinite, since reactivation of the genetic apparatus of highly variable differentiated cells, including neurons, may end in the exhaustion of their ability to synthesize RNA and proteins [162, 189]. As indicated above, this local wear of differentiated cells achieving lengthy hyperfunction constitutes the basis of a number of diseases.

One can imagine that this depreciation is most probable when the genetic apparatus activation essential to restore the basis of adaptation recurs repeatedly under aging. Thus, loss of long-term adaptation, which might possibly be needed in the future, presents a major risk particularly in old age. *Support of the structural foundations of the adaptation needed in future through relatively low periodic loads on responsible functional systems is much more economical than repeated deadaptation–readaptation cycles, each such cycle being characterized by quite high structural cost.*

To sum up, it must be emphasized that deadaptation expresses the reversibility of any long-term adaptation and constitutes the necessary prerequisite for forming new adaptive responses and changing the whole phenotype in accordance with environmental requirements. This process is provided by reducing the systemic structural trace of adaptation by decreasing nucleic acid and protein synthesis on the basis of the $F \rightleftarrows G$ interrelation mechanism in the cells that form the functional system responsible for adaptation, and also at the expense of the above-cited molecular mechanisms of accelerated structural degradation. In this case, the short-lived cell membranous structures, whose potencies constitute the essence of adaptation and adaptive increase of functional possibilities, atrophy earliest and to the greatest extent. With physiologic deadaptation, the process provides a return from adaptation to a certain conventional norm, and with further progress it leads to the selective atrophy of the key cell structures, to deeply impaired functions, and may be identified as pathologic deadaptation. Thus, like adaptation, deadaptation has only relative biologic expediency.

Chapter 2
Role of Stress Reaction in Urgent and Long-term Adaptation

> "Nature is simple and is not rich with excessive causes of things."
>
> Newton

Urgent and long-term adaptive responses involve decisive participation of the neurohormonal mechanism of the whole organism. In its most distinctive form, this participation is seen in the form of a stress reaction. The stress reaction was originally discovered by Selye [506], who with remarkable insight identified it as a general 'adaptive syndrome.' This term is essentially based on the fact that stress does not simply preceed adaptation to a new factor or environmental situation, but plays an effective role in adaptation, namely in forming a functional system effectively responding to a given factor as well as a structural trace enhancing the system's power. At the same time, an effective functional system with a systemic structural trace are known not to form in so-called hopeless situations, when the factor acting on the organism is too strong or when the situation in the environment is so complex as to render the required adaptive response unfeasible. As a result, initial homeostatic disorders persist, and the stress response stimulated as a result becomes exceedingly intensive and prolonged. Precisely in this situation, the response may turn from a link of adaptation into a link of pathogenesis and cause numerous stress diseases, from ulcerous lesions of the stomach and serious injuries of the heart to blastomatous growth.

From this general concept of the role of stress response in adaptation to disease, this chapter will examine the principal positive effects of stress response – its contribution to the development of urgent and long-term adaptation – and only chart the possibility of transforming each of those effects into an injurious factor in so-called stress diseases, whose pathogenesis and possible prevention are discussed in the chapters that follow.

Stress Mobilization and Redistribution of Resources

The general adaptational syndrome is known to consist chiefly of considerable excitation of the higher vegetative centers and, as a result, of the adrenergic and hypophysial-adrenal systems. This causes the effect of high concentrations of catecholamines and glucocorticoids, both these factors possessing a high action range whose principal feature is to mobilize the energy and structural resources of the organism. Indeed, catecholamines naturally increase the cardiac output and mobilize the glycogen reserve of the liver and induce hyperglycemia, lipolysis, and increased fatty acid content in the blood. This results in an enhanced flow of oxygen and oxidation substrates to the tissues. Acting on a genetic level, glucocorticoids activate gluconeogenesis and transamination and, hence, conversion of amino acids into glucose, i.e., the structural reserve of the organism into its energy reserve. Generally speaking, a stress induced by any environmental factor outwardly appears to be a generalized response of mobilization involving the entire organism.

However, an elementary analysis shows that the role of stress in adaptation is not restricted to simple mobilization of structural and energy resources. There are at least three stress effects that exceed the limits of this mobilization. The first effect is that mobilization of energy resources during stress is really coupled with their nonuniform distribution. For example, the state of readiness to struggle and, particularly, of the struggle itself are known to be characterized in animals not only by generally mobilized respiration and circulation, but also by the considerable constriction of vessels in the organs of the abdominal cavity and of nonactive muscles in the simultaneous dilation of active muscle vessels [1, 556]. This fact unambiguously shows that the excess oxygen, glucose, and fatty acids resulting from the mobilization response is selectively directed to the systems that perform enhanced functions.

With respect to the effects of glucocorticoids, this uneven distribution of resources freed during stress is present even more unambiguously. The catabolic effect of glucocorticoids is known to be characterized by a situation in which these hormones inhibit protein and nucleic acid synthesis and activate protein degradation in the muscular, connective, limphoid, and fatty tissues of the skin; as a result, the amount of free amino acids in the blood considerably increases. It was noted long ago that against a background of this generalized catabolic effect glucocorticoids produce an anabolic effect in the liver, i.e., they activate the synthesis of a system of proteins – enzymes specifically responsible for neoglucogenesis, transamination, and for the synthesis of blood plasma albumin [593]. In this case, the amino acids released during the structural degradation of other organs and which abound in the blood are apparently used to activate protein synthesis.

At present, there are facts that permit one to assume that this effect of redistribution of structural resources in the form of amino acids and, possibly, nucleotides and other compounds, constitutes a characteristic feature of the stress reaction and, perhaps, the biologic essence of this nonspecific component of adaptation. Indeed, it is well known that animals develop the stress reaction during initial considerable physical loads: The corticosterone concentration in the blood increases; the blood shows a negative nitrogen balance, eosinopenia, and thymus involution; and the stomach is occasionally affected by ulcerous lesions. At the same time, nucleic acid and protein synthesis becomes pronouncedly activated in the heart [54] and skeletal muscles [515] to result in a greater mass of mitochondrial structures in the myocytes and a pronouncedly developed general hypertrophy.

Similarly, surgically created considerable stenosis of the aortic ostium is accompanied in animals by the development of the stress reaction with pronounced atrophy of the thymus, which removes thymocytes almost completely. Against this background, the nucleic acid and protein synthesis in the myocardium becomes highly activated, and the heart rapidly hypertrophies [318].

The role of stress in redistributing structural resources and, moreover, their concentrations in systems responsible for adaptation, are apparent in the familiar evidence that when the organism finds itself in a new dangerous situation, or when it is injected with a new heterologous antigen against a background of a more or less pronounced stress reaction, it develops new defensive conditioned reflexes or specific immunity to the new antigen. An essential link in both these processes is the activation of nucleic acid and protein synthesis in a definite system of neurons or in specific immunogenetic systems. This activation is apparently potentiated by an excess of oxidation and biosynthesis substrates, an excess that occurs on account of the catabolic effect of stress hormones in most tissues.

These and other similar factors lead one to the idea that the stress reaction involves not just mobilization of energy and structural resources but also their discriminative redistribution or, more simply, transfer from systems that do not participate in adaptation to a given concrete factor to systems specifically responsible for that adaptation. Due to this redistribution, the oxygen, oxidation substrates, and precursors used for synthesizing nucleic acids and proteins are supplied unidirectionally to the systems that achieve adaptive hyperfunction, i.e., to systems that generate the structural trace which in future would form the basis for long-term adaptation, the foundation of stable enhancement of the organism's resistance to the environment.

Stress-enhanced Resistance to Hypoxia

The fact that a short-term, a fortiori adaptogenic stress may urgently enhance the resistance of a vitally important organ to hypoxia and, very probably, be provided by increased glycolysis, was established relatively recently in our laboratory [339]. Experiments were performed on Wistar male rats, weight 220–250 g. The animals were divided into three series: (a) control; (b) those subjected to lenthy emotional painful stress (EPS); and (c) those subjected to short-term EPS. Emotional painful stress was reproduced in the form of anxious neurosis by the known Desiderato method [80]. The essence of this stress action is that the rats are placed into a special chamber to which a 4-mA current is supplied through the floor; the animals may avoid the painful effect by climbing onto a platform in the center of the chamber. Approximately 30 min after the rats had developed this conditioned reflex, they were subjected to short intense current charges (6 mA for 2 s) via the platform floor at random intervals for 5.5 h; the total duration of stress was 6 h for the second series of animals, and only 1 h for the third.

The principal features of this EPS model were that first, a lengthy conflict developed between the existing conditioned reflex of avoiding the current by climbing onto the platform and the unconditioned pain stimulant on that platform and, second, in the protracted waiting for the painful effect because the shocks were inflicted at random intervals.

One hour after the termination of stress in experimental and control animals, their hearts were extracted under urethane anesthesia and studied for contraction in isovolumetric conditions by the Fallen method [107].

Table 7. Contractile function of isolated hearts of rats exposed to short- and long-term EPS [M ± m]

Indices	Experimental series	Frequency of heart contractions per min			
		120		300	
Developing pressure, mm Hg	1. Control, $n = 10$	109.8	± 9.1	116.0	± 10.4
	2. EPS 6 h, $n = 10$	69.1	± 3.9	86.2	± 4.7
	3. EPS 1 h, $n = 9$	105.4	± 3.8	130.1	± 11.2
	p_{1-2}	<0.001		<0.001	
	p_{1-3}	>0.1		>0.1	
Diastolic pressure, mm Hg	1. Control, $n = 10$	3.0	± 0.8	6.1	± 1.2
	2. EPS 6 h, $n = 10$	8.1	± 0.7	9.8	± 1.1
	3. EPS 1 h, $n = 9$	4.6	± 0.8	5.8	± 1.0
	p_{1-2}	>0.1		>0.1	
	p_{1-3}	>0.1		>0.1	
Maximum pressure elevation rate, mm Hg/s	1. Control, $n = 10$	2,050	± 140	2,875	± 175
	2. EPS 6 h, $n = 10$	1,380	± 110	2,240	± 150
	3. EPS 1 h, $n = 9$	2,100	± 180	3,385	± 270
	p_{1-2}	<0.01		<0.05	
	p_{1-3}	>0.1		>0.2	
Maximum pressure drop rate, mm Hg/s	1. Control, $n = 10$	1,125	± 72	1,600	± 105
	2. EPS 6 h, $n = 10$	890	± 83	1,310	± 80
	3. EPS 1 h, $n = 9$	1,461	± 91	2,014	± 120
	p_{1-2}	<0.05		<0.05	
	p_{1-3}	<0.05		<0.05	
Relaxation index, s^{-1}	1. Control, $n = 10$	10.3	± 1.1	13.8	± 1.2
	2. EPS 6 h, $n = 10$	12.9	± 1.2	15.2	± 1.4
	3. EPS 1 h, $n = 9$	13.9	± 1.2	15.5	± 1.5
	p_{1-2}	>0.1		>0.1	
	p_{1-3}	>0.1		>0.1	

[a] *n*, number of animals in group.

The data in Table 7 indicate that at low contraction frequency (120 per min) the contractile function of animal hearts that had sustaines protracted EPS was found to be depressed. This was seen in decreased pressure and in the rise and drop of pressure rates by about one-third. Diastolic pressure, which in the given experimental conditions reflects the dynamics of the myocardial diastolic tension, proved to have increased 2.5-fold. At high contraction frequency (300 per min), when increased calcium flow to cardiomyocytes activates the SR calcium pump and the myofibril contractile apparatus, these stress-induced disturbances of contraction and relaxation were less pronounced, but generally retained.

The effect of short-term EPS on the contractile function of the heart proved essentially different. At 120 beats/min it was not apparent. At 300 beats/min, it tended to increase the developing pressure and its rate and also to show a pronouncedly increased pressure drop, i.e., a relaxation rate amounting to 40% ($p < 0.05$).

Therefore, unlike protracted EPS, our short-term stress caused no disturbances in the contractile function of the heart. The principal effect of this action was evident in the second stage of the experiment, when the hearts were placed in conditions of hypoxia. Table 8 shows that long-term stress-exposed animals responded to hypoxia with about the same depression of the principal contraction and relaxation paramters as in controls, the only difference being that in the 20th min of hypoxia, the hypoxic contracture in the former was slightly more pronounced than in the latter. With subsequent reoxygeneration, hypoxic contracture, characterized by the diastolic pressure level, disappeared more slowly in rats that had sustained long-term stress than in controls.

The resistance to hypoxia of the hearts of animals that had endured short-term stress was essentially elevated. Table 8 also shows that prior to beginning of hypoxia in hearts more or less 'exhausted' by the preceeding stage of the experiment, i.e., by induced high contraction rate, the endured short-term stress effects caused two significant differences, namely an approximately 40% increase in the myocardium relaxation index and coronary flow. With equal perfusion pressure in all the experimental series, the last shift is indicative of reduced resistance in the coronary bed.

Under hypoxia, the effect of short-term stress was seen even more distinctly. For example, at the 20th min of hypoxia, the developing heart pressure and pressure elevation velocity were higher than in control rats by 69% and 119%, respectively. The pressure drop velocity (relaxation velocity) appeared to be increased 2.7 times and, consequently, the relaxation index increased by almost 60%. As a result of such high resistance of relaxation to hypoxia, the hypoxic contracture proved to be 2.6 times less. The coronary flow in the hearts of rats exposed to short-term stress proved to be almost 90% higher under hypoxia.

With reoxygenation, all the contraction and relaxation parameters in the hearts of animals exposed to short-term stress were completely restored in the 1st min; specifically, hypoxic contracture was found to disappear in the 1st min. In control rats in the same reoxygenation stage, the developing pressure and pressure rise and drop velocities remained more than twofold diminished. Twenty minutes later, in the late reoxygenation stage, these differences remained considerable and significant for the parameters characterizing relaxation, i.e., for the pressure drop rate and the diastolic pressure level. On the whole, this result signifies that short-term stress increased the resistance of the contractile function of the heart to hypoxia. As a result, the preferentially defended process proved to be relaxation, in which the functioning of the membrane sarcolemmic calcium pump and the cardiomyocyte SR are known to play a decisive role in the timely removal of calcium from the myofibrils.

Table 8. Development of hypoxic contracture of isolated hearts of rats exposed to long- and short-term emotional painful stress (EPS)

Developing pressure, mm Hg	1. Control, n = 11	105.8 ± 3.8	57.8 ± 3.5	22.9 ± 2.2	49.0 ± 3.5	76.1 ± 4.6
	2. EPS 6 h, n = 10	68.4 ± 4.9[c]	48.0 ± 4.3	14.9 ± 1.2[b]	44.6 ± 3.8	65.2 ± 3.0
	3. EPS 1 h, n = 9	102.2 ± 3.9	59.2 ± 4.2	38.6 ± 2.1[b]	109.9 ± 6.9[b]	88.0 ± 7.1
Diastolic pressure, mm Hg	1. Control, n = 11	5.4 ± 0.6	9.4 ± 1.2	39.6 ± 3.2	27.2 ± 3.5	17.3 ± 1.3
	2. EPS 6 h, n = 10	8.1 ± 0.8[b]	9.1 ± 0.1	48.5 ± 4.5[b]	33.2 ± 3.5	23.4 ± 3.0[b]
	3. EPS 1 h, n = 9	4.4 ± 2.1	6.5 ± 1.5	15.0 ± 1.4[c]	6.1 ± 1.4[c]	6.0 ± 1.2[c]
Contraction velocity, mm Hg	1. Control, n = 11	1,842 ± 75	919 ± 77	424 ± 50	904 ± 70	1,220 ± 71
	2. EPS 6 h, n = 10	1,380 ± 110[c]	900 ± 87	286 ± 27[b]	586 ± 80[c]	1,125 ± 70[c]
	3. EPS 1 h, n = 9	2,150 ± 120[c]	1,014 ± 101	875 ± 92[c]	2,016 ± 190[c]	1,740 ± 158[b]
Relaxation velocity, mm Hg/s	1. Control, n = 11	1,163 ± 25	404 ± 46	230 ± 21	421 ± 67	692 ± 42
	2. EPS 6 h, n = 10	840 ± 45[c]	415 ± 60	162 ± 20[b]	375 ± 66	685 ± 40
	3. EPS 1 h, n = 9	1,451 ± 120[c]	557 ± 48	508 ± 44[c]	1,125 ± 108[c]	1,100 ± 96[c]
Relaxation index, s^{-1}	1. Control, n = 10	10.0 ± 1.0	8.0 ± 0.6	8.2 ± 0.6	10.1 ± 0.9	8.9 ± 0.9
	2. EPS 6 h, n = 10	12.3 ± 1.1	8.6 ± 0.6	10.8 ± 1.1	8.3 ± 1.1	10.6 ± 1.1
	3. EPS 1 h, n = 9	14.3 ± 1.2[b]	9.4 ± 0.8	13.0 ± 0.9[c]	12.3 ± 0.9	13.4 ± 1.0[b]
Coronary flow, ml/min	1. Control, n = 10	8.0 ± 1.4	6.3 ± 0.5	6.0 ± 0.8	6.5 ± 0.5	7.0 ± 0.7
	2. EPS 6 h, n = 10	7.5 ± 0.5	7.5 ± 1.4	4.9 ± 0.3	6.5 ± 0.8	6.7 ± 1.1
	3. EPS 1 h, n = 9	11.0 ± 1.2	10.5 ± 0.8	11.3 ± 1.1	11.2 ± 0.7	11.6 ± 1.1

n, number of animals in series.
[b] $p < 0.05$.
[c] $p < 0.001$.

Table 9. Loss of CPK to perfusate by isolated heart of animals exposed to short- and long-term stress

	Initial aerobic conditions	20 min hypoxia	20 min reoxygenation
1. Control, $n=12$	9.0 ± 1.8	34.0 ± 2.6	49.5 ± 2.8
2. EPS 6h, $n=10$	11.4 ± 1.2	51.9 ± 5.0	60.3 ± 4.1
3. EPS 1h, $n=19$	9.8 ± 0.9	21.5 ± 3.5	26.4 ± 3.9
	$p_{1-2} > 0.1$	$p_{1-2} < 0.001$	$p_{1-2} < 0.001$
	$p_{1-3} > 0.1$	$p_{1-3} < 0.001$	$p_{1-3} < 0.001$

It would appear that the increase of coronary flow and, correspondingly, the decrease in the resistance of the coronary bed in the hearts of animals that had undergone short-term stress should be assessed from the same viewpoint since such a shift cannot be caused simply by increased relaxation of the smooth muscles of the coronary arteries and arterioles.

The data in Table 9 characterizes the CPK release from an isolated heart into the perfusate under aerobic conditions, hypoxia, and reoxygenation. Table 9 shows that with normal oxygenation the CPK loss is minimum, and with hypoxia it is increased. Reoxygenation involves further CPK release into the perfusate. It is significant that both with hypoxia and reoxygenation the CPK release from the hearts of animals that had endured long-term stress is increased compared with controls by 52.8% and 21.8%, respectively, and in the hearts of animals that had endured short-term stress the CPK release, conversely decreased compared with controls by 38.8% and 26.5%, respectively. Inasmuch as enzyme loss by the heart is commonly regarded to be the result of energy-shortage injury of the cardiomyocyte sarcolemmal membrane and quantitatively correlates with the degree of ATP decline in the myocardium [531], these data show that preliminary short stress enhanced the resistance of the cardiomyocyte membrane apparatus to hypoxic injury.

On the whole, the experimental results show that intensive but short-term (just 1 h) EPS caused certain quite stable changes in the heart, which definitely remained after the organ was isolated from under the spectrum of the neurohormonal effects of the whole organism. The changes were present in the form of two phenomena. First, relaxation in the myocardium and the coronary bed became activated; this was evident in the increased myocardium relaxation index and in the decreased coronary bed resistance in aerobic conditions. Second, the heart's resistance to hypoxia grew considerably; this was seen in the increased stability of relaxation to hypoxia, in the abrupt decrease in the degree of hypoxic contracture, and also in the reduced hypoxic loss of CPK by the heart. It is significant that with continued stress (prolonged sixfold) both these positive shifts disappeared completely even to be replaced by a certain decline in the heart resistance to hypoxia.

There are two most probable mechanisms for activating the relaxation system and enhancing cardiac resistance to hypoxia in short-term stress. First, catecholamine-determined activation of glycolysis, which plays a role both in providing energy for relaxation and in increasing resistance to hypoxia. Second, the effect of stress hormones on the membrane lipid bilayer, which may enhance the activity of membrane enzymes, specifically those of the calcium pump.

Lipotropic Effect of Stress in Biomembranes

The determining role of biomembranes in the functioning of cells signifies that the adaptive or injurious effects of any factor, especially such a complex set of factors as those involved in stress, are manifest in indirectly in the whole organism, via cell membrane systems. The fulfillment by biomembranes of their principal functions, namely the barrier function and the protein-enzyme matrix, ion transport channel, and receptor functions, is known to be feasible due to the specific biomembranous structure. The main feature of this structure is the presence of a hydrophobic zone formed by two phospholipid layers, which incorporate functionally active membrane proteins. Figure 11 reflects this basic property of the biomembrane liquid-mosaic model [265] and shows that in respect to localization in the phospholipid bilayer membrane proteins are classified in two classes. The lesser class (about 30%) includes peripheral proteins weakly bound with the membrane, chiefly at the expense of electrostatic interactions, and integral proteins that form globules, which may include sugars and specific lipids, with the formation of glyco- and lipoproteins. These proteins are immersed in the lipid bilayer, and some may permeate the membrane.

In terms of our presentation, it is significant that the principal vitally important biomembrane proteins, e.g., Na-K-ATPase, Ca-ATPase, receptor-enzyme adenyl cyclase complex, and proteins that form channels of slow calcium conduction, are integral proteins whose activity very strongly depends on the lipid microsurrounding, i.e., on the composition and physical condition of the lipid bilayer. The significance of the lipid bilayer in the activity of these proteins and, hence, its role as a factor largely determining the functions of whole organs and systems may be illustrated by numerous examples, one of which is cited below.

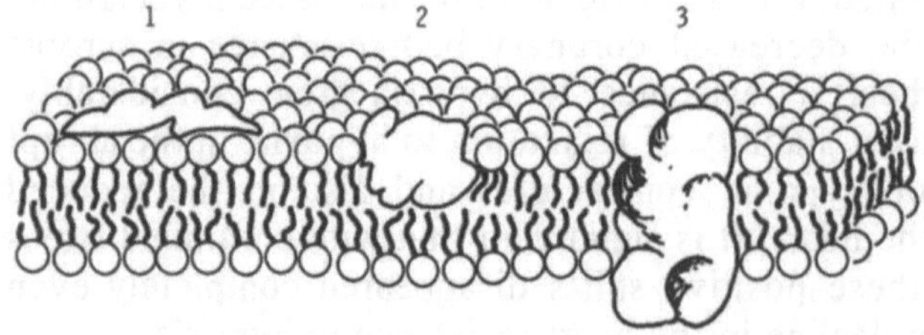

Fig. 11. Structure of biologic membrane; phospholipid bilayer and main types of membrane proteins localized therein. *1*, pheripheral protein; *2*, integral protein localized on one side of membrane; *3*, membrane-permeating proteins

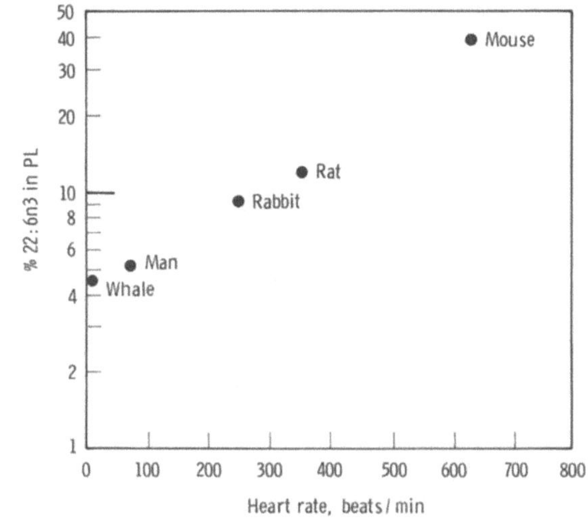

Fig. 12. Interdependence of decosohexogenic acid in phospholipids (*22:6n3 in PL*) of myocardial membranes and heart contraction frequency in various animal species. (After Gudbjarnasson et al. 1978)

In Fig. 12, the curves characterize the dependence between the content in the membranous phospholipid bilayer of cardiomyocytes of one of the fatty acids, viz dacasohexaene acid (22:63), and the heart rate in various animal species [152]. As is apparent from Fig. 12, the content of this polyunsaturated fatty acid with six double bonds in its hydrocarbon 'tail' is almost tenfold higher for the mouse myocardium than for the human myocardium, and the mouse heart rate is correspondingly about ten times higher than the human heart rate. One can understand this important fact of evolutionary physiology only through the principle that the fatty acid composition of the phospholipids from cardiomyocyte membranes in general and its conduction system in particular largely determine the functioning of the ion conduction channels and cation pump enzymes, and thereby the functional possibilities of the heart and its automatism.

At present, it has been proven that in the course of individual life and the organism's adaptation to the environment, the fatty acid composition in the organs and tissues may be modified by diet and by administrating nicotine and catecholamines [152]. In principle, catecholamines and other stress hormones may modify not only fatty acid composition but also other lipid bilayer parameters; hence, to understand the contribution of stress to adaptation and pathology, it is essential to examine generally the basic mechanisms at the expense of which stress response can modify the membranous lipid bilayer. At present, at least four processes are known to provide membrane renovation in the organism and to possess the ability to change their phospholipid and fatty acid composition, fluidity, and thus the lipid surrounding of vitally important integral proteins and the functions of the organism's organs and systems. All these processes are essentially activated under the action of the principal stress factor, that is catecholamines and some other stress hormones.

The first process of this kind is the phospholipid biosynthesis that takes

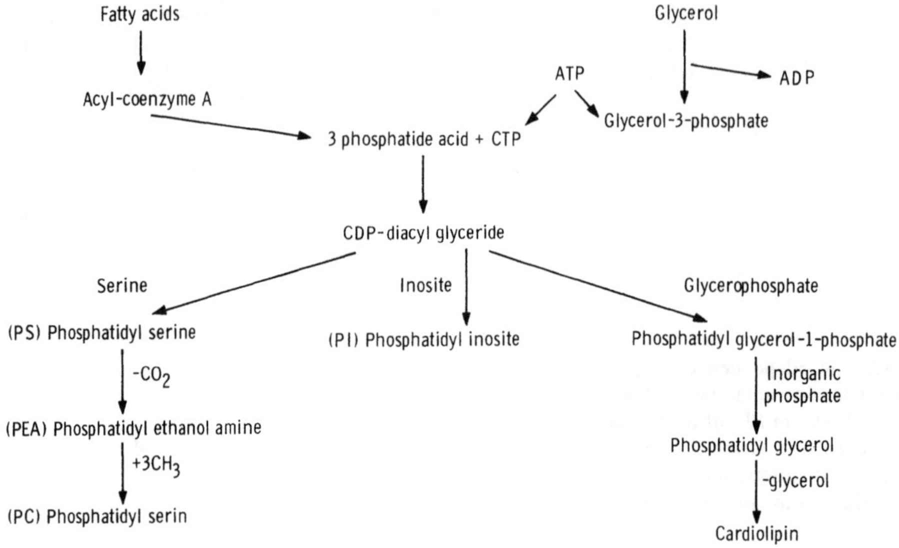

Fig 13. Synthesis of cell membrane phospholipids

place in the endoplasmic reticulum (Fig. 13). The ACTH and catecholamines released in increased amounts during stress were found to activate phospholipid synthesis. For instance, ACTH increases inclusion in the hydrocarbon chains of phospholipid molecules of labeled precursors, ^{14}C-arachidonate and ^{14}C-acetate, the maximum amount of label being contained in phosphatidyl choline, and then in reducing order in phosphatidyl ethanolamine, phosphatidyl serine, and phosphatidyl inositol [257].

In a similar way, membrane phospholipid synthesis is activated by catecholamines by the criterion of inclusion of labeled phosphorus [393]. Among other things, it was shown that the introduction into the perfusate of a working heart of physiologic doses of noradrenaline increases the inclusion of labeled palmitic acid by more than 70%, while removal of catecholamines from the myocardium by reserpinization conversely decreased this index twice. Further, isoproterenol, a synthetic analog of noradrenaline, was found to stimulate the enzymatic synthesis of phosphatidyl monomethyl ethanolamine and phosphatidyl choline in erythrocyte shadows in the presence of methyl group donators, CH_3-adenosyl-CH_3-methionine. This stimulation was inhibited by propranolol [177] resulting in increased fluidity of membranes and higher activity of membrane-bound enzymic systems.

The second process is phospholipid degradation by phospholipases, whose application points are shown in Fig. 14. Phospholipase A specifically hydrolyzes the split-off of the acyl residue in position 2, the given reaction product being called lysophosphatide. Phospholipase B catalyzes the split-off of the second or both fatty acids. Phospholipase C causes the bond between phosphoric acid and glycerol to hydrolyze, and phospholipase D splits off the XO group to form phosphatidic acid. The principal mass of phospholipases is concentrated in cell lysosomes; however, there are also phospholipases

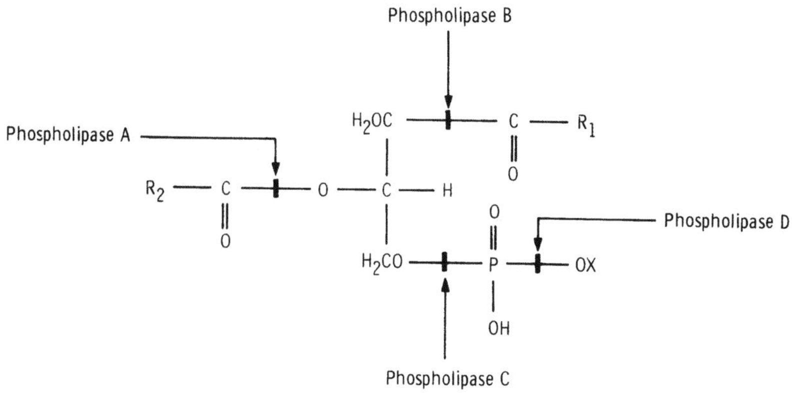

Fig. 14. Phospholipase action. Explanation in text

directly contained in various cell membranes. It should be noted that in the aqueous phase phospholipases do not show catalytic activity, active conformation being created only upon adsorption and immobilization of the enzyme at definite surfaces, i.e., in the case of membranes, at the penetration of the enzyme into the lipid hydrophobic phase, this process taking place with the participation of Ca^{2+} [460]. At the same time, calcium plays the role of a phospholipase activator. In addition to phospholipase, the membranes also include specific acyl transferases, which attach acyl groups, fatty acid residues, to corresponding lysophosphatides. Coordinated functioning of multienzyme complexes consisting of phospholipases and acyl transferases provide continous renovation of lipids in the membranes and quite fine responses to various changes in environmental conditions by including in the given phospholipids different fatty acid residues with definite chain lengths, nonsaturation degrees, double bond positions, etc.

It appears significant to us that catecholamines, the concentration of which in the blood naturally grows with stress, activate phospholipases in at least three ways and, consequently, they activate the degradation and renovation of membrane phospholipids. This takes place, first, at the expense of direct activation of phospholipases by catecholamines [291] and, second, indirectly via a system of kinins. In this case, bradikinin, which forms in increased amounts, specifically activates phospholipase A_2 [183]. Finally, the effect of catecholamines on the cell naturally increases the influx therein of calcium, an activator of phospholipases in general and in mitochondria in particular [460]. Thus, there can be no doubt that during stress even a physiologic adrenergic effect with limited intensity activates phospholipolysis and, consequently, renovation and actual change of the membrane fatty acid composition and this may, in turn, cause adaptive changes in the activity of membrane-bound proteins.

The third process to play a role in renovating the membrane lipid composition is peroxidation of unsaturated phospholipid fatty acid residues. Its essence, examined in greater detail below (Chap. 6), is determined by the fact that in addition to oxydase four-electron oxygen reduction over the cytochrome oxydase of the respiratory chain an oxygenase pathway is continuously present in the organism resulting in the one-, two-, and three-electron reduction of O_2 and, correspondingly, in its active free-radical forms, viz superoxide (O_2) anion radical, hydrogen peroxide (H_2O_2), and hydroxyl radical (OH). These active oxygen forms attack unsaturated phospholipid fatty acid residues, the oxygen being included in the molecule of the oxidized substrate to form phospholipid hydroperoxides, by which the entire process is identified as lipid peroxidation (LP).

Phospholipid hydroperoxides are unstable compounds, which when degrading in physiologic conditions produce effects that modify the membranous lipid bilayer, specifically decreasing the content of unsaturated fatty acids in the lipid surrounding of vitally important membrane-bound proteins; this may involve the forming of additional channels of calcium permeation, the so-called peroxide clusters [200]; the effect of LP products may labilize lysosomes and from then release phospholipases [258], and also produce the above-cited action of these enzymes.

With regard to stress response, the concrete situation of interest here, it is significant that EPS was unambiguously shown to involve LP activation and the accumulation of LP products in the brain, heart, muscles, and other organs [370]. In accordance with modern concepts, this activation is caused by catecholamines through two mechanisms. First, it is brought about via a beta-receptor – adenyl cyclase system, i.e., at the expense of cAMP and increased Ca^{2+} influx, which probably activates LP. Second, LP activation occurs as a result of the oxidation of stress-induced excess adrenaline to adrenochrome; this produces adrenaline semiquinone, which 'discards' the electron onto oxygen resulting in the above-mentioned superoxide radical (O_2). Thus, LP is undoubtedly a mechanism via which the stress response modifies the lipid bilayer of the cell membrane to be able to determine the adaptive changes or functional and structural disorders in the organs and systems.

The *fourth* and last *mechanism* which in this presentation may play a role in modifying the membrane lipid bilayer during stress is associated with the well-known fact that catecholamines activate lipases in the adipose and other tissues causing, consequently, a highly increased concentration of fatty acids in the blood plasma and, *possibly,* the *incorporation of various fatty acids in the lipid bilayer of membranes of vitally important organs.* Fatty acids are known to possess a dual effect. First, they are substrates, whose oxidation results in ATP aerobic resynthesis. Second, like other amphiphiles, they interact with cell membranes. This was thoroughly reviewed by Katz and Messineo [209], who showed that the fatty acid effect depends on its concentration. At relatively low concentration, membrane permeability is noted to decrease due to membrane stabilization. High fatty acid concentration results in a detergent-like effect, which deprives the membranes of the ability to fulfil their barrier function. The above-mentioned lysophosphatides possess a still more pronounced injurious effect, the ability to disrupt cell membranes [581]. It is important that in this case, which involves moderate or short-term stress, a physiologic rise in the concentrations of fatty acids transported in the blood plasma by albumins can stabilize membranes or cause adaptive changes.

All the examined processes that compose the lipotropic stress effect in biomembranes can play a role in adaptation only as long as the activities of the enzymes, ion transport channels, and receptors in the membranes depend on their lipid surrounding. The available evidence indicates that the requirements of these proteins to their lipid surrounding may vary. In some cases, they may be highly nonspecific and be satisfied by any set of phospholipids; in other cases, there is the need for the presence of a strictly specific set of phospholipids for a given enzyme system to be active. For example, Goldman and Albers [138] used Na^+-K^+-ATPases from the electric organ of a skate to show the need for phosphatidyl serine to form a phosphorylated enzyme in the stage of Na^+-dependent phosphorylation and its K^+-dependent dephosphorylation. Magnesium-dependent conformational changes in Na^+-K^+-ATPase depend on phosphatidyl choline.

Phospholipids play an important part in the functioning of adenylate cyclase, via which the effect of catecholamines, the principal stress factor, comes into being. Adenylate cyclase is known to represent a sophisticated multimolecular complex comprising at least three functional sites: (a) the receptor (control) site located on the external surface of the membrane, the site that selectively binds various active substances; (b) the catalytic site, located on the membrane's internal surface, which is responsible for generating cAMP; and (c) the communicative site connecting the first two [474]. Solubilized adenylate cyclase in the myocardium is known to lose sensibility to catecholamines and hormones, but to retain the ability to bind them [263]. When various phospholipids are added to this adenylate cyclase, it restores its specific sensibility to different agents; for example, phosphatidyl serine regains its sensibility to glucagon [270] and phosphatidyl inositol regains its sensibility to noradrenaline [271]. The same selectivity in the reducing effect of various phospholipids was determined for the thyroid gland [590] and liver [469]. Inasmuch as phospholipids have no direct relationship to receptor sites, it would be natural to assume that they constitute an important link in the activity of the communicator site. It is also possible that various phospholipids determine the specific activation of adenylate cyclase when corresponding biologically active substances act on the receptor site [176].

Presently existing evidence permits the assertion that other cell enzyme systems are also dependent on lipids. For example, removal of lipids inactivates the Ca ATPase of the SR; as a matter of fact, attempts to reactivate the enzyme by phospholipids proved ineffective since lipid removal from Ca ATPase is apparently accompanied by an irreversible loss of its hydrophobic surfaces [165].

In studying the polyenzymic systems of the mitochondrial respiratory chain, phospholipids were found to play a substantial role in the activity of cytochrome-C-reductase, CoQ-cytochrome oxydoreductase, NAD-dehydrogenase, and succinate dehydrogenase [287, 587], and hence in ATP anaerobic resynthesis, which constitutes the basis for energy supply to various functions.

Thus, the lipid microsurrounding of proteins, which, as has now been shown, can be relatively quickly modified by the principal stress hormones, is a real determinant of the activity of the principal membrane-bound enzymes. It is highly significant that this type of lipotropic stress may come about in the cell membranes of the systems responsible for adaptation so as to enhance rapidly the physiologic possibilities of these systems before structural synthesis occurs therein de novo and forms a systemic structural trace. This signifies that in addition to mobilization and the selective redistribution of resources, glycolysis activation and higher resistance to hypoxia, the lipotropic stress effect may play a role in forming urgent adaptation and creating the prerequisites for long-term adaptation.

Antioxidant Effect of Glucocorticoids and the Influence of Synthetic Antioxidants on Urgent Adaptation to Physical Loads

Excitation of adrenergic control and the effect of its mediators, catecholamines, plays such an important role in the adaptive and harmful effects of stress that many researchers regard the administration of adrenaline, noradrenaline, and isoproterenol, the synthetic analog of noradrenaline, as a stress model. Indeed, the injuries caused by catecholamines reveal very strong similarities to stress injuries. Administration of noradrenaline or isoproterenol naturally activates LP [249] and phospholipases [155, 291], causes excess fatty acids to occur in the blood [535], increases calcium influx into the cells [470], there is a pronounced decline in ATP concentration [572], focal contracture-type injuries occur in the myocardium [475], and there are subsequent necrobiotic foci, which, in merging, may occupy a considerable area in the myocardium, for instance [448].

As will be shown below, similar qualitative changes may be revealed in the hearts of animals that had endured serious EPS (Chap. 6). Yet, quantitatively they are considerably less pronounced: The significant decrease in ATP concentration is not observed and necrobiotic foci are small. Naturally, this striking quantitative difference may be explained by the fact that with exogenous administration of catecholamines their acting concentration is higher than with stress. However, alongside this questionable explanation there is another possibility that stress reaction is provided by the coordinated excitation of a considerably more complex system in which the hypophysial-adrenal complex, which determines steroid hormone release, plays the principal role, not by the isolated excitation of adrenergic regulation. Besides their specific effect at the genetic apparatus level, these hormones are known to be membrane stabilizers [75] and antioxidants [509] and, consequently, capable of modulating the action of catecholamines during stress by limiting their injurious effect.

Indeed, the investigations of Gukasov et al. [154] from Y. A. Vladimirov's laboratory showed that LP activation induced in a system of isolated mitochondria can be effectively suppressed by steroid hormones. In this case, female sexual hormones, i.e., estrogens, and also glucocorticoids possess the strongest antioxidant effect. This permits the inference that catecholamine-induced LP activation may also be restricted by the antioxidant effect of steroid hormones.

This view is in agreement with the results of Gladkova and Rykova [134], who showed that in female rats experimental myocardial infarction, in the pathogenesis of which stress plays an important role along with ischemia, is smaller and generally less serious than in males. At the same time, the estrogen content in the blood of females increases and in males, conversely, decreases with infarction. The protective effect of steroid hormones during ischemic injuries was also demonstrated by other researchers [262, 272]; it agrees with the fact that highly different antioxidants not only prevent stress injuries but also suppress changes in catecholamine content and metabolism characteristic of stress response (Chap. 8). This possibly takes place due to suppression by antioxidants of catecholamine synthesis since dopamine β-hydroxylation, the key reaction in this process, occurs through the influx into the molecule of hydroxyl dopamine and has proved to need O_2^-, an anion of superoxyde-radical, to be achieved, i.e., it is an oxygenase reaction of peroxidation [85]. Generally speaking, the available evidence makes probable the assumption that steroid hormone hyperproduction is the component of stress response that through LP suppression limits the adrenergic component of the same response to create ultimately an optimal adaptive situation.

Inasmuch as the stress response is achieved instantly during the effect of any considerable environmental change upon a nonadapted organism, this assumption signifies that by simulating natural regulatory mechanisms we can, by means of sufficiently strong natural and synthetic antioxidants, intervene in the development of initial urgent adaptation to highly different environmental factors. It may be assumed that with reactions characterized by the availability of oxygen, activation of aerobic metabolism, pronounced excitation of adrenergic regulation, and activation of catecholamine synthesis, one may expect excess LP activation and a positive adaptational effect of the antioxidants that eliminate that activation. A typical reaction of this kind occurs in untrained animals and humans in response to maximal life-compatible hypoxic hypoxia.

From these general arguments, we compared the effects of the powerful antioxidant M-1, which, like estrogens, is a phenol-type LP inhibitor, for animal resistance to maximal physical loads and hypoxic hypoxia [386]. The experiments were performed on Wistar male rats, weight about 200 g. In a new series of experiments, the animals were divided into four groups. The first was the control group; the second was injected daily with doses (20 mg/kg) of M-1 intraperitoneally for 3 days; the third group was subjected to a single maximally long physical load in the form of running on a treadmill (16 m/min) until 'breakdown.' Finally, the animals from the fourth group were first injected M-1 (20 mg/kg) and then subjected to the same physical load. Running time was measured in minutes; after that, all the rats from all four groups were killed to determine the alanine aminotransferase (AlAT) activity in the blood plasma, and the lactate dehydrogenase (LDH) activity.

In the second series of experiments, we distinguished the same four groups of animals as in the first series; however, these tests differed in that the animals that were preliminarily given M-1 were made to run in a treadmill not until 'breakdown,' but for exactly the period of time that proved maximal for animals not administered M-1. Immediately after being subjected to physical load, all the rats were killed and determined for lactic acid concentration in the blood plasma.

From Table 10 it follows that the maximal running time in a treadmill until 'breakdown' (average 67 min) was characteristic of animals that did not receive M-1, and those that did receive M-1 could run for 108 min. In other words, the antioxidant was instrumental in enhancing endurance, measured by maximal running time, by about 60%. Table 10 also shows that rats not administered M-1 showed pronounced fermentemia under maximal load: AlAT activity increased 2.8 times, and LDH almost twofold. In animals preliminarily administered the peroxidation inhibitor, the degree of fermentemia after more protracted loads was much lower: AlAT activity grew by only 65%, and LDH by only 40%.

From Table 11, it follows that the antioxidant did not affect the lactate concentration in the blood of inactive animals; however, after loads of equal duration, the lactate concentration in animals that received the antioxidant proved many times lower than in those not administered the LP inhibitor.

Table 10. Effect of antioxidant on maximal running time in treadmill and on fermentemia

	Maximal running time in treadmill, min	Enzyme activity in blood plasma, μmol/ml	
		AlAT	LDH
1. Control, $n = 10$	–	1.34 ± 0.10	21.48 ± 1.99
2. Control + antioxidant, $n = 9$	–	1.28 ± 0.09	20.09 ± 1.55
3. Maximal load, $n = 10$	66.9 ± 1.1	3.83 ± 0.09	34.96 ± 2.21
4. Maximal load + antioxidant, $n = 9$	108.1 ± 5.7	2.10 ± 0.10	27.96 ± 1.59
p_{3-4}	< 0.001	< 0.05	< 0.001

Table 11. Effect of antioxidant on lactate content in blood plasma of rats subjected to maximal physical load (76 min)

	Lactate content in plasma, μmol/ml
1. Control, $n = 10$	7.4 ± 0.05
2. Antioxidant, $n = 10$	7.3 ± 0.06
3. Maximal load, $n = 10$	19.5 ± 2.2
4. Maximal load + antioxidant, $n = 10$	11.08 ± 0.9
Significance	p_{1-2} not significant
	$p_{1-3} < 0.001$
	$p_{1-4} < 0.001$
	$p_{3-4} < 0.001$

Inasmuch as according to modern sufficiently well-grounded concepts lactate concentration in the blood restricts the duration of physical work, it is understandable that in our experimental situation the control animals stopped running because of fatigue, and those protected by the antioxidant could continue to run for a long time. On the whole, the data obtained show that antioxidants may, in principle, enhance the resistance of an untrained organism to physical loads and deserve further relevant scrutiny.

In terms of our presentation, it is significant that the effect of antioxidants on animal resistance to hypoxic hypoxia proved to be opposite. In special experiments, M-1 (100 mg/kg) and another antioxidant, oxypyridine (80 mg/kg), were injected intraperitoneally once a day for 3 days to Wistar male rats. Two hours following the last injection of control rats, those that were given M-1 and oxypyridine were 'lifted' to 9,000 m in an altitude chamber. The 'ascent' lasted 30 min; after that, the animals were kept at that altitude for 1 h. Their mortality was calculated for each group.

Table 12. Effect of synthetic antioxidants on mortality of rats subjected to hypoxic hypoxia

	Mortality, %
1. Control, $n = 18$	25.0 ± 2.1
2. M-1, $n = 18$	58.3 ± 4.3
3. Oxypyridine, $n = 18$	41.7 ± 3.7
Significance of differences	$p_{1-2} < 0.001$ $p_{1-3} < 0.001$

From Table 12 it follows that at hypobaric hypoxia the antioxidants increased the rat death rate by 1.5–2 times, i.e., decreased animal resistance to hypoxia. In our view, this was presumably caused by the fact that under oxygen shortage the suppression of catecholamine synthesis in adrenergic neurons of vitally important brain centers restricts the possibility of retaining respiration and life itself. Additional inhibition of catecholamine synthesis with antioxidants narrows this restriction and increases mortality. Yet, both this and other possible explanations of the results undoubtedly require many-sided verification. In this case, it is significant that exogeneously administered antioxidants and possibly endogenous antioxidants, i.e., the antioxidant status of the organism, may play an important role in achieving urgent adaptation to environmental factors.

Post-stress Activation of Nucleic Acid and Protein Synthesis and Its Role in Forming Long-term Adaptation

The above-cited stress-reaction effects, viz mobilization and enhanced resistance to hypoxia associated with glycolysis activation, redistribution of resources, and lipotropic and antioxidant effects, play a direct role in urgent adaptation and create the prerequisites for forming a systemic structural trace constituting the basis of subsequent long-term adaptation. At the same time, the stress reaction includes components that play an exclusive and preferential role precisely in long-term adaptation.

The first component of this kind concerns catecholamines, glucocorticoids and, possibly, other hormones, whose secretion grows with stress; they have a direct effect on metabolism and the forming of structures in the cells of systems responsible for adaptation. This direct effect of stress-realizing systems on the forming of the systemic structural trace is further examined in detail in Chap. 5, using the formation of temporary relationships and higher adaptive responses as examples.

Another component is the so-called post-stress activation of nucleic acid and protein synthesis or the anabolic stress phase, which develops soon after the completion of any single stress effect; this phase plays a major role in forming the systemic structural trace and is discussed below.

A still obscure question in the role of stress response in developing adaptation is how a well-known catabolic stress is coordinated with pronounced activation of nucleic acid and protein synthesis, which is invariably present in systems responsible for adaptation and constitutes the basis of the systemic structural trace. For example, it is not clear how this catabolic effect is coupled with the activation of nucleic acid and protein synthesis in cerebral neurons in the formation of temporary relationships, or with the activation of nucleic acid and protein synthesis and proliferation of lymphocytes in immunocompetent organs during the initial immune response, etc.

One possibility is that with single stress action the catabolic phase is followed by the anabolic phase, i.e., by generalized activation of nucleic acid and protein synthesis. This phase can potentiate an interrelation between function and the genetic apparatus and thereby also the structural trace forming in the systems specifically responsible for adaptation to definite environmental factors. To evaluate the probability of this assumption, we shall below examine facts that permit: first of all, the quantitative assessment of the degradation of nucleic acids and proteins in the initial catabolic phase of serious EPS; second, the demonstration of a generalized post-stress activation of nucleic acid and protein synthesis, i.e., the anabolic phase of stress response; and third, the demonstration of the potentiating effect of the anabolic phase of a nonspecific stress response to the forming of such a fortiori varying specific adaptive responses as the primary immune response and the defensive conditioned reflex [388].

The experiments were performed on white Wistar male rats, weight 120–150 g. The EPS was reproduced by the method described above [80]. In studying the catabolic effect of EPS in the first stage, we determined RNA and protein degradation. To that end, the animals were injected with labeled precursors and then some of them after 24 h were examined for initial radioactivity of tissue RNA and protein. After that, EPS was induced in half of the remaining rats. Subsequently, following definite periods, we determined the radioactivity of RNA and protein in control and EPS-exposed rats.

Specific protein radioactivity in the tissues of these series of animals was determined 1, 3, 5, and 7 days following the injection of a labeled precursor. Three indices characterizing RNA and protein degradation were examined: the half-life of these molecules ($t^{1}/_{2}$), their turnover time (T), and the share of the RNA and protein replaced during 24 h. In the next investigational stage, RNA degradation was assessed by the release with urine of pseudouridine, which is known to be contained preferentially in transfer RNA [100].

It turned out that after EPS the ribosomal RNA degradation rate and, still more, the protein degradation rate increased in all the organs. Indeed, the RNA half-life in the liver, heart, and brain decreased by 28%, 25%, and 14%, respectively; corresponding figures for protein half-life in these organs were 36%, 25% and 34%, respectively.

In terms of our presentation, it is significant that the increased RNA and protein degradation revealed in the experiments could have been unambivalently reduced or prevented by introducing a central-action inhibitor, namely gamohydroxybutyric acid (GHBA), which was found to suppress or decrease excitation of stress-realizing systems (Chap. 6). Further, it was shown that in the day following termination of EPS excretion with urine of nucleotides that form part of the RNA composition, viz uracyl and pseudouridine, increases by 60%. This signifies that RNA degradation in the organism was substantially increased under the influence of EPS. Approximately three-fifths of the pseudouridine excreted with urine is known to be due to the decomposition of transfer RNA and two-fifths due to the degradation of ribosomal RNA [100]. Thus, abrupt increase in pseudouridine corresponds to the evidence that EPS causes considerably accelerated degradation not only of ribosomal but also transfer RNA (Table 13).

From Table 13 it follows that increased pseudouridine and uracyl excretion with EPS may be prevented by injecting GHBA. Together with the above-mentioned evidence on the effect of GHBA, this means that increased degradation rate and decreased RNA and protein life in EPS-exposed animals are

Table 13. Effect of EPS and GHBA upon pseudouridine and uracyl excretion with rat urine

	Control	Time from termination of stress action			
		EPS	EPS + GHBA	EPS	EPS + GHBA
Excretion of pseudouridine, mg/day	0.773 ± 0.030	1.243 ± 0.053	0.856 ± 0.062	0.602 ± 0.039	0.809 ± 0.041
n	24	8	8	8	8
p		< 0.001		< 0.01	
Excretion of uracyl, mg/day	0.136 ± 0.006	0.221 ± 0.26	0.127 ± 0.014	0.156 ± 0.012	0.141 ± 0.008
n	24	8	8	8	8
		< 0.001		< 0.01	

[a] *n*, number of animals

indeed caused by excitation of the stress-realizing systems and the action of stress hormones, for suppression of the stress response either prevents or decreases these events. Thus, shortened life is a concrete quantitative expression of the stress catabolic effect; this shift characterizes the structural cost of the stress response.

In studying the anabolic effect of EPS, we examined the RNA and protein biosynthesis dynamics in different periods after that action. In this case, the animals were killed 2–160 h after EPS. The RNA content and biosynthesis, and also the protein biosynthesis, were examined in the tissues of the brain, heart, liver, spleen, thymus, and testis. To that end, labeled precursors were injected intraperitoneally 1 h prior to killing the animals when studying the RNA and 4 h before scrutinizing the protein. The intensity of RNA synthesis was studied by inclusion of ^{14}C-orotate (specific activity 26 mCi/mmol), 20 μCi/100 g body weight. The RNA was extracted by the known method of Schmidt and Thannhauser. Its content was determined spectrophotometrically and expressed in mg per g wet tissue, and also in mg per organ. The protein synthesis rate was investigated by inclusion of radioactive amino acids: ^{35}S-methionine, ^{14}C-alanine, and ^{14}C-glycin, all injected at 20 μCi/100 g.

In all the experiments, the RNA and protein radioactivities were determined by means of a gas meter and expressed in impulses/min/1 mg (specific radioactivity). For RNA, we also calculated radioactivity for the whole organ equal to the product of the specific activity by the RNA content in the organ.

The investigation results showed that the structural cost of stress response is not exhausted by the cited increase in RNA and protein degradation. It turned out that directly following EPS degradation growth is combined with suppression of RNA and protein biosynthesis. This combination is sufficiently well known and constitutes the basis for a negative nitrogen balance and the decrease in animal body weight during serious stress actions; persisting for a long time, it may lead to disorders in the adaptive possibilities of the organism. However, the situation proved otherwise with our single EPS. Following the catabolic effect, all the organs studied naturally developed activated biosynthesis and increased RNA content, which in turn led to increased protein biosynthesis and a certain, not always significant, increase in the weight of given organs. The curves in Fig. 15 and 16 illustrate this. An analysis of the graphs reveals the following three points.

First, the dynamics of the indices studied is basically the same for all the six organs examined; 2 h after termination of stress, the RNA biosynthesis and content and the protein biosynthesis were lower than in the initial level, subsequently to increase. In addition, in conformity with the logic of the process, the RNA and protein biosynthesis attains maximal levels after 24 h, and the increment of the total RNA content in the organ and the organ weight proves to be at a maximum after 48 h. The RNA and protein biosynthesis and content tend to increase only after 5 days, but after 1 week they do not differ from the initial level.

Thus, the post-stress activation stage of RNA and protein biosynthesis is many times longer than the biosynthesis suppression stage. The former provides a sufficiently lengthy period of 'readiness,' during which the organs

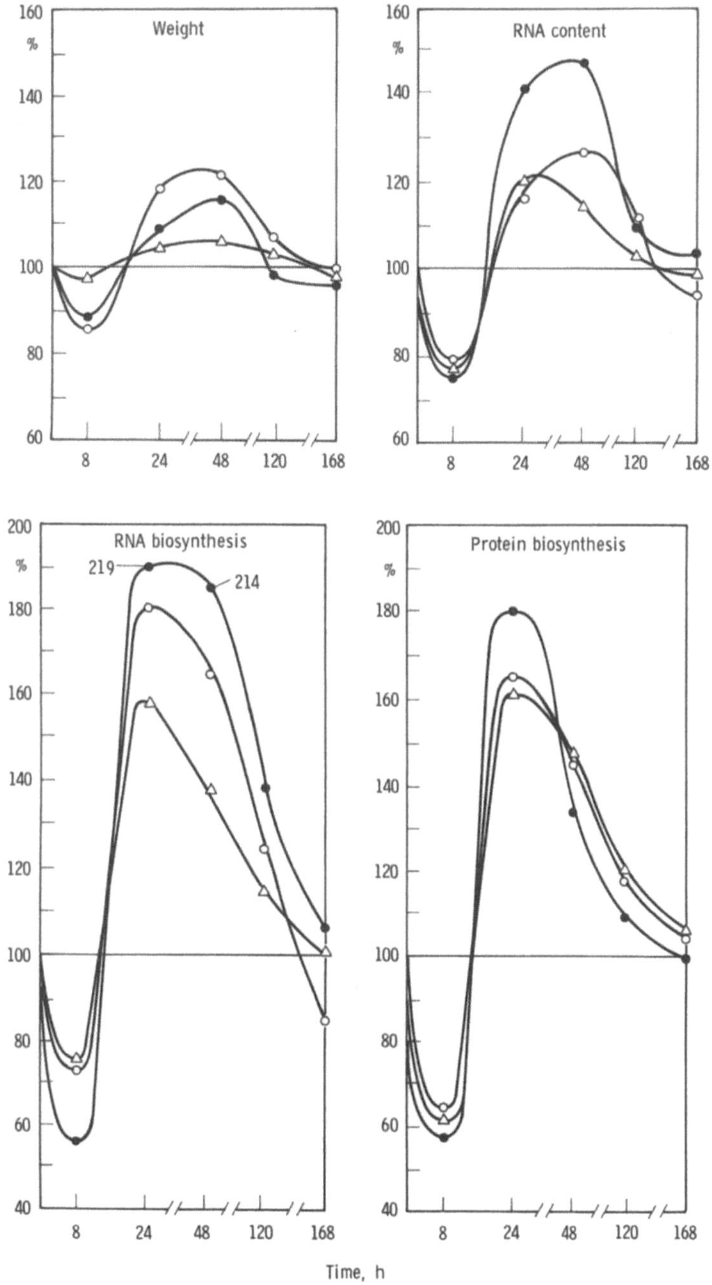

Fig. 15. **Changes in organ weight, RNA content, RNA biosynthesis, and protein biosythesis in the heart (○), brain (△), and liver (●) after stress.** ***Ordinate axis*****, percentage of initial levels**

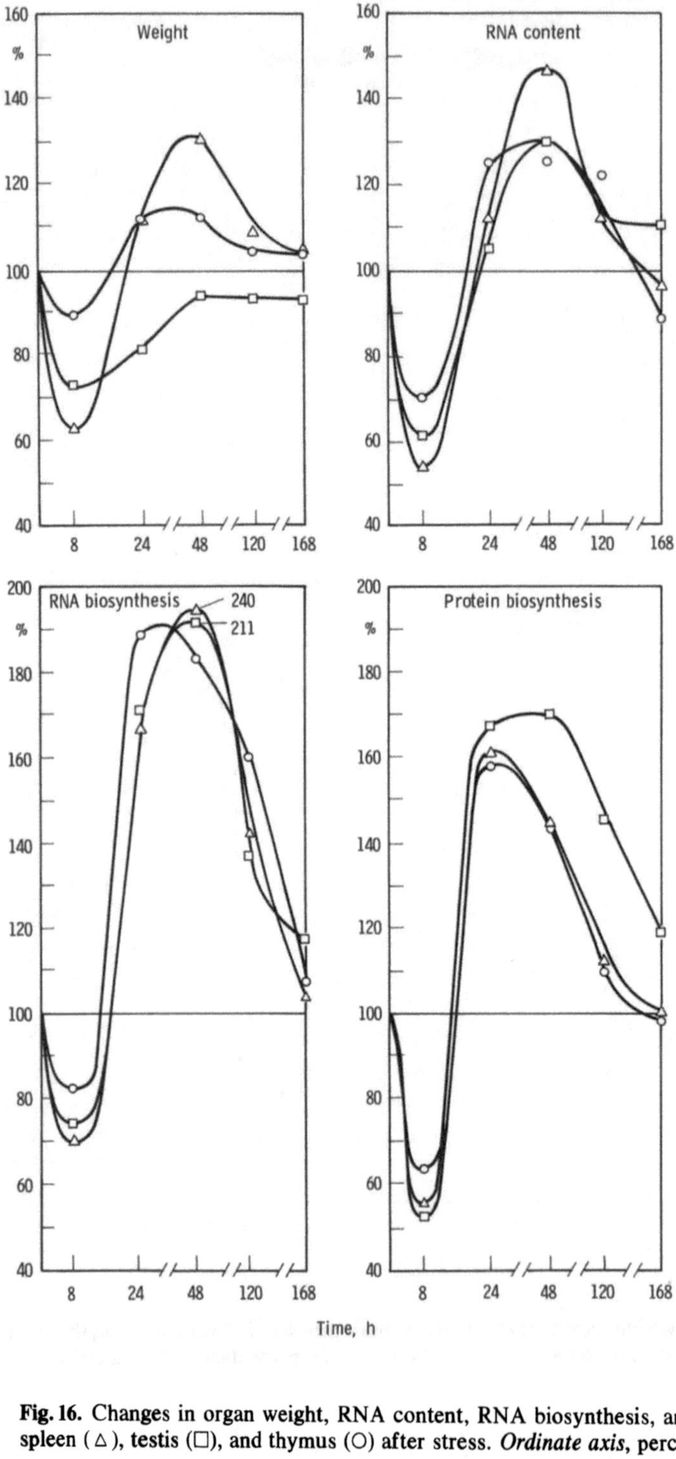

Fig. 16. Changes in organ weight, RNA content, RNA biosynthesis, and protein biosynthesis in spleen (△), testis (□), and thymus (○) after stress. *Ordinate axis*, percentage of initial levels

and systems can respond to physiologic load by a more rapid and greater activation of biosynthesis and, correspondingly, by a quicker formation of the necessary structures than would have occurred under a sudden load in the absence of an increased pool of ready ribosomes and higher activity of the entire protein-synthesizing apparatus.

Second, in the group of organs where cell migration is unobserved, the catabolic and anabolic effects are more pronounced in the liver than in the heart and brain, where both effects are relatively low (see graphs in Fig. 15). For example, judging from RNA biosynthesis activation, the maximal catabolic effects for these three organs constitute −44%, −27%, and −25%, and the maximal anabolic effects 119%, 79%, and 57%, respectively. This 'gradient' in the stress suppression of biosynthesis and its post-stress activation apparently corresponds to the physiologic role of the organs in realizing the stress-reaction.

And, third, the experiments for studying biosynthesis stipulated a relatively short interval between injecting radioactive precursors and killing the animals. Correspondingly, the results obtained in studying the spleen, thymus, and testis depended relatively little on cell migration. The fact that the maximal post-stress increase of RNA biosynthesis in those organs constituted 90%–95%, the RNA content 25%–45%, and the organ masses 10%–30% of the total affords grounds for assuming that the stage of post-stress activation of biosynthesis occurs in a sufficiently conspicuous form in immunocompetent and generative organs, and this may in turn affect the organism's adaptational responses achieved with the participation of these organs.

The generalized post-stress activation of nucleic acid and protein biosynthesis apparently represents a regulation-caused, i.e., hormone-induced, phenomenon. In assessing the possible mechanism of this phenomenon, one should keep in mind that the abrupt increase in glucocorticoid and catecholamine concentrations during the stress reaction is preceded by an increased content in the blood af ACTH and proximate opiate-like hormones [153, 481], the activating effect of these hormones on protein synthesis having been proven for some organs [521]. Subsequently, an increase was noted in the secretion and concentration in the blood of thyroid hormones [106, 441], which together with glucocorticoids lead to increased secretion of the somatotropic hormone, that is of the growth hormone [302, 435]. The anabolic effect of physiologic concentrations of thyroid hormones and the growth hormone causes no doubt. The hyperglycemia observed during the stress reaction could apparently cause hypersecretion of insulin, which activates protein synthesis in the muscle tissue [449]. Evidence shows that activation of the insulin apparatus during stress [21] is accompanied by a decrease in insulin concentration in the blood [96, 126] and, hence, is combined with enhanced insulin utilization in the tissues.

On the whole, the question of the dynamics of hormonal shifts after stoppage of stress effects and, correspondingly, of the hormonal mechanism of post-stress activation of nucleic acid and protein synthesis requires further investigation, which would go beyond the bounds of this account.

In terms of our study, it was advisable to elucidate a question essentially important for understanding individual adaptation, namely how stress-induced generalized activation of nucleic acid and protein biosynthesis affects specialized adaptive responses occurring in the organism in the post-stress period and based on the development of structural traces in systems specifically responsible for adaptation. The effects of sustained EPS on forming the initial immune response caused by the injection of sheep erythrocytes, and on conditioned defense reflexes, will be examined in concrete examples.

Experiments on 130 August rats, weight 120–150 g, were performed to study the influence of previously endured EPS on the development of initial immune response. The EPS was reproduced by the above method. The animals were immunized with sheep erythrocytes (SE) 2 h after terminating the effect of the stress situation by a single intraperitoneal injection of an a fortiori small dose of antigen (AG), viz 1% SE suspension (1 ml). To characterize the primary immune response, we determined the number of straight antibody-forming cells (ABFC) in the spleen by the method of Jerne and Nordin [196] after preliminarily enriching the splenocyte population with viable cells to as much as 85%–92% by the Boehmer and Shortman [40] technique. Concurrently, we studied the number of hemagglutinins in the blood serum and the B-lymphocyte content in the spleen. The latter were indicated by indirect methods on living cells sécured to glass by means of a Fernvar film. Rabbit serum against August rat globulins and fluorescent ass serum against rabbit globulins were adsorbed twice with acetonized rat liver powder. These indices were determined in dynamics on days 3, 5, 8, and 14 day following immunization.

From Table 14 it follows that EPS causes the initial immune response to increase in August rats at all stages. This was evident by the fact that after the injection of the antigen into EPS-exposed animals, the antibody (hemagglutinin) titer increased more steeply to remain higher in all stages of the response. As follows from Table 14, this post-stress increased ability to synthesize antibodies is provided by a multiple increase in ABFC, reckoned both in millions of spleen cells and for the entire spleen; at the same time, the total number of spleen cells also increased with a constant percentage of B lymphocytes.

It is also interesting that an increase in the ABFC number in the course of initial immune response is more pronounced than the antibody titer in animals previously exposed to EPS. For example, on day 5 of the immune response, the hemagglutinin titer increased by less than 150% in stress-exposed animals, while the number of ABFC increased by five times more for the spleen; correspondingly, the number of antibodies formed by 1 million ABFC was decreased.

This correlation possibly expresses increased functional potency and the large reserve of the system responsible for adaptation, in which additional stress effect was instrumental in forming an excessive structural trace, the ABFC pool, far from always entirely utilized when administrating a known low antigen dose. As for the mechanism of post-stress increase in the number of ABFC, one may assume that it is caused by accelerated differentiation of B lymphocytes into plasmic cells in EPS-exposed animals.

This phenomenon may be based on changes occurring under stress in a

Table 14. Effect of EPS on dynamics of initial immune response in August rats

	Period following immunization			
	3 days		6 days	
	Experimental	Control	Experimental	Control
ABFC/million	20.4 ± 3.4	18.04 ± 1.96	170.8 ± 36.3[a]	26.08 ± 3.3
	($n = 9$)	($n = 14$)	($n = 15$)	($n = 22$)
ABFC/spleen	10,199 ± 841	12,272 ± 1,906	103,206 ± 31,263[a]	19,902 ± 5,028
Number of cells in spleen, million	542.5 ± 38.5	531 ± 38.7	705 ± 77.4[a]	490 ± 29.0
Number of B-cells in spleen, %	44.5 ± 3.62	38.6 ± 1.82	40.6 ± 1.33	41.04 ± 1.59
	($n = 5$)	($n = 6$)	($n = 11$)	($n = 15$)
Antibody titer	0.26 ± 0.04[a]	0.11 ± 0.02	1.37 ± 0.04[a]	0.97 ± 0.07[a]
	($n = 15$)	($n = 12$)	($n = 16$)	($n = 26$)

	Period following immunization			
	8 days		14 days	
	Experimental	Control	Experimental	Control
ABFC/million	39.7 ± 10.6	22.8 ± 3.1	29.8 ± 8.0	12.4 ± 1.9
	($n = 7$)	($n = 14$)	($n = 5$)	($n = 11$)
ABFC/spleen	40,068 ± 9,613[a]	14,930 ± 3,348	33,979 ± 5,924[a]	8,460 ± 1,440
Number of cells in spleen, million	1,074 ± 159	531.6 ± 64.5	1,050 ± 163.5[a]	548 ± 55.5
Number of B cells in spleen, %	40.7 ± 1.92	39.9 ± 2.6	43.2 ± 2.01	39.8 ± 2.34
	($n = 6$)	($n = 9$)	($n = 5$)	($n = 6$)
Antibody titer	1.43 ± 0.08	1.31 ± 0.048	1.54 ± 0.21	1.32 ± 0.11
	($n = 8$)	($n = 14$)	($n = 7$)	($n = 12$)

[a] differences from control significant.
[b] n, number of animals.

cooperation of cells interacting during the immune response – B lymphocytes and T cells (helpers and suppressors). In addition, EPS may also cause a decline in the suppressor function of T lymphocytes so as to enhance the intensity of the H-lymphocyte response to SE, a thymus-dependent antigen, as was observed in the experiments of other investigators [600]. The mechanism of stress effect on immunogenesis is a complex issue requiring special inquiry. Nonetheless, it is apparent that the final EPS effect in our experiments was in a greater number of spleen cells and an even greater increase in ABFC in the course of initial immune response.

In principle, this evidence on the post-stress activation of immunogenesis is in agreement with the results of previously published studies [141] and permits one to underline that under certain conditions the nonspecific stress reaction comes out as a factor that potentiates the development in immunocompetent organs of a systemic structural trace, which constitutes the foundation of highly specific immunogenetic responses.

The next stage of our research was concerned with the *influence of EPS on the development and retention of conditioned defense reflexes.* The assumption that the activation of RNA and protein synthesis constitutes an essential link in the mechanism of forming brain memory and thereby of temporary connections and new habits has been well grounded in relevant studies of the last several decades. For example, a specific cerebral cortex function, i.e., locking and fixation of temporary connections in cortical neurons, is known to entail naturally the activation of nucleic acid and protein synthesis, which is absent during ready, previously formed conditioned reflexes and with isolated actions of conditioned and unconditioned stimuli [295, 517]. In accordance with this, inhibition of nucleic acid and, even more, protein synthesis either delays or makes impossible the fixation of temporary connections [344], while any factors that cause activation of nucleic acid and protein synthesis in the cerebral cortex possess an opposite effect [308, 347].

From these facts arose the idea that in developing temporary connections, environmental stimuli, acting in definite sequence on the receptors, generate a constellation of intensely functioning neurons in the brain, a constellation that includes both neurons responsible for the reception of a previously indifferent signal and neurons responsible for unconditioned-reflex adaptational activity. In the first neurodynamic stage, or the stage of short-term memory, the physiologic hyperfunction of each of the excited neurons leads to the increased synthesis of nucleic acids and proteins therein by means of intraneuronal connections between the function and the genetic apparatus. This phenomenon, constituting the essence of the consolidation stage, leads to increasing migration of the newly synthesized neuron proteins to its processes and, as a result, to transformation of synaptic connections between neurons. This causes the initially functional process to obtain its structural provision, and the group of excited neurons transforms into a certain integral and structurally secured functional system that is retained for a long time after terminating excitation and, therefore, constitutes the structural basis of memory [317].

This hypothesis and the post-stress activation of biosynthesis mentioned

Table 15. Effect of EPS on development and retention of conditioned defense reflex of active avoidance (CRAA)

	Control	Time elapsed after terminating stress actions				
		2 h ($n = 15$)	1 day ($n = 12$)	6 days ($n = 10$)	12 days ($n = 12$)	20 days ($n = 12$)
Number of combinations needed to develop contitioned reflex	16 ± 2.25	11.2 ± 1.29 $p > 0.1$	9.3 ± 1.23 $p < 0.01$	13.5 ± 1.99 $p < 0.01$	18.8 ± 1.49 $p > 0.1$	20.3 ± 2.35 nonsignificant
Number of combinations needed to restore reflex when checking retention	3 ± 1.54	3.7 ± 0.16 $p < 0.01$	3.0 ± 0.67 $p < 0.01$	4.07 ± 0.74 $p < 0.01$	10.5 ± 0.9 $p > 0.1$	11.5 ± 1.25 nonsignificant
Decrease in conditioned reflex, %	42 ± 5.85	83 ± 3.35 $p < 0.01$	70 ± 6.73 $p < 0.05$	71.3 ± 4.15 $p < 0.01$	47.3 ± 4.07 $p > 0.1$	41.2 ± 5.01 $p > 0.1$

above, a phenomenon quite pronounced in the brain, served as the foundations for studying the development and retention of conditioned reflexes during various intervals after EPS. The experiments [388] involved Wistar male rats (weight 180–200 g), in which a conditioned defense reflex of active avoidance (CRAA) in a Y-shaped labyrinth was developed. The conditioned stimulus was light and the unconditioned stimulus was electric current. The reflex was developed until a criterion of four faultless responses in five consecutive trials was attained; retention was checked 24 h after development, prior to which control animals were not subjected to any actions, while the rest were subjected to 6 h EPS using the above technique.

A conditioned reflex was developed in various animal groups 2 h and 1, 6, 12, and 20 days after EPS termination. The results shown in Table 15 indicate that even 2 h and 1 day after EPS the conditioned reflex was observed significantly to develop more rapidly and to become sharply retained. Indeed, the number of combinations needed for the reflex to develop decreased under EPS from 16 to 11 and 9, and the number of combinations necessary to restore the reflex when checking its retainment decreasement from 9 to 1.7 and 3, i.e., three to five times. Accelerated development of the conditioned reflex subsequently stopped, while increased retention, on the other hand, was distinctly evident 6 days after EPS, but not after 12 and 20 days.

Thus, the principal result is that after EPS the brain retains a certain shift, which promotes an increase in the degree of retention of a temporary connection, its fixation. Inasmuch as fixation is largely caused by the activation of nucleic acid and protein synthesis, post-stress activation of nucleic acid and protein synthesis may very probably operate here. This post-stress activation is really observed in the brain and, as has been shown in recent years, may occur as a result of direct action by at least two stress hormones, viz ACTH [94] and glucocorticoids [23]. However, Table 15 shows that the most pronounced increase in retention of the conditioned reflex is observed at its development 2 h after EPS, i.e., in conditions when RNA and protein biosynthesis in the brain have still not increased. Hence, one may assume that apart from the post-stress activation of biosynthesis, post-stress higher rates of development and retention of the conditioned defense reflex are apparently caused by a stronger activating effect of the reticular formation, which occurs via ascending adrenergic brain systems and can play a role during the fixation of temporary connections [216].

On the whole, the presented facts unambiguously show that the post-stress activation of nucleic acid and protein synthesis, the stress anabolic phase, may potentiate long-term specialized adaptational responses to specific factors.

General Mechanism and Principal Stages of Phenotypic Adaptation

The above provides grounds for assuming that the systemic structural trace constituting the basis of specific adaptation and the nonspecific stress reaction are inseparably interrelated, i.e., coordinated within the framework of a single

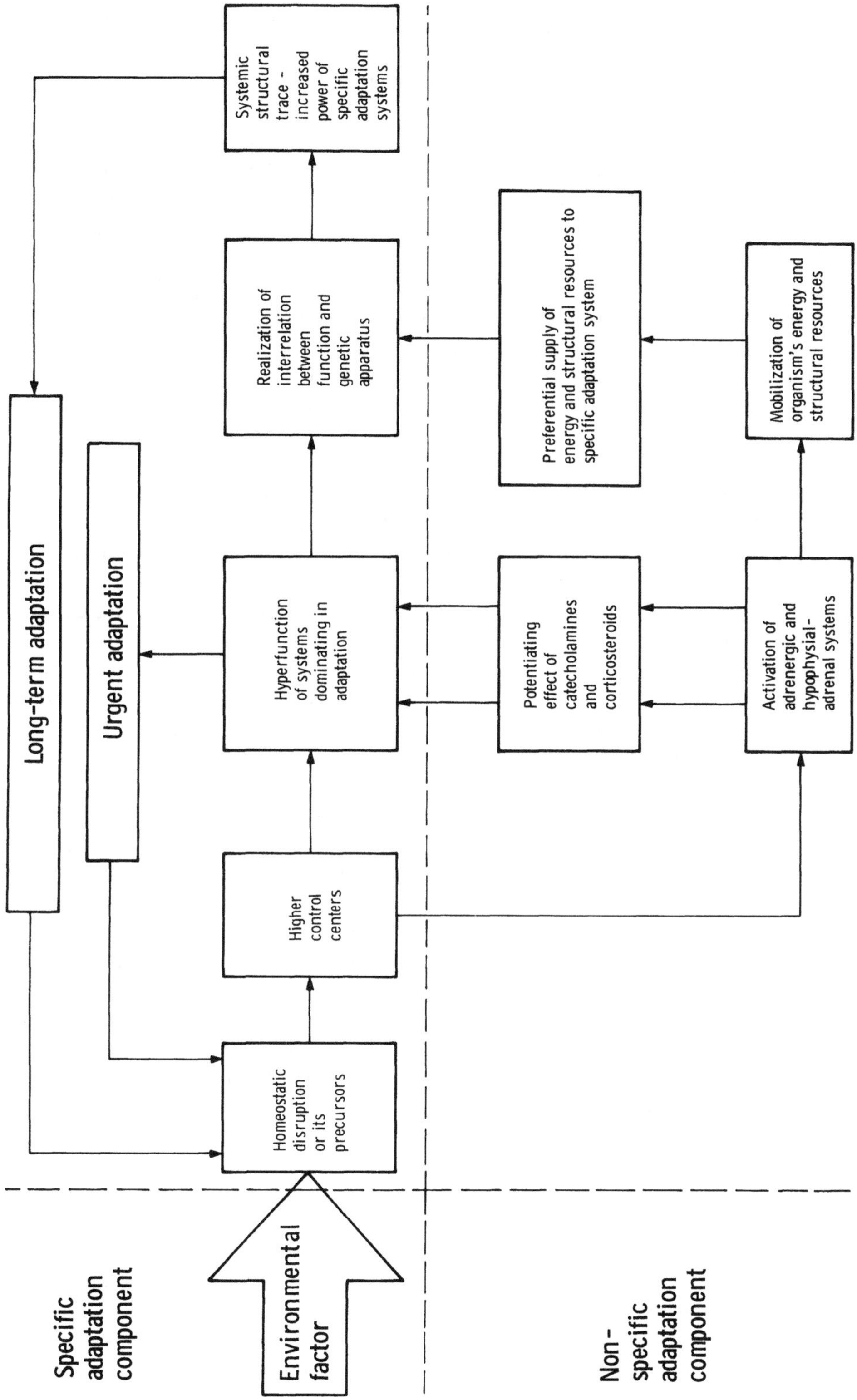

Fig. 17. Mechanism of phenotypic adaptation. Explanation in text

adaptational response. This idea of the mechanism of phenotypic adaptation is summarized by the scheme in Fig. 17, which shows that disruption of homeostasis by environmental factors or a signal that such disruptions are possible in future, activate the systems responsible for adaptation via regulatory mechanisms. As a result, the following two phenomena occur: First, mobilization of the functional system specifically responsible for adaptation to a given factor and, second, a nonspecific stress reaction. The system of specific adaptation subsequently fulfils an enhanced function and dominates in the vital activity of the organism. On the basis of a relationship between function and the genetic apparatus and upon the potentiating effect of the stress reaction, the cells of the dominant system develop a structural trace, which substantially enhances its potency. This results in reliable and long-term adaptation, which eliminates the initial disruption of homeostasis and renders the stress response superfluous. The increased functional possibilities of the system responsible for adaptation are based on a change in the ratio of cell structures, namely on the selective increase in the mass and potency of the structures responsible for control, ion transport, and energy supply. As was shown in detail above, this very shift provides not only high potency but also economy of the system responsible for adaptation and high reactivity of its links for controlling signals.

In the course of individual life, the organism encounters a whole aggregate of environmental factors; from the mechanism just examined this results not in one structural trace but in a complex aggregate of systemic structural traces, which deeply change the appearance of the organism and form its phenotype in conformity with environmental requirements. In addition, once formed the systemic structural trace and adaptation to a definite environmental factor are not eternal. Environmental changes are accompanied by changes in the spectrum of adaptational responses: A new set of long-term adaptational responses replaces more or less gradually the existing set of long-term adaptational responses. As was shown above, this occurs by deadaptation, in the course of which previously established systemic structural traces disappear through the decrease in nucleic acid and protein synthesis and the activation of structural degradation. The stress reaction, which constitutes the essential initial link in any new adaptation, should, due to its catabolic effect, obviously simultaneously potentiate deadaptation, the elimination of old structural traces that have lost their biologic significance. In other words, stress does not just provide the transfer of resources from the inactive systems to the active ones, but is an apparatus that simultaneously catalyzes the forming of some systemic structural traces and the disappearance of others. A general adaptation syndrome thus develops as a factor that plays an important role in reprograming the adaptive possibilities of the organism to solve new problems posed by the environment, and thereby the role of a factor that catalyzes the phenotypic changes in the course of life.

The suggested concept of phenotypic adaptation allows one to chart the basic stages of that process, which are examined in greater detail below, using concrete long-term adaptive responses as examples. The first stage of urgent adaptation is characterized by mobilization of the preexisting adaptive mechanisms, viz hyperfunction or initiation of a functional system responsible for adaptation. This hyperfunction lacking structure, despite its insufficient perfection, makes it possible for the organism to 'hold out' until the development of long-term adaptation. In the behavioral sphere, it is a stage of wasteful and only occasionally successful responses; in adaptation to physical loads, it is a stage of extreme work with maximal circulation and respiration minute volume, with lacticemia close to critical level, and so on. This initial stage is essentially based on the triad involving (a) hyperfunction of the system specifically responsible for adaptation to a given factor; (b) a stress reaction; and (c) violation of functions caused by shifts in homeostasis.

When inherited or acquired deficiency of any link in the systems is involved in response to a new environmental factor, urgent adaptation may apparently prove so incomplete that the organism either becomes sick or avoids solving the problem dictated by the environment; diseases caused by unrealized adaptation may correspond to these situations.

The second stage of transition from urgent to long-term adaptation is characterized by activation of nucleic acid and protein synthesis in the cells of the system specifically responsible for adaptation, i.e., by enhanced power of that dominant system and gradual decline of the stress reaction. Subsequently, stable adaptation develops in most cases. However, the transition stage may be protracted in so-called hopeless situations, when the factor that acts on the organism is too strong or when the environmental situation too complex, so that the required adaptational response proves unfeasible. In such situations, the functional system fails to be effective, and the systemic structural trace does not form therein. As a result, initial homeostatic disruptions persist and the stress reaction thereby stimulated become exceedingly intense and extended. Precisely in this situation, the stress reaction may be transformed from an instrument of adaptation into an instrument of injury and cause numerous stress diseases like the ones examined below (see Chap. 6).

The third stage of long-term adaptation is characterized by the presence of a systemic structural trace, perfect adaptation, and the absence of a stress reaction to a definite factor or situation. With excessively protracted or directed adaptation of this kind, domination of a given system in accordance with the basic principle of the dominant may lead to a one-sided development of the organism. For example, animal adaptation to physical loads has proven to be accompanied by a decrease in liver and kidney structural reserves, and adaptation to hypoxia may be accompanied by immunodeficiency. Such 'disarmament' of specific systems may be a prerequisite of diseases of one-sided development (see Chap. 5).

The fourth stage of wear and functional insufficiency is not indispensible. It develops only with excessively tense adaptation and is characterized by a situation in which high loads on systems dominating in adaptation lead to excessive hypertrophy of their cells and, as a result, to structural and functional disorders. This has been shown to occur possibly on account of two mechanisms. The first one is that with high levels of hypertrophy cell growth proves to be imbalanced and accompanied by the selective retardation of the structural masses responsible for receiving control signals, ion transport, and energy supply. In a sense, this situation the opposite of the structural complex of optimal adaptation and has been studied in detail for large cardiac hypertrophy, in which it leads to depression of the contractile function (see Chap. 4).

The second mechanism involves, after lengthy hyperfunction and hypertrophy, the nerve centers, endocrine glands, and executive organs, which may develop a peculiar complex of 'local wear,' evident as dereased nucleic acid and protein synthesis, disrupted structural renovation, death of a certain number of cells, and the development of organ sclerosis (see above).

Structural Cost of Adaptation

In this chapter, I have frequently used the expression 'adaptation cost,' introduced in one of my previous works [320]. The structural cost of adaptation is the number of structures, i.e., the nucleic acid and protein molecules, which the organism must additionally synthesize to achieve a given adaptational response. The 'structural cost', thus, consists, first, of molecules that are additionally degraded in the initial, urgent, i.e., stress phase, of adaptational response and must be resynthesized as stable adaptation forms; second, of the mass structure increment (i.e., essentially from the molecule population) normally observed in the transition and stable stages involving organ hypertrophy; and third, of accelerated renovation of the structure responsible for adaptation of the system that may be retained during the entire adaptation period. Indeed, the experiments outlined above showed that following a single 6-h stress action, most internal organs were found to have substantially shortened RNA and protein half-lives during the subsequent 11 days.

Thus, even a single stress action considerably decreases the life of cell structures and requires their rapid resynthesis for a lengthy period of time, i.e., it has a high structural cost; it is precisely for this reason that such repeated effects play a role in wearing the organism in the course of aging.

The structural cost of adaptation may be disclosed with even greater definition directly in the organs, in which a systemic structural trace forms through the activation of nucleic acid and protein synthesis. For example, when studying heart adaptation to continuous load, its compensatory hyperfunction (see above), it was shown that from the very outset RNA degradation in the myocardium is sharply accelerated; however, its total content nonetheless grows since RNA biosynthesis is more increased than its degradation. As a result, during the initial 11 days of compensatory hyperfunction caused by narrowing of the aorta, i.e., during a period in which heart hypertrophy developed, the amount of RNA synthesized in the myocardium proved to be twofold greater than in controls [321]. A similar situation was observed with the protracted and considerably increased functions of any other organ or system; in all cases, long-term adaptation is provided through increased degradation and even greater nucleic acid and protein biosynthesis, i.e., it has a definite structural cost. The amount of RNA that may be transcribed on the RNA matrices of undividing differentiated cells is not infinite; neither is the number of protein molecules that may be translated from each messenger RNA matrix. For this reason, excessively protracted and tense long-term adaptation achieved at too expensive structural cost may be the cause of local wear or local aging, the basis for exhaustion of the organism's adaptive possibilities (see above).

To determine approximately the structural cost of each specific adaptational situation today and to do this more exactly in the future is of major practical significance. One can distinguish at least five situations when the structural cost of adaptation is particularly high.

As was just shown, the first situation, which has tremendously high structural cost, involves heavy and lengthy stress.

The second situation is the initial or emergency stage of adaptation to the effect of any factor. As was shown above, in this stage the direct shortage of energy-rich phosphorous compounds in cells and the catabolic effect of stress hormones lead to the considerable damage of membranes and to the degradation of other structures, this being evident in fermentemia, negative nitrogen balance, and other similar shifts.

The third situation is most strikingly apparent in the transitional stage of adaptation, when pronounced activation of nucleic acid and protein synthesis and increase in the mass of cell structures develop in the form of hypertrophy of individual organs and whole functional systems in response to load, hypoxia, cold, or any other factor causing energy shortage. Understandably, if the resultant stable adaptation proves essential to the organism during its entire subsequent life, or for a long period of time, then the question concerns economical utilization of the structural reserve. In this case, when the environmental factor to which adaptation occurred had subsequently ceased to act for a lengthy period of time to start acting again after deadaptation, the systemic structural trace forms anew. Correspondingly, the structural cost of adaptation proves extremely high.

The fourth situation is connected with the previously examined role of the dominant principle in long-term adaptation. The forming of a pronounced systemic structural trace in the dominant functional system responsible for adaptation (structural prevalence of that system) may be accompanied by various degrees of atrophy or functional deficiency of other systems. Then, under environmental changes and abrupt loads on these deficient systems, they would be found to have a large energy shortage and damaged structures, followed by major activation of nucleic acid and protein synthesis. In other words, a change in the organism's adaptational program, i.e., in its vital vector, would have considerable structural cost.

Finally, in the fifth situation the system protractedly dominant in tense long-term adaptational response may be subjected to direct wear. In other words, this may result because the genetic apparatus of differentiated cells had exhausted its ability to generate new RNA and, correspondingly, protein portions, which form cell structures. The outcome of this process is the development of sclerosis in a given organ or system and functional insufficiency, the phenomenon of local aging of overloaded systems cited above. This outcome appears highly probable when the first three of the above situations are strikingly pronounced in adaptation, namely, with a great load on an untrained organism, repeated loss and subsequent recovery of adaptation, and increasing prevalence and relevant one-sided development. The evolutionally determined desire of the organism to respond harmoniously to environmental factors naturally prevents this outcome. It is significant that even if all these factors were reduced to a minimum, the adaptation occurring in optimal conditions would nonetheless be combined with definite additional degradation and the forming of cell structures, i.e., would have a specific structural cost. In this case, the structural cost of adaptation may be different at equal intensity of the environmental factor under equal load conditions. Indeed, as was shown above, an environmentally determined increase in physiologic function activates RNA transcription and protein synthesis through the occurrence in the cell of a relative ATP shortage, which again is a factor that determines structural degradation. Under equal additional load and equally increased ATP utilization in the myofibrils, membrane pumps, and secretory granules, the ATP concentration may decrease by different degrees depending on the potency of its resynthesis system in the cell, primarily on the power of the mitochondrial system and transport thereto of substrates and oxygen. For example, with a powerful ATP resynthesis system as in trained animals, the ATP and CP concentrations under maximal loads on the heart decrease less than in untrained animals [356] and, correspondingly, RNA and protein synthesis is not activated (see Fig. 2), and heart hypertrophy should also be less pronounced.

The opposite situation would arise if simultaneously with additional load on the organ the researcher were to limit the potency of the ATP resynthesis system in its cells by oxygen shortage. Such experiments were performed by Hubner [184], who, prior to removing one kidney, subjected the other kidney to anoxia of varying duration. As a result, the compensatory hyperfunction of the remaining kidney led to unusually high activation of nucleic acid and protein synthesis. In other words, the lower the potency of the ATP resynthesizing apparatus, the higher the ATP shortage under load, and the higher the activation of nucleic acid and protein synthesis and the resultingly higher hypertrophy with which the organism has to 'pay' for the developing long-term adaptation, the higher the structural cost of that adaptation. With a highly potent ATP resynthesizing apparatus, the hypertrophy of a given organ and, correspondingly, the structural cost of adaptation prove to be less. *This means that when two organs are subjected to an equal load the one whose unit mass has the ability to generate more energy in ATP form hypertrophies less. It copes with the load primarily by using the large aerobic reserve of available tissue and secondly by means of the increasing tissue mass. The highly potent and effective systems of energy supply prevent uneconomical utilization of the organism's structural resources.*

This concept corresponds to the results of the familiar experiments performed by Beikert [25], who for 12 weeks trained guinea pigs by swimming and then compelled them to swim until complete exhaustion. As a result, animals with highest heart hypertrophy were found to become exhausted earlier than all the rest. In our view, the fundamentally important results of the clinicophysiologic studies performed by Dembo et al. [77] should be understood in the same way;

these authors showed that among the most highly trained people, namely masters of sport and first-class athletes, the number of persons with pronounced heart hypertrophy was fewer than among second- and third-class sportsmen. According to those authors, the number of persons with clinically determinable heart hypertrophy among athletes who had trained for over 10 years and, consequently, were highly resistant to physical loads, was less than among those who had been training for shorter periods of time.

In assessing this evidence, one should keep in mind that under natural conditions the potency of the mitochondrial system per cell unit mass depends not only on the intensity of previous training sessions, but is also considerably predetermined genetically. In full conformity with this, the so-called aerobic power (the amount of oxygen consumed per kilogram of body weight during maximal work) varies widely in untrained or equally trained persons. The degree of hypertrophy should fluctuate in the same measure under equal load, which is highest at the lowest potency of the system of oxidative ATP resynthesis in the muscle tissue.

Generally speaking, the above facts may be summarized in the following form: Optimal adaptation develops with the highest potency of the system of oxidative ATP resynthesis and entails that high functional adaptation results are combined with moderately hypertrophied cells, i.e., prove to be achieved at minimal structural cost. In this case, more or less economical adaptation is predetermined not only by the intensity and conditions of environmental factors, but by the genetically determined properties of the organism.

Significantly enough, the transitions of adaptation into a disease cited above provide no grounds to assess adaptation negatively as a whole, for the organism manages to adapt to most real environmental situations. This perfection of evolutionally determined responses is due, first of all, to the fact that the organism possesses a system of behavioral responses that permit it to avoid excessively lengthy contact with physical and chemical environmental factors creating direct loads on the functional systems, and it thus evades excessively tense and protracted adaptation. Second, even serious and protracted stress eventually weakens in 'hopeless situations,' since the organism possesses multiform systems that restrict stress and prevent stress injuries (see Chap. 7). And third, even tense long-term adaptation to a continuously active environmental factor far from always results in disease since the duration of the above-mentioned stage of stable adaptation in many cases approaches specific life expectancy. These powerful adaptation mechanisms long ago resulted in purposefully developed adaptation being used empirically to prevent diseases. Further study and control of adaptation mechanisms will reveal substantial prospects in this respect. Adaptation to altitude hypoxia, which we shall examine next, allows us to concretize the developed concept on adaptation stages and assess the possibilities for using adaptation for preventive purposes.

Chapter 3
Development of Adaptation to Hypoxia and Its Preventive Use

"We bitters drink
to escape poison..."
William Shakespeare

In recent decades, adaptation to altitude hypoxia has become an increasingly important issue in biology and medicine. Its urgency and the current extensive research in many countries have been determined by the need to solve several pressing and vital problems, chiefly: (a) to adapt numerous groups of people who have just settled in mountainous regions to altitude hypoxia; (b) to provide adaptation to the combined effects of altitude hypoxia and physical loads; (c) to make possible high-altitude complex intellectual activity, essential for operating modern scientific instruments; and (d) to overcome the Andes disease and deadaptation after descending from mountains. A no less important problem is the use of adaptation to hypoxia for preventing diseases of the circulatory organs and the brain, and also for enhancing resistance to all kinds of unfavorable factors.

Relevant investigations by Haldane and Priestly [158], Barkroft [22], Sirotinin [524], Gazenko et al. [131], Barbashova [19, 20] and the present author [320] have permitted the distinction of at least four levels of mutually coordinated adaptational mechanisms within the complex architecture of adaptation to hypoxia.

1) Mechanisms whose mobilization can provide a sufficient supply of oxygen to the organism despite its shortage in the environment; hyper-

Table 16. Effect of actinomycin 2,703 on oxygen absorption by intact rats and those adapted to altitude hypoxia

	Number of animals	Oxygen absorption, ml/kg/h		
		sea level	altitude 7,000 m	p
Control	20	2,505 ± 101	1,687.5 ± 101	<0.01
Control + actinomycin 2,703	12	2,290 ± 186	1,438.0 ± 130	<0.01
Adaptation to hypoxia for 10 days	12	2,285.9 ± 87	1,847.9 ± 180	<0.01
Adaptation to hypoxia for 25 days	12	2,274.6 ± 56	2,043.3 ± 54	<0.02
Adaptation to hypoxia for 25 days + actinomycin 2,703	12	1,741.4 ± 147	1,278.2 ± 161	<0.05

Lesser oxygen utilization in the above experiments did not depend on the extent of hypothermia since decline of rectal temperature in animals administered and not administered actinomycin was about the same.

ventilation; heart hyperfunction providing the transport of increased amounts of blood from the lungs to the tissues; and polycytemia and the corresponding increase in oxygen capacity.

2) Mechanisms that make possible a sufficient supply of oxygen to the brain, heart, and other vitally important organs in spite of hypoxemia, viz dilatation of the arteries and capillaries of the brain, heart, etc.; decrease in the diffusion distance for oxygen between the capillary wall and the cell mitochondria due to the formation of new capillaries and changes in the cell membrane properties; and increased ability of the cells to utilize oxygen due to increase in myoglobin concentration.

3) Greater ability of the cells and tissues to utilize blood oxygen to generate ATP notwithstanding oxygen shortage. Some investigators assume that this possibility may be realized either through the increased affinity of cytochrom-oxydase, the end enzyme in the respiratory chain, to oxygen, through a simple increase in the number of mitochondria per cell unit mass, or through the increased coupling of oxidation with phosphorylation.

4) Increased ATP anaerobic resynthesis owing to activated glycolysis, which many researchers assess as a significant adaptive mechanism.

A simple analysis shows that these basic adaptation factors occur through the activation of nucleic acid and protein synthesis and the development of structural changes in the systems responsible for adaptation. This activation develops during adaptation to hypoxia in systems responsible for oxygen transport, i.e., in the blood system [34, 409, 430], lungs [42, 325], heart [354, 355], and also in organs that do not take part in oxygen transport, primarily in the brain [13, 349, 357].

It should be emphasized that the activation of nucleic acid and protein synthesis does not just accompany adaptation, but plays the role of an essential link therein. This follows from the results of our joint work with Malkin and Annanurova et al. [353], which showed that the elimination of synthesis activation by means of actinomycin, an inhibitor of RNA synthesis, impairs adaptation, suppresses oxygen utilization by the organism, and results in the death of most of the animals. Of adapted rats, 65% died when actinomycin was administered for 5 days between days 20 and 25 of adaptation in a dose of 10 μg/kg, which did not cause death in controls. Oxygen utilization by survived rats was impaired.

Table 16 shows that when unadapted animals were first taken to a height of 7,000 m, they began to consume about 30% less O_2. In the course of adaptation, this shortage in oxygen absorption gradually decreased so that by day 25 the animals adsorbed almost the same amount of oxygen as at sea level. This discovery was made by Kreps et al. [242] and represents a striking expression of adaptation to oxygen shortage in the environment. After actinomycin was administered for 5 days, the oxygen utilization shortage was again equal to 26%. In addition, yet another important phenomenon was evident in adapted animals under the influence of actinomycin: Oxygen absorption dropped not only in the altitude chamber but at sea level as well. Thus, shortage in oxygen utilization during inhibition of RNA synthesis is also observed when there is no oxygen shortage in the environment, without increased load on the functions of external respiration and circulation. This impaired utilization can hardly be explained by injured transport systems. In this connection, one should keep in mind that in suppressing RNA synthesis, actinomycin injures the structures where this synthesis is most intense. For example, actinomycin suppresses the thyroxin-induced formation of mitochondria. Possibly, therefore, the above process concerns disorders in the biosynthesis of mitochondrial structures and formation of mitochondria, i.e., inhibited formation of organelles, which in the end determine utilization of O_2 by the organism and ATP resynthesis and play a key role in adaptation to hypoxia.

On the whole, this means that activation of nucleic acid and protein synthesis constitutes an essential link in adaptation to hypoxia. This activation brings about the formation of a systemic structural trace which constitutes the basis of long-term adaptation to that factor. A large number of publications, including several monographs in Russian [290, 320, 398–400] analyze various aspects of adaptation to hypoxia.

In line with our task, we shall not duplicate these reviews but we shall restrict ourselves to examining the dynamics of adaptation to hypoxia, the main features that characterize its systemic structural trace and utilization for preventive purposes.

Systemic Structural Trace and Principal Stages of Adaptation to Hypoxia

In its early stage, adaptation to hypoxia develops essentially differently from adaptation to other environmental factors. The difference is primarily that contrary to factors causing motor activity, oxygen shortage in the environment initially does not affect exteroceptors but imperceptibly and gradually invades the internal medium causing hypoxemia and thereby disrupting homeostasis. Only after the occurrence of hypoxemia does oxygen shortage begin to act as a stimulant of chemoceptors of the aortal-carotid zone, as well as directly on centers that control respiration and circulation, and also on other organs thus causing at least three interrelated series of events.

First, hypoxemia causes an increase in function of systems specifically responsible for oxygen transport from the environment into the organism and its distribution therein, i.e., lung hyperventilation, increased cardiac output, dilatation of brain and heart vessels, narrowing of vessels in the abdominal organs and in muscles, and resultant rise in arterial pressure.

Second, it is characterized by activation of the adrenergic and hypophysial-adrenal systems, i.e., by a stress reaction. This nonspecific component of adaptation plays a role in mobilizing the circulatory and external respiratory apparatuses; however, it is simultaneously evident in a sharply pronounced catabolic effect, i.e., in negative nitrogen balance, loss in weight, atrophied fatty tissue.

Thirs, in restricting ATP resynthesis in the mitochondria, acute hypoxia directly depresses the functions of a number of systems, primarily in the higher sections of the brain, this being apparent in impaired intellectual and motor activities.

This combined mobilization of systems specifically responsible for oxygen transport and nonspecific stress reaction with impaired functioning of the apparatus that controls behavior and motion constitutes a syndrome characterizing the initial stage of urgent, yet largely incomplete adaptation to hypoxia. In the second transitional stage of adaptation, a shortage in the energy-rich phosphorous compounds in the cells of systems that perform increased functions and are subjected to hypoxemia causes activation of nucleic acid and protein synthesis. In the course of adaptation to hypoxia, this activation of biosynthesis involves a wide range of organs and systems and results in an extensive systemic structural trace with ramified architecture. For instance, activation of nucleic acid and protein biosynthesis in the bone marrow becomes the basis for erythroid cell proliferation, which results in the development of adaptive polycytemia and increased oxygen capacity. Activated synthesis of nucleic acids and proteins in lung tissue cells leads to lung hypertrophy and an increased respiratory surface, and in the myocardium it results in a set of changes highly similar to those observed with physical loads, i.e., to increasingly potent adrenergic heart control, considerably increased myoglobin concentration and transmissive capacity of the coronary bed, and on the whole an increasingly powerful system of energy supply to the heart.

In the brain, nucleic acid and protein synthesis is also naturally activated during adaptation to hypoxia and apparently plays a role in the structural

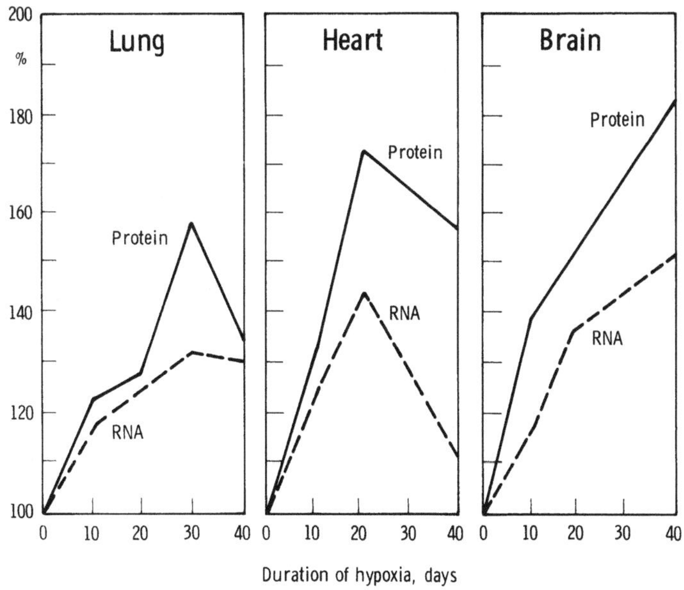

Fig. 18. Dynamics of RNA concentration and protein synthesis (percentage of initial values) in the lungs, heart, and brain during adaptation to hypoxia

changes that make it possible for the brain to extract the needed amount of oxygen from hypoxemic blood.

The curves in Fig. 18 characterize the dynamics of nucleic acid and protein synthesis in the lungs, heart, and brain of animals during adaptation to periodic hypoxic effects in an altitude chamber [325, 345, 354, 355]. Basically similar activation was seen in mountains under natural altitude hypoxia. It is significant that apart from other structures this activation results in mitochondrial and multienzymic complexes of glycolysis, i.e., structures directly responsible for the transformation of energy into an ATP form that can be utilized for cell functions. A characteristic feature of the transitional stage is that the systemic activation of biosynthesis is combined with declining but still persisting manifestations of urgent adaptation.

The third stage of stable adaptation is characterized by the complete formation of the systemic structural trace, whose architecture is characterized by several features that play a decisive role both in adaptation to hypoxia and in the use of that adaptation for preventive purposes. The first feature is the increased potency and simultaneous economy with which the external respiratory and circulatory systems function. Increase in the lung respiratory surface, total chest capacity, and power of the breathing musculature enhance the inspiratory volume and the coefficient of oxygen utilization from the inhaled air. As a result, the initial hyperventilation decreases and the organism receives the necessary oxygen at a relatively lower respiration rate.

During adaptation to hypoxia, the heart grows in efficiency with a 1.5–2-fold increase in coronary bed capacity, higher myoglobin concentration, and a greater number of mitochondria in the myocardium, i.e., with increasing potency of the energy supply system [214, 428, 542]. At the same time, the polycytemia and increased oxygen capacity observed during adaptation result in a situation in which the necessary oxygen can be supplied to the tissues at a lower minute volume; increased tissue capacity to extract oxygen from passing blood may have the same economizing significance during adaptation. As an outcome, a stronger heart functions in an efficient mode and, therefore, has a higher reserve.

The second feature of the systemic structural trace during adaptation to hypoxia is that it provides a decrease in the basal metabolism and economical oxygen utilization by tissues. Indeed, it has been established that long-term adaptation to hypoxia involves a slightly decreased release of the thyreotropic hormone from the hypophysis, diminished secretion of thyreoid hormones, and structural changes in the thyroid gland indicative of hyperfunction [205, 301, 520]. At the same time, humans and animals were noted to have a decreased basal metabolism [144, 229], which was proven for the whole organism but was also seen in lower oxygen utilization by individual organs, by the heart, for instance.

The fact that the heart absorbs less oxygen during the same cardiac output is highly significant. It shows that utilization of oxygen becomes more effective for the contractile function during adaptation. In other words, we have here a heart that utilizes 30%–40% less oxygen, but performs its usual work. This phenomenon was discovered by Moret et al. [404] who used direct methods to measure oxygen utilization by the heart function in highlanders living at an altitude of 4,000 m. We later showed that the same increase in effective oxygen utilization may be obtained in the hearts of animals adapted to hypoxia in altitude chambers [329]. Thus, the increase in the effectivity of oxygen utilization has been proven at least for the heart.

The third and highly important feature of adaptation to altitude hypoxia is evident in deep changes in neural control, which are pronounced both for the higher regions of the nervous system and for circulation control. Activated nucleic acid and protein synthesis, retained in the brain at a definite stage of stable adaptation to hypoxia, is accompanied both by increasingly retained conditioned reflexes and accelerated transition of short-term memory into long-term memory and, what is particularly important, by the increased stability of extreme brain stimulants, conflicting situations, epileptogens, hallucinogens, etc. (see Chap. 4).

Changes also develop at the level of vegetative and, specifically, adrenergic control during adaptation to hypoxia. Increasingly powerful adrenergic control of the work of the heart, which is apparent in hypertrophy of sympathetic neurons [454], increased numbers of sympathetic fibers in the myocardium, and as previously noted more intense and less prolonged inotropic response of the heart to noradrenaline, are important changes of this kind [320, 371, 454]. This is accompanied by a decrease in the myogenic tonus of resistive vessels and by their lesser response to noradrenaline [332]. Such changes in adrenergic regulation of the heart and the vascular bed provide a situation in which an increase in cardiac output during behavioral responses, first, begins and ends more quickly and, second, is accompanied by less elevated arterial pressure, i.e., is on the whole more economical.

The fourth feature of the systemic structural trace during adaptation to hypoxia predetermines major changes in the control of water-salt metabolism and the vascular tone. This phenomenon is essentially as follows: During adaptation to hypoxia, the partial atrophy of the hypothalamic supraoptical nucleus responsible for the forming of vasopressin, an antidiuretic hormone [238], and the quite pronounced atrophy of the aldosterone-secreting adrenal glomerular zone quite naturally develop at the expense of a still rather obscure mechanism [402]. This adaptive hypofunction of the supraoptical nucleus and adaptive hypoaldosteronism are naturally accompanied by a stable decline in the organism's sodium and water reserves. In other words, the sodium and water content in hypoxia-adapted humans and animals is stably reduced, and the K/Na ratio in the plasma, erythrocytes, and vascular wall conversely somewhat increased [363].

In accordance with contemporary concepts on ion transport, such shifts in water-salt metabolism may be combined with a decrease in the calcium content of the vessel cells and should in that case lead to a decline in the myogenic component of the vascular tone. A decrease in the calcium content of vascular muscles upon adaptation to hypoxia is still to be confirmed; however, as was just mentioned, a decrease in the myogenic component of the vascular tone during adaptation even to moderate hypoxia does exist. Thus, adaptive hyposecretion of the antidiuretic hormone, hypoaldosteronism, decrease in the sodium and water reserves, and a diminishing of the myogenic component of the vascular tone are significant features of adaptation to hypoxia.

The fifth feature of the systemic structural trace may presently be formulated only hypothetically as constituting an increase in the power of the inhibitory and modulatory systems, which at the brain level synthesize such inhibitory mediators as GABA, glycin, dophamine and, possibly, encephalins, and at the level of other tissues are represented by prostaglandin and antioxidant systems. This question is discussed in greater detail below.

The systemic structural trace of adaptation to hypoxia with its previously mentioned distinctive features lowers hypoxemia, reduces hyperventilation to a very modest degree, eliminates mobilization of the circulatory system, completely eliminates the stress reaction, reduces to a certain minimum activated nucleic acid and protein synthesis in oxygen-transport systems, and helps eliminate disorders in intellectual and motor activities. *This stage of stable adaptation may be retained for many years and pass into the fourth stage of exhaustion,* normally assessed as chronic mountain disease, this taking place only at too strong hypoxia and in the case of certain other factors.

Hurtado [186, 187] aptly defined the chronic mountain disease as a loss of natural adaptation by people who were born and had always lived in the mountains. Numerous cases of this disease are chiefly seen in abruptly intensified adaptational shifts at the stage of oxygen transport from the lungs to the tissues, namely – additional activation of erythropoesis revealed by bone marrow biopsy, and progressing polycytemia and hyperfunction of the heart, which is evident in additionally increased cardiac output, progressing hypertension in the lesser circulation, and hypertrophy of the right ventricle, uncommonly large even for highlanders. Simultaneously, arterial hypotension and sharp fatigability develop.

The subsequently pathologic mobilization of hemopoiesis and circulation, and event inherent in the chronic mountain disease, may presumably be caused by two different initial injuries. First, it may be due to impaired external respiration (hypoventilation), to some measure compensated by growth in oxygen capacity and minute volume. Second, it may be caused by impaired oxygen utilization by tissue mitochondria and the resultant additional mobilization of transport mechanisms, i.e., by a unique return to the emergency stage of the process.

Hurtado's investigations show that with chronic mountain disease events actually develop in conformity with the first possibility: The patients are noted for hypoventilation, which is compensated by stronger hemopoiesis and circulation. In studying the mechanism of this hypoventilation, Hurtado showed that patients afflicted by this disease have substantially reduced hyperventilation, which develops in response to the administration of carbon dioxide, i.e., there is reduced sensitivity of the respiratory center to CO_2.

From this, Hurtado assumed that wear of neurons in the respiratory center constitutes the principal pathogenic mechanism of the chronic mountain disease. During adaptation to hypoxia, the respiratory center achieves moderate but continuous hyperfunction and is an object of hypoxemic action. As is shown in detail below, these factors depend on the same mechanism to activate the genetic apparatus of differentiated cells in general and that of neurons in particular; in effect, they activate nucleic acid and protein synthesis therein. Such activation is not infinite and may terminate when the neurons exhaust their ability to resynthesize RNA and proteins, i.e., it may end in exhaustion [162, 190].

In terms of our presentation, it is significant that adaptation to hypoxia develops through the same four stages that were postulated above as the principal stages of any phenotypic adaptation. The stable adaptation stage was found to be characterized by a more economical functioning of the system responsible for adaptation apparently to consummate the final stage of wear through the pathogenic mechanism of local wear common to many adaptations.

Adaptation to Hypoxia, a Preventive Factor

A ramified structural trace, which to one extent or another involves all organs and tissues, in the reason why adaptation to hypoxia is concomitant with an increased resistance to many other factors. At present, it has been shown that stable adaptation to hypoxia, a stage that occurs due to a lengthy stay at moderate altitudes or because of periodic hypoxic effects in an altitude chamber, may be used for preventing experimental circulatory and brain diseases. The evidence obtained by our laboratory on the effect of prior adaptation to hypoxia on the development of heart insufficiency during experimental heart failure and myocardial infarction should be assessed precisely from this viewpoint [320].

Figure 19 shows that 2 days after aortic coarctation in emergency compensatory hyperfunction and acute cardiac insufficiency, the CP concentration in the rat myocardium dropped by more than twofold; however, it hardly declined in rats preadapted to altitude hypoxia and possessing a powerful energy-supply system. Upon emergency compensatory hyperfunction of the heart, the noradrenaline content in the myocardium was 2.5 times lower, whereas, in animals preadapted to hypoxia and in possession of a powerful apparatus of adrenergic control, the decrease was much less pronounced and on the verge of significance.

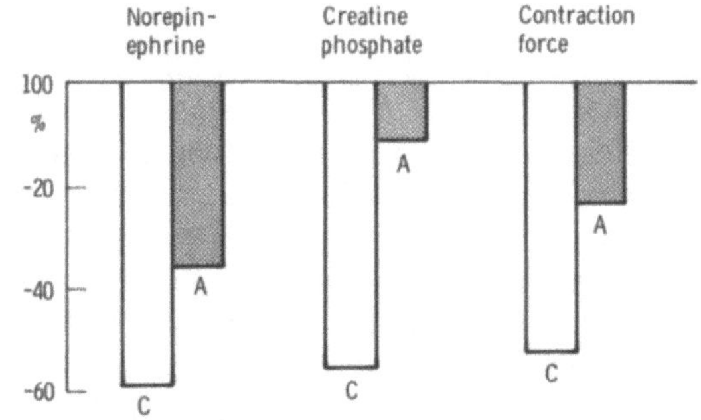

Fig. 19. Effect of preliminary adaptation to altitude hypoxia on norepinephrine and creatine phosphate concentrations in the myocardium and on the state of the heart contractile force during CHH emergency stage. *C*, control; *A*, adaptation. *Ordinate axis*, percentage of initial values of the above indicators

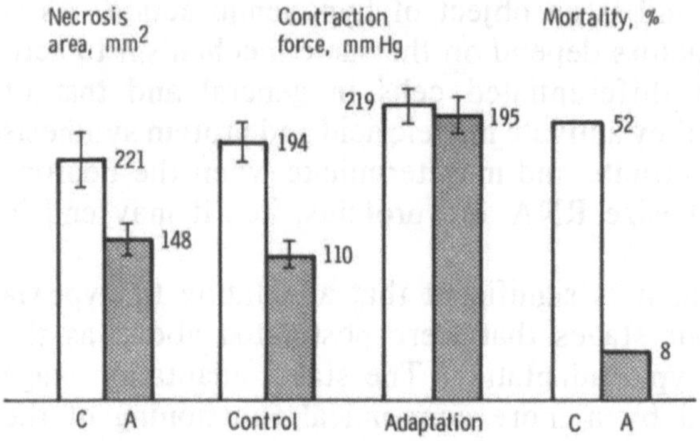

Fig. 20. Effect of preliminary adaptation to altitude hypoxia on necrotic area, myocardial contraction strength, and mortality in experimental infarction

In accordance with this evidence, it was noted that in control Wistar rats subjected to these experiments, the emergency stage of compensatory hyperfunction of the heart was in more than 50% of the cases complicated by dyspnea, hydrothorax and edema of the subcutaneous fat in the neck and chest. *Prior adaptation to hypoxia prevented this syndrome of acute stagnant circulatory insufficiency.*

With ischemic necrosis of the myocardium, the preventive effect of adaptation to hypoxia is also quite definitely pronounced. Figure 20 compares the area of the necrotic zone in preadapted and control animals 2 days after ligating the left coronary artery. Prior adaptation is seen to have diminished the area of the ischemic necrosis by about one-third. The effect of prior adaptation on defects of contractile function on rat mortality in ischemic necrosis proved to be much more pronounced. Indeed, Fig. 20 shows that 1 day after the occurrence of ischemia, the maximal contraction power of the left ventricle in unadapted animals was 44% lower than in controls; in preadapted rats, this defect constituted only 11%, i.e., was four times less frequent. One day after myocardial necrosis had been induced, the mortality among unadapted rats was 50%, and among adapted animals only 8% or six times less. These data indicate that the *preventive effect of prior adaptation during ischemic necrosis of the myocardium was evident not only in a decrease in necrotic area but also in the more perfect compensatory hyperfunction of the unaffected regions of the myocardium and, possibly, via such an extracardial factor as more economical utilization of oxygen by the organism.* In other words, the principal part in this preventive effect was played jointly by the increasing power of the cardiac muscle and by the economical utilization of oxygen, both factors developing during adaptation to hypoxia.

Other features of the systemic structural trace, for example, changes providing higher brain resistance to extreme stimuli and increased power of the modulatory and inhibitory systems, can be no less important in preventing diseases caused by adaptation to hypoxia. These changes are most significant in preventing stress injuries.

Until very recently, disorders in metabolism and heart contraction during emotional stress were readily postulated by everyone but studied by no one. The existing evidence was confined to numerous indications of impaired

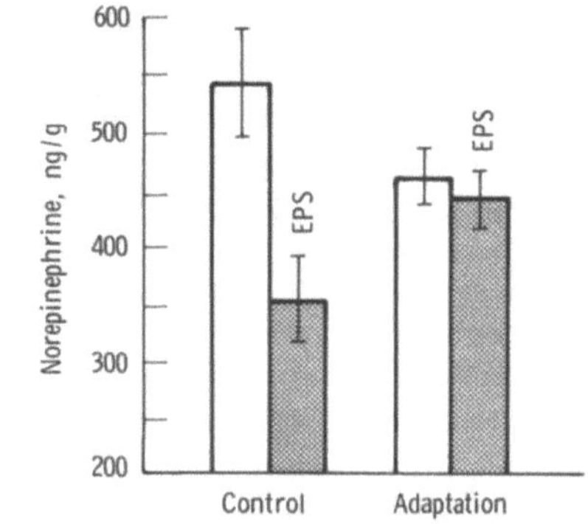

Fig. 21. Effect of preliminary adaptation to altitude hypoxia on norepinephrine content in the myocardium of stress-exposed animals. ***Undashed columns,*** animals not exposed to stress; ***dashed columns,*** stress-exposed animals

cardiac rhythm during neurotic states of animals and humans. At present, it is clear that endured EPS is followed by activation in the myocardium of free-radical oxidation of lipids and by labilization of lysosomes. The effects of lipid peroxides and lysosomal enzymes damage membrane structures and this is seen in disrupted calcium transport. At the same time, the catecholamine content in the myocardium becomes increasingly low and heart contraction is impaired. These data are examined in greater detail in Chap. 6, which characterizes EPS and the disturbances in metabolism and heart structures occurring therewith. At this point, it appears desirable to discuss the evidence that shows that prior adaptation to hypoxia largely prevents stress injuries of the heart [364].

Figure 21 shows that after EPS the noradrenaline content in the left ventricular myocardium proved to be lower by about one-third. In adapted animals, the noradrenaline content did not decrease under EPS: Adaptation prevented the catecholamine reserve in the myocardium from decreasing. This may have possibly been due to an adaptive increase in brain resistance to extreme stimuli and, consequently, to a lower adrenergic effect upon the heart and lower catecholamine demand. At the same time, presumably reducing the adrenergic effect, adaptation prevents lipid peroxidation in the myocardium from activating as it does under stress.

The data in Table 17 show that in unadapted animals exposed to EPS, the peroxide content in the myocardial lipids, according to chemiluminescence,

Table 17. Effect of prior adaptation to hypoxia on the intensity of myocardium lipid peroxidation under EPS

Animal group	Number of animals	Intensity of luminescence, counts/s/mg lipids	p
1. Control	7	16.0 ± 1.5	$p_{1-2} < 0.001$
2. EPS	13	41.5 ± 4.0	$p_{1-3} < 0.05$
3. Hypoxia	8	24.2 ± 2.7	$p_{3-4} > 0.1$
4. Hypoxia + EPS	6	21.0 ± 1.6	

grew three times compared with controls; in adapted animals, the peroxide content prior to EPS was insignificantly increased, but did not change under the influence of EPS. Thus, adaptation prevented a major injury factor from occurring. Finally, Fig. 22 shows that after EPS the contractile function indices proved to be 40%–60% lower when an increased contraction rate was

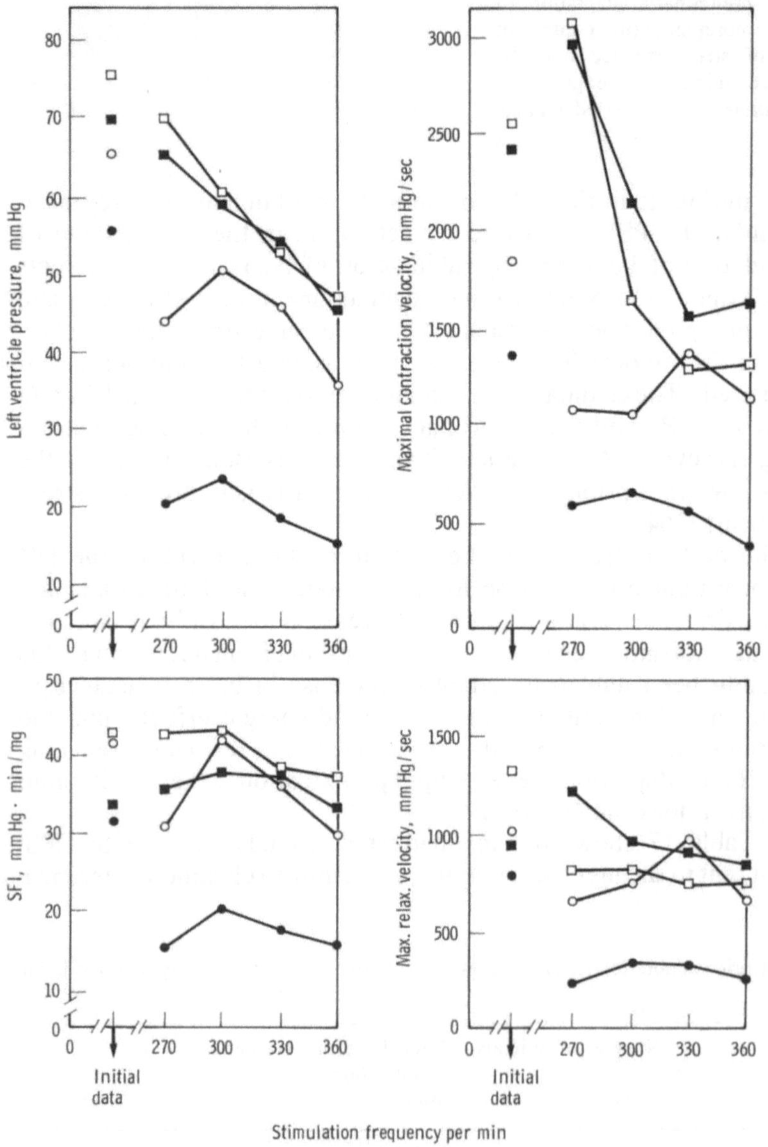

Fig. 22. Effect of preliminary adaptation to hypoxia on contractile function of hearts of stress-exposed animals during gradual increase in heart contraction frequency. ○ control; ● stress; □ adaptation to hypoxia; ■ adaptation to hypoxia and stress

imposed on the heart, while in preadapted animals this impairment in the contractile function did not occur.

In assessing the mechanism of this phenomenon, one should keep in mind that adaptation to hypoxia protects the heart not only from stress injuries but also from high doses of isoproterenol, a synthetic analog of noradrenaline [448]. This permits the assumption that factors localized in the heart itself may play a major role in enhancing its resistance to catecholamines, specifically in the mechanism preventing the adrenergic activation of peroxidation under stress. Some such factors are the following antioxidant systems of cell membranes: Superoxide dismutase, glutatione peroxidase, and catalase, and also the presence of tocopherol, and so on. Having established the absence of peroxidation with stress in hypoxia-adapted animals [364], one could on this basis assume that adaptation to hypoxia enhanced the power of the organism's antioxidant systems. Very recently, this assumption was directly confirmed by Gerasimov et al. [133], who showed that animal adaptation to hypoxia caused a statistically significant increase in antioxidant enzymes in the brain and liver. Reverting to the prevention of stress injuries of the heart by means of adaptation, one should keep in mind that the absence in adapted animals of the activation of peroxidation under the effect of stress should logically prevent stress injuries of organs.

The assumption that prior adaptation to hypoxia enhances animal heart resistance to stress and oxygen shortage has been revealed even more definitely in experiments in which an isolated heart was the subject of research. In our laboratory, these experiments were performed on control animals, animals adapted to hypoxia in an altitude chamber, animals exposed to EPS, and animals adapted to hypoxia and subjected against such a background to the same stress action [552].

The contractile function of the heart for all four groups of animals was studied by the method of Fallen et al. [107], in which the heart contracts isometrically around a latex bulb that replaces the left ventricular cavity. Parallel with recording the contractile function, we took samples of perfusate that had passed via the coronary bed to determine the CPK activity by the Bergmeyer [27] method. The CPK release from the myocardium is known to be one of the most reliable criteria of myocardial cell injury, and with hypoxia it naturally increases proportionally to the decline in ATP concentration in the myocardium [531]. Correspondingly, the program of experiments performed on an isovolumic heart stipulated a hypoxic test, during which the pO_2 in the perfusion solution decreased for 20 min from 600 to 150 mm Hg. The hypoxic test naturally involved depression of the contractile function and considerably enhanced the release of CPK from the myocardium. As a result of comparisons of data obtained for the four groups of animals (Table 17), it was possible to solve two separate problems: First, the effect of hypoxia-adaptation on heart resistance to stress injury, and second, the effect of adaptation on heart resistance to hypoxic injury.

Table 18 presents the data on CPK release from an isolated heart into the perfusate. From Table 18 it follows that 35 min after the start of perfusion, the hearts of control animals released CPK into the perfusate and after

Table 18. Effect of acute hypoxia on CPK activity in the perfusate of isolated rat hearts during adaptation to hypoxia and EPS

	Number of animals	CPK activity		
		35 min perfusion	95 min perfusion	20 min hypoxia
1. Control	8	46.87 ± 6.7	19.65 ± 2.15	31.64 ± 3.78
2. EPS	7	86.55 ± 10.52	29.56 ± 1.18	47.26 ± 5.25
3. Adaptation	6	24.31 ± 3.34	13.1 ± 2.64	18.68 ± 1.25
4. Adaptation + EPS	6	20.3 ± 3.06	12.24 ± 1.3	24.96 ± 2.72
Significance of differences				
p_{1-2}		<0.01	<0.01	<0.05
p_{1-3}		<0.01	>0.05	<0.01
p_{1-4}		<0.01	<0.02	>0.05
p_{2-4}		<0.01	<0.01	<0.02

95 min the amount of released CPK was reduced more than twofold. This dynamics of CPK release is familiar and is determined by the fact that when the heart is isolated the myocardium is injured by hypoxia; subsequently, with prolonged heart function under moderate oxygenation, this injurious factor ceases to act. A hypoxic test performed in these conditions again led to considerable CPK release.

For rat hearts adapted to hypoxia, the CPK dynamics remained the same as in controls; however, the absolute values of the CPK released activity proved to be 2–1.5 times less in all experimental stages; prior adaptation to hypoxia had enhanced the resistance of an isolated heart to this factor. In stress-exposed animals, the dynamics of CPK release also remained unchanged; however, the absolute values of its activity in the perfusate at all experimental stages proved to be 2–1.5 times higher than in controls. As was to be expected, the endured stress had reduced rat heart resistance to hypoxia.

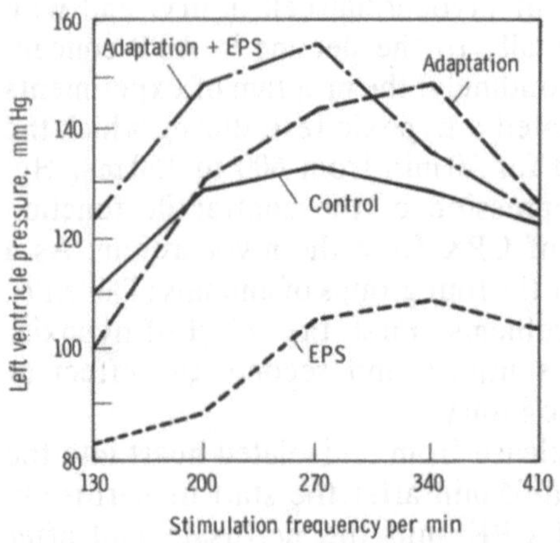

Fig. 23. Effect of stress and adaptation to altitude hypoxia on contraction force of left ventricle of isolated rat heart during increasing pacing frequency

The chief experimental result was that prior adaptation to hypoxia fully prevented the injurious effect of stress by the enzyme release criterion (see Table 18, bottom line). Creatine phosphokinase release from the myocardium of adapted animals that had endured stress does not essentially differ from that in animals adapted to hypoxia without having been exposed to stress. In other words, adaptation prevented both the injurious effect of hypoxia and the potentiating effect of stress on that injury.

Figure 23 shows the contractile function of the heart contracting isovolumically (i.e., that of the same heart whose CPK release was studied). As is apparent from Fig. 23, with equal frequency of contractions induced by an electrostimulator, the pressure developed by the hearts of stress-exposed animals was lower than in controls by about one-third ($p < 0.05$). In animals that had endured stress against the background of adaptation, this defect in contractile function was absent; in intact animals adapted to hypoxia, the pressure developed was likewise the same as in controls.

The curves in Fig. 24 show the response of an isolated heart to an hypoxic test. The left part of the figure shows that after the occurrence of hypoxia, the developing systolic pressure was sharply reduced. From the 10th min of hypoxia, it became extremely low and virtually the same for the hearts of animals from all four groups, namely 8–9 mm Hg. Restoration of the contractile function following elimination of hypoxia occurs differently for the hearts of animals from different groups. Stress-exposed animals are characterized by the slowest restoration of the contractile function and hypoxia-adapted animals have the fullest restoration. The two other groups of animals occupy an intermediary position: In those that had endured stress involving adaptation to hypoxia, the contractile function of the heart under reoxygenation is restored more quickly and to a fuller extent than in control animals not subjected to stress.

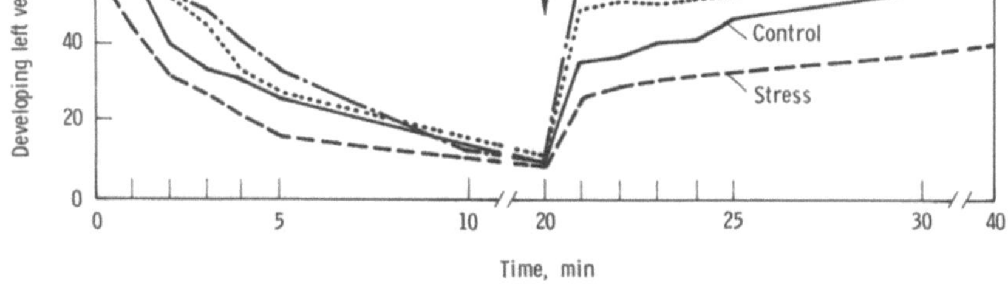

Fig. 24. Effect of preliminary adaptation to altitude hypoxia on contraction force of left ventricle of isolated rat heart exposed to acute hypoxia

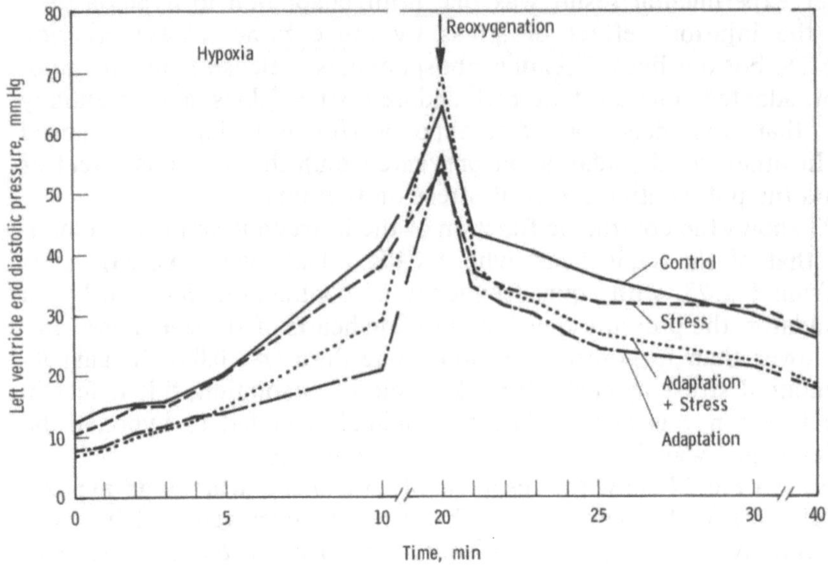

Fig. 25. Effect of preliminary adaptation to altitude hypoxia on dynamics of hypoxic heart contracture in stress-exposed animals

The curves in Fig. 25 characterize the development of hypoxic contracture of the hearts of animals of all four groups and the gradual elimination of this contracture under reoxygenation. The only major difference in these data is that in the course of reoxygenation, hypoxic contracture in animals previously adapted to hypoxia is eliminated more quickly and to a greater extent, irrespective of whether they had been exposed to stress or not. *Thus, prior adaptation to hypoxia prevents stress injuries of the heart by the criterion of enzyme release from the myocardium and, correspondingly, prevents depression of the contractile function of the heart after stress.*

An analysis of the mechanisms of increasing resistance to stress injuries is the subject of the last chapter in this book; however, the very fact that adaptation may be used to prevent stress injuries appears unquestionable.

An important role in the use of adaptation to hypoxia for preventing various injuries is played by the adaptational component cited above, which involves hyposecretion of the antidiuretic hormone, hypoaldosteronism, a stable decrease in the sodium and water reserve, and a decline in the arterial myogenic tone and adrenoreactivity. At one time it was noted that this set of shifts is in many respects opposite to that observed in human hypertension and all forms of experimental hypertension in animals. Hence, adaptation to the periodic action of hypoxia in an altitude chamber was used to prevent experimental hypertension. At first, we obtained evidence showing that prior adaptation to hypoxia inhibits renal and DOCA hypertension [363]. However, the pathogenesis and course of these experimental diseases are quite remote from human hypertension.

Spontaneous hereditary hypertension of Okamoto rats (SHR), which is pathogenically quite close to human hypertension and develops through prehypertensive stages, early hypertensive and developed hypertensive stages, is presently being intensively studied. Evidently, an assessment of the preventive effect of hypoxia-adaptation on this model is quite significant for cardiology. This assessment was recently made in our joint experiments with Barbarash and Dvurechenskaya [365]. The curves shown in Fig. 26 perimit one to trace the dynamics of arterial pressure in control Wistar rats and in Okamoto SHR, which, beginning from the age of 20 days, underwent a course of hypoxic effects in an altitude chamber. As is apparent from Fig. 26, during 3 months of observation, the arterial pressure in control animals increased 10–15 mm Hg to reach 118 mm Hg, which corresponds to the normal age dynamics. In SHR, the blood pressure during the same period increased by 30 mm Hg to reach a group average of 155 mm Hg, which corresponds to the usual dynamics of hereditary hypertension. Finally, in animals of the same SHR strain that 'visited' the altitude chamber the disease did not develop; their systolic pressure by the end of the observation period was 121 mm Hg, i.e., did not differ from that of controls. Significantly, this preventive effect shown in Fig. 26 by the shaded zone was especially pronounced at the age of 3–4 months, when prior adaptation of animals was maintained by means of only two hypoxic actions a week.

In estimating the probable mechanism of the preventive effect of adaptation to hypoxia in SHR, one should keep in mind that at least three links were detected in the pathogenesis of this form of hypertension.

1) Hereditary hypertension involves increasing activity of adrenergic control, which is evident in increased concentration of catecholamines in the adrenals and in their greater spontaneous release. An increased thyroid gland function [417] proved in SHR was an outcome of this activation of the sympatho-adrenal system. The substantial role of this set of shifts in hereditary hypertension pathogenesis stems from the fact that immunologic sympathectomy

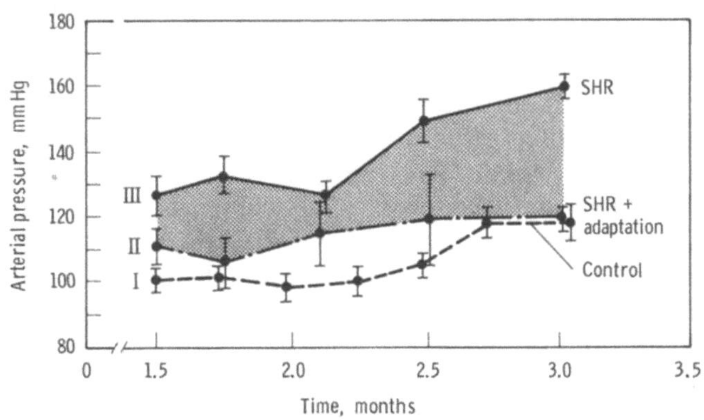

Fig. 26. Effect of adaptation to altitude hypoxia on dynamics of arterial pressure in spontaneously hypertensive rats (SHR). *Dashed zone* indicates preventive effect of adaptation to hypoxia

[72] or thyroidectomy [417] in early ontogenesis prevents spontaneous hereditary hypertension (SHH) from developing.

2) Hereditary hypertension entails a hyperaldosteronism syndrome expressed by increased aldosterone secretion [463] and enhanced appetite for salt [73]. At the same time, it is also found to involve function activation and hypertrophy of the hypothalamic nucleic responsible for the secretion of the antidiuretigenic hormone. In accordance with this, SHH is noted to result in enhanced Na^+ and water content in the organism [125], while the addition of NaCl to food potentiates SHH development [73].

3) Hereditary hypertension involves decreased ATPase activity of the SR in the vascular smooth muscle cells [8]. This is naturally accompanied by increased calcium content in the vascular smooth muscle cells [8, 125], this being the direct cause of increased tonic contraction of the smooth muscles and elevated arterial pressure in SHR.

The interconnection between these three links in the pathogenetic chain of hereditary hypertension is obscure. In addition to the traditional view that stimulation of the higher regions of the nervous system activates adrenergic regulation and sequentially actuates all the three links, the opposite possibility has been postulated, namely, that the membrane genetic defect and disruption in the Na-K ATPase function entails as a compensatory response changes in vascular tone control, which constitute the basis of hypertension.

In the first mechanism, the preventive effect of adaptation to hypoxia during SHH and other forms of hypertension may be explained prior to resolving the central issue of its pathogenesis. This possibility is determined by the fact that adaptation to hypoxia naturally leads to structural changes in regulatory systems. These changes are the opposite of those in hereditary hypertension. For instance, developed adaptation to hypoxia reduces the activity of the sympathetic nervous system and the thyroid gland function. Simultaneously, hypoaldosteronism develops, evident in the decrease in aldosterone synthesis and partial atrophy of the adrenal glomerular zones; the size of the hypothalamic supraoptical nuclei diminishes. These changes in adapted animals were found to result in a stable decline of the Na^+ and water reserve and a lower Na^+ concentration in the tissues [363].

This shift in water-salt metabolism control is the most probable cause of the preventive effect of hypoxia and adaptation thereto in hereditary hypertension in animals. This view agrees with the results of experiments in which Barbarash and ourselves [374] compared the effect of hypoxia on diuresis and natriuresis in control rats and SHR, the hypoxia having been reproduced in an altitude chamber. Urea was taken from all the animals twice – under normal oxygen conditions outside the altitude chamber and under hypoxia in an altitude chamber at a 'height' corresponding to 4,400 m above sea level. The interval between the first and second urine sampling was 2 days. Table 19 shows that the effect of hypoxia on Wistar rats caused considerably reduced excretion of water, Na^+, and K^+, chiefly due to the lower rate of glomerular filtration. Decline in natriuresis was less pronounced than that in K^+ excretion, and was statistically nonsignificant. Changes in electrolyte excretion were

Table 19. Effect of acute hypoxia on diuresis and natriuresis in SHR and Wistar rats

Indication[b]	Wistar rats		SHR	
	Prior to hypoxia	During hypoxia	Prior to hypoxia	During hypoxia
Water excretion, ml	1.7 ± 0.2	1.0 ± 0.3[a]	1.0 ± 0.1	1.8 ± 0.6
Water filtration, ml	124.4 ± 1.2	56.1 ± 14.0	91.2 ± 17.7	101.0 ± 12.2
Water reabsorption, ml	122.7 ± 1.3	57.1 ± 13.9[a]	90.2 ± 17.8	99.2 ± 12.3
Water reabsorption, % of filtration	98.21 ± 0.3	99.0 ± 0.29	98.9 ± 0.21	98.2 ± 0.6[a]
K^+ concentration, mEq/l	49.5 ± 6.2	38.5 ± 4.1[a]	51.3 ± 3.1	47.0 ± 4.7
Na^+ concentration, mEq/l	84.9 ± 13.2	66.6 ± 0.3[a]	121.2 ± 16.8	82.5 ± 5.1[a]
Na/K factor	1.74 ± 0.21	1.79 ± 0.24	2.28 ± 0.09	1.77 ± 0.09
K^+ excretion, mEq	0.085 ± 0.003	0.047 ± 0.015[a]	0.035 ± 0.002	0.081 ± 0.018[a]
Na^+ excretion, mEq	0.147 ± 0.009	0.120 ± 0.039	0.126 ± 0.024	0.150 ± 0.039
Na^+ filtration charge	14.22 ± 1.4	6.42 ± 1.6[a]	9.51 ± 1.85	10.53 ± 1.37
Na^+ reabsorption, mEq	14.07 ± 1.4	6.30 ± 1.60[a]	9.38 ± 1.85	10.38 ± 1.27
Na^+ reabsorption, % of filtration charge	98.96 ± 0.32	98.10 ± 0.30	98.67 ± 0.23	98.57 ± 0.38

[a] Significance of differences: < 0.05.
[b] Calculated for 12-h period and 100 g body weight.

greater during than after hypoxia. This hypoxic effect observed in adult animals in their 5th month of life corresponds to the evidence contained in the literature on the inhibition of water and electrolyte excretion under the action of acute hypoxia and allows one to assume that the delay of sodium and water can play a specific role in the increase of arterial pressure normally observed in humans and animals directly after ascent to high altitude.

In SHR, the effect of acute hypoxia was qualitatively the reverse of that observed in Wistar rats and was evident in the considerably increased water excretion due to a lower reabsorption percentage and slightly increased filtration. At the same time, potassium excretion was found to be higher than that of sodium. Unlike normotensive animals of the same age, in adult rats with hereditary hypertension acute moderate hypoxic hypoxia possesses pronounced diuretic and also natriuretic effects, which may play a role in the hypotensive effect of hypoxia.

Thus, increased sensibility of SHH rats to the natriuretic effect of hypoxia may play the role of a preventive effect of lengthy hypoxia during hereditary hypertension. This interpretation of the antihypertensive effect of adaptation to hypoxia is in agreement with the notion that natriuresis and restricted consumption of sodium with food are successful in treating human hypertension.

Adaptation to hypoxia blocks the principal pathogenetic SHH link and prevents an increase in arterial pressure. A genetically determined disease proves to be essentially inhibited by timely purposefully directed changes in the habitat.

In assessing the above evidence on the preventive effect of adaptation to hypoxia during animal circulatory diseases, it should be emphasized that they mostly coincide with the results of Soviet and foreign epidemiologic studies. These studies show that in highlanders the blood arterial pressure [300] and cholesterol content [478] are lower; hypertension [185] and myocardial infarction [300] are much less frequent and smoother. This aspect of the issue has been the subject of lengthy studies by Mirrhakhimov [398–400]. From the above data, we have recommended the use of adaptation to the periodic action of hypoxia in an altitude chamber for preventing certain human circulatory diseases [320].

It has become clear that adaptation to hypoxia enhances resistance not only to factors causing diseases of the circulatory system. In recent years, this adaptation both in experiments and in clinical studies has proved capable of inhibiting certain allergic diseases and related immunodeficient states. Contrary to acute hypoxia which suppresses immunogenesis, adaptation to hypoxia is known to activate the latter. For example, permanent highland residents and animals adapted to periodic moderate hypoxia show a stronger immune response to bacterial antigens and an elevated level of serum immunoglobulins [63, 95]. Together with Evseev, Davydova, and Giber [376], we studied the effect of adaptation to periodic hypoxia on certain indices of nonspecific resistance and on the immune response in healthy animals and those suffering from adjuvant polyarthritis.

White strainless male rats (250–300 g) divided into four groups were used in the experiments. Group 1 included intact animals and served as controls; in group 2 animals adjuvant arthritis was reproduced by introducing Freund's complete adjuvant (0.2 ml) into the hind pads; groups 3 and 4 were adapted to altitude hypoxia by a daily 6-h stay in an altitude chamber with atmospheric pressure equivalent to that at 5,000 m, the procedure lasting 1.5 months. On day 30 after the beginning of adaptation, adjuvant arthritis was reproduced in group 4 animals. On day 52 after the beginning of adaptation, all the animals were subjected to a single immunization by intraperitoneal injection of sheep's erythrocytes. The hemagglutinin production was examined on days 7, 14, and 21 after immunization by the classic method. The dynamics of the nonspecific resistance indices (lysozyme and complement) was studied on days 30, 39, 45, and 52 after the beginning of adaptation. The lysozyme content in the serum was determined by the nephelometric method for *Micrococcus lysodeiticus* [88]. The complement level in the serum was determined photometrically by 50% hemolysis and expressed in conventional units. The degree of arthritis was estimated by the edema index reflecting the increase in volume (percentage) of the affected joints. The dynamics of the spreading inflammatory changes was also recorded in the animals.

Rats adapted to altitude hypoxia were found to have a more intense immune response to sheep's erythrocytes. The hemagglutinin titers in adapted animals were significantly higher in control rats at all observation times following immunization (Fig. 27). The most pronounced difference between experiment and control was noted on day 14 after immunization, when the hemagglutinin levels in adapted rats were more than three times higher than in intact

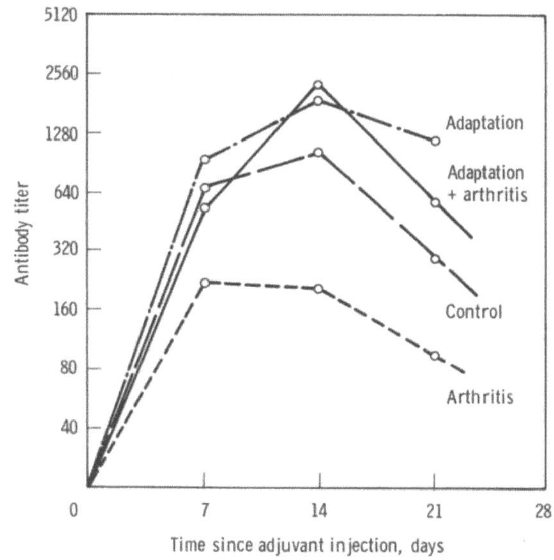

Fig. 27. Effect of preliminary adaptation to altitude hypoxia on hemagglutinin titers

animals. In the course of adaptation, the rats showed higher lysozyme activity in the serum (Table 20). For instance, the lysozyme level in control animals was 30.1 ± 0.53; on day 30 after the beginning of adaptation, it was 30.5 ± 1.05; and on day 45 it was 39.8 ± 0.6 ($p < 0.001$). The complementary activity of the blood serum in rats adapted to altitude hypoxia remained essentially the same.

Adaptation to altitude hypoxia substantially weakened the extent of adjuvant polyarthritis, estimated by the edema index and the inflammatory changes spread factor (Table 20). For example, inflammatory changes advancing to the joints of all paws and the tail was observed only in one of seven adapted rats and in five of seven control animals ($p < 0.001$). At the same time, adaptation affected changes in nonspecific resistance and immune response, changes normally observed with adjuvant arthritis. From Table 20 and Fig. 27 it follows that in nonadapted animals with adjuvant arthritis the complement and lysozyme content in the serum proved to have increased by 50% and 30%, respectively, compared with the initial level. The immune response is, on the other hand, sharply decreased at all times after administration of the antigen. Adaptation considerably reduced these deflections inherent in adjuvant arthritis: In the course of arthritis development in adapted animals, the complement level was 30% lower than in unadapted animals and the immune response was close to that in adapted animals lacking arthritis. Correspondingly, on day 14 after immunization, the hemagglutinin titer in animals afflicted by the adjuvant disease and unadapted to hypoxia was eight times lower than in adapted rats with arthritis.

Thus, adaptation to altitude hypoxia leads, first, to higher lysozyme activity in the serum; second, to enhanced immune response to sheep's erythrocytes; and third, prevents immune response suppression inherent in adjuvant arthritis

Table 20. Effect of adaptation to altitude hypoxia on development of the adjuvant disease in rats

Group	Number of animals	Edema index			Lysozyme			
		Day 38	Day 45	Day 54	Day 30	Day 38	Day 45	Day 52
Control	9	–	–	–	31.4 ± 1.0	30.1 ± 0.53	30.6 ± 0.7	31.2 ± 1.3
Adjuvant arthritis	7	136.1 ± 14.7	147.1 ± 20.0	162.0 ± 22.0	28.8 ± 1.0	38.1 ± 0.5[a]	38.3 ± 0.5[a]	39.6 ± 1.05[a]
Adaptation to altitude hypoxia	9	–	–	–	30.5 ± 1.0	33.7 ± 0.15[a]	39.8 ± 0.6[a]	39.2 ± 0.45[a]
Adaptation to altitude hypoxia + adjuvant arthritis	7	101.3 ± 8.2	110.8 ± 13.6	103.5 ± 8.1	31.2 ± 1.0	38.5 ± 0.3[a]	38.7 ± 0.3[a]	37.4 ± 1.05[a]

Group	Number of animals	Complement			
		Day 30	Day 38	Day 45	Day 52
Control	9	121.6 ± 6.8	110.0 ± 2.8	120.0 ± 6.2	114.0 ± 2.8
Adjuvant arthritis	7	121.6 ± 6.8	186.6 ± 6.8[a]	179.0 ± 14.5[a]	204.6 ± 5.9[a]
Adaptation to altitude hypoxia	9	138.2 ± 8.9	125.3 ± 6.9	130.7 ± 2.8	124.5 ± 9.9
Adaptation to altitude hypoxia + adjuvant arthritis	7	118.1 ± 5.6	139.0 ± 2.6[a]	134.6 ± 6.9[a]	159.2 ± 8.5[a]

[a] Differences within each group are significant compared with initial values.

and lessens inflammatory changes in the joints. The chief experimental result is that *adaptation to periodic hypoxia activates antibody genesis in normal animals, prevents the antibody-genesis suppression normally observed with adjuvant arthritis, and renders the arthritis itself less pronounced. In explanation of this fact, one should keep in mind that B lymphocytes play a decisive role in antibody genesis, while T lymphocytes are essentially involved in allergic inflammation during adjuvant arthritis. Hence, one may assume that in increasing antibody generation by B cells, adaptation to the periodic effect of hypoxia reduces T-lymphocyte activation.*

On the whole, the evidence obtained during experimental prophylaxis creates a prerequisite for a clinicophysiologic study of the question of using adaptation to periodic hypoxia with a view to preventing human diseases. When conducting such research, one should keep in mind the major significance of the dose of hypoxic effect, for the issue is concerned with a strongly acting factor which, if overdosed, may possibly cause negative consequences.

One of the hypoxia dosing methods easily achievable with an altitude chamber involves such a gradual increase of 'altitude' and duration of 'ascent' as to reduce to a minimum the phenomena characterizing the initial, emergency stage of adaptation, to extend the transition stage, and to obtain a set of adaptive changes characteristic of the third stage of stable adaptation and favorable for the organism. This would essentially involve a gradual ascent within 1 week from sea level 'height' to an 'altitude' of 2,000–2,500 m with subsequent daily ascents to the same 'altitude' for 1–1.5 months. This kind of adaptation to the periodic action of hypoxia involves training not only for hypoxia, but for ascents and descents; it does not lead to deadaptation of the organism from normal living conditions at sea level and the relatively brief duration of the action (4–5 h) permits at a certain stage of adaptation the combination of adaptation to hypoxia with adaptation to other factors. Finally, after adaptation has formed, the need for daily effects becomes superfluous; the result may be retained by means of two to three actions per week. These positive elements of adaptation to periodic hypoxic action provide grounds for a clinicophysiologic study of this approach in using phenotypic adaptation of the organism for preventive purposes. At the same time, it is obvious that the prevention of diseases, specifically circulatory diseases, by the directed, scientifically based development of a phenotype cannot be achieved through adaptation to any single factor. The issue ultimately stipulates that adaptation to a well-dosed and individually selected set of factors occurs and the major task of modern medicine is to ground scientifically and introduce into practice this combined adaptation.

Chapter 4
Compensation, a Form of Adaptational Response of an Injured Organism[1]

"Every evil is somehow compensated."

Seneca

Compensatory adaptations are important responses of the organism to injury; they are evident when the organs and systems directly unaffected by the action of the injurious agent take on the functions of the degraded structures by way of substituting hyperfunction or qualitatively changed function. The compensatory process to one extent or another eliminates functional disorders caused by injury and compensation, thus, becomes a major factor of clinical recovery. Besides compensation, other adaptational responses of the injured organism play an important role in recovery. They provide either elimination or restriction of the injurious factor (generation of antibodies, phagocytosis, inflammation), temporary immobilization of organs and systems (protective inhibition), and direct restoration of the degraded organ or tissue (regeneration). Thus, compensation should not be identified with the entire set of adaptational responses of an injured organism; the question is concerned with one of its adaptational responses. The compensatory process may take place at intracellular, intraorganic, intrasystemic, and intersystemic levels. For example, upon degradation of a polyploid cell genome, intracellular compensation is achieved at the expense of increased RNA synthesis over the survived genomes. When some renal nephrons are destroyed during nephrosclerosis, the intraorganic compensation occurs because the nephrons that survive enhance their function and hypertrophy. When the stomach is removed or its function impaired for a long time, intrasystemic compensation is to some extent provided by the increased secretory function of underlying regions of the digestive system. Intersystemic compensation is observed in many clinical situations, specifically during anemia, when injury of the hemapoietic system, decreased number of erythrocytes, and reduced ability of the blood to transport oxygen are lengthily compensated by greater cardiac output, i.e., by compensatory hyperfunction of the heart and the circulatory system as a whole.

Compensatory adaptations occur in the same general manner as the other adaptational responses mentioned. In line with this, we shall consecutively examine the concepts of urgent and long-term compensation and the concept of the systemic structural trace as the foundation of compensation. Like any other adaptational response, compensation develops in two stages, namely urgent and long-term. For example, when his right hand is injured, a man

1 Two special monographs were devoted to the problem, and this chapter summarizes those investigations.

instantly starts to use his left hand to perform the functions of the disabled extremity. This urgent compensation is very important in extreme situations; however, it is a fortiori imperfect. Subsequently, training and formation in the brain of a system of new structurally secured temporary connections results in the development of habits providing long-term compensation – relatively perfect performance by the left hand of operations normally performed by the right hand.

After one kidney is either removed or switched out by a pathologic process, urgent compensation occurs in the form of compensatory hyperfunction of the remaining kidney. However, this compensation is imperfect since its excretory function during the first few days is lower than that of two kidneys, despite the fact that all the nephrons become simultaneously involved in the kidney function and the functional reserve of the single kidney is mobilized. Subsequently, the mass of the remaining kidney and its formative nephrons increases; compensatory hypertrophy of the organ develops and, as a result, the function of one kidney begins to correspond to that of two kidneys. The organ restores its functional reserve and establishes perfect long-term compensation which, in the absence of additional injuries, may persist throughout the whole life [38, 228]. Compensation caused by pathologic switch-off or removal of any one of a pair of organs develops in a similar way.

Similarly, surgical removal of the upper half of the small intestine results in urgent compensation in the form of increased secretion by the stomach and pancreas; this is evident in increased trypsin, lipase, and amylase activity. The compensatory hyperfunction of the glandular apparatus of the stomach and the excretory region of the pancreas is followed by their compensatory hypertrophy, which constitutes the basis of stable long-term compensation [110].

The degree to which urgent compensation is pronounced and the dynamics of its transition into long-term compensation are highly dependent on how rapidly injury develops. For instance, with acutely occurring heart failure due to abruption of the bicuspid valve, trauma, infarction of the papillary muscle, or experimental narrowing of the aorta, the heart urgently develops compensatory hyperfunction in response to the increased load. This occurs at the expense of the Starling mechanism, the interval-force mechanism, and a positive inotropic catecholamine effect. In the beginning, it is not accompanied by myocardial hypertrophy but is complicated by a set of shifts characteristic of cardiac insufficiency. Subsequently, along with rapidly developing heart hypertrophy and enhanced heart power, signs of cardiac insufficiency prove to be eliminated, and urgent compensation is replaced by stable long-term compensation. Yet, in most cases, injury of the heart valve apparatus in rheumatic patients develops gradually; correspondingly, the load on the heart grows gradually – cardiac hyperfunction is less pronounced and the hypertrophy-induced hyperfunction of the heart also develops slowly. In this situation, transition of urgent compensation to long-term compensation is extended in time to complicate its clinical determination. However, this does not change the essence of the process; hence, the forming of urgent but insufficiently perfect compensation and its subsequent transition to stable long-term compensation are general regularities in the development of compensation.

To understand the nature of the compensatory process, one must first reveal the mechanism of urgent compensation and then the mechanism of transition of urgent compensation into long-term compensation. When studying urgent compensation, one should keep in mind that structural degradation caused by injury and subsequent functional disorder ultimately change the constancy of the organism's internal medium, its homeostasis. Disruption of homeostasis affects receptors, nerve centers, and executive organs to cause the organism's response. In this response, as in any other adaptive response, one should distinguish the activation of systems specifically responsible for the compensation of the injury-induced functional defect from the activation of the adrenergic and hypophysial-adrenal systems nonspecifically involved in any injury in the organism. For example, when the function of one lung is shut off by croupous inflammation, the oxygen shortage and carbon dioxide excess in the blood stimulate the chemoceptors in the aortal-carotid zone as well as directly the respiratory and vasomotor centers in the medulla oblongata. These are the same or, possibly, different stimulators engendered by substances that during the inflammatory process excite definite hypothalamic centers. This causes a response in the form of a compensatory hyperfunction of the external respiratory and circulatory apparatus in conjunction with pronounced activation of the adrenergic and hypophysial-adrenergic systems, i.e., with the stress reaction.

Anokhin's [6] studies on the decisive role of afferent signaling in generating functional systems that specifically compensate a given injury are highly significant in understanding how these systems develop. His studies of compensation during disturbed motor functions and functions of the internal organs resulted in the formation of several general principles characterizing the development of a system that compensates an injury-caused functional defect. (1) The principle of signaling the defect, whereby the first impulse to start compensatory adaptations arises. (2) The principle of progressive mobilization of reserve compensatory mechanisms, which enables us to understand the actuation sequence of compensatory mechanisms. (3) The principle of inverse afferentation of individual fractional stages of recovery of impaired functions. (4) The principle of sanctioning afferentations, in which the last excitation that had determined the success of functional recovery on the periphery is secured in the nerve centers and, especially, in the cerebral cortex. (5) The principle of relative instability of compensated function, which permits the assessment of the strength of each final compensation.

The above principles are applicable to highly variable forms of compensation. For instance, injury of the lower extremity impairs equilibrium and walking. This entails changed information from vestibular receptors, muscle proprioceptors, cutaneous receptors, and visual receptors. The processing of this information by analyzers in the central nervous system causes the function of definite motor centers and muscle groups to change so as to restore to some measure equilibrium and to preserve, in changed form, the act of walking (signaling of defect). As the injury becomes more severe, defect signaling may increase and then involve new nerve centers and corresponding muscle groups in the compensating functional system (progressive mobilization of reserve compensatory mechanisms). Subsequently, as the compensation becomes effective or the injury itself is eliminated, the composition of the afferent pulse flow entering the higher region of the nervous system would change and certain regions of the functional system previously involved in compensatory activity (inverse afferentation of stages of recovery of impaired functions) would correspondingly shut off. Ultimately, a quite stable anatomic defect retained after recovery would persist through a definite set of afferentations entering the higher regions of the nervous system, afferentations that on this basis would form a stable combination of temporary connections providing optimal compensation, e.g., minimal limping during a given injury.

After hemorrhage, anemia and oligemia lead in a similar manner to tissue hypoxia, and this set of shifts is the stimulator of the chemoceptors of distension of the aortal-carotid zone and the bulbar centers themselves. In response to this functional defect signaling, the cardiac output increases and the vessels of the abdominal cavity organs and muscles narrow. These significant responses reduce the hypoxia syndrome in vitally important organs but do not fully eliminate this functional defect. Hence, impulses from volumoceptors localized in the region of the right auricle involve a major link in the compensatory functional system, namely increased secretion of an antidiuretic hormone by the supraoptical nucleus and anterior section of the hypophysis and of aldosterone by the adrenal glomerular zone. These hormones block excretion of Na^{+} and water from the organism; as a result of such progressive mobilization of reserve compensatory mechanisms, oligemia is rapidly eliminated and tissue hypoxia is also partially eliminated. The tissue hypoxia, which was effective from the very start and was not eliminated by the described responses, stimulates the formation of erythropoietins in the kidneys, which in turn stimulate erythropoiesis. Regeneration of erythrocytes eliminates anemia and, consequently, hypoxia. Removal of this functional defect terminates increased impulses from the above-mentioned receptor zone (inverse afferentation on restoration of functions), and the compensatory functional system, having played its role, is eliminated.

Unlike restoration of blood after hemorrhage, most injuries in humans and higher animals are not followed by complete restoration of the degraded organs and the functional system that had formed in urgent compensation is preserved in subsequent life. This is specifically observed in the previous examples of major injuries of extremities, paired internal organs, heart valvular apparatus, and gastroenteric tract; in addition, in all these situations, incomplete urgent compensation becomes reliable long-term compensation.

As with other adaptational responses, the phenomenon directing the transition from urgent to long-term compensation is the increased cell function of systems responsible for compensation, which causes an increased transcripton rate of ribonucleic acids over the DNA structural genes in the nuclei. The resultant increase in the polysome number enables protein synthesis and additional structures to increase, being evident in the form of cell hypertrophia or hyperplasia. In other words, we have an intracellular control mechanism connecting the cell function level with the cell's genetic apparatus; this mechanism was exhaustively examined in the above examples of heart compensatory hypertrophy and other adaptational responses.

The significance of the interrelation between function and the genetic apparatus as an essential link in long-term adaptational responses, specifically in compensation, has been proven at all levels of the organism's regulatory hierarchy. For example, when during retraining experiments the leading right paw has to 'cede' its status to the left paw, the development of this habit is accompanied by activation of RNA and protein synthesis in specific cerebral cortex neurons responsible for the forming of corresponding temporary connections and habits (see Chap. 5, pp. 201–205). With compensatory hyperfunction of the heart, kidney, lung, and part of the liver that had remained after hepatectomy, activation of nucleic acid and protein synthesis is also the basis for the hypertrophy of those organs.

It is significant that during compensation nucleic acid and protein synthesis is activated not only in the cells of the executive organ sustaining the increased load but also develops and leads to definite structural changes in all the links of the compensating functional system, i.e., the development of a systemic structural trace forming the basis of long-term compensation. For example, congenital heart diseases accompanied by a flow from the right heart to the left heart and by pronounced hypoxemia were found, besides heart hypertrophy, to result in activated nucleic acid and protein synthesis in the bone marrow and in pronounced polycytemia, providing increased oxygen capacity and a greater number of mitochondria in the muscles with their resultant ability to capture oxygen from the blood.

When a person loses his right hand, the systemic structural trace is evident in new connections in the cerebral cortex, which permit the left hand to perform acts normally performed by the right hand, in hypertrophy of the motor neurons innervating the retained extremity, in hypertrophy of its muscles, and also in changes in the shoulder girdle bones. A combination of these structural changes, which develop at different levels, forms a systemic structural trace constituting the basis of compensation.

Evidently, with different injuries, the compensating functional system and the systemic structural trace forming therein differ in architecture. With damage of the nervous system and subsequent compensation of motor and intellectual disorders, the architecture is most complex; when individual internal organs are shut off, it is relatively simple. However, in all cases, the interrelation between function and the cell's genetic apparatus is the key link in the mechanism that engenders the systemic structural trace.

It should be emphasized that the operation of a compensatory functional system does not in itself provide stable compensation. The basis of such compensation is the development in that system of a structural trace. This is particularly evident in situations in which the compensatory functional system is relatively simple and arises directly after injury. For example, after considerable narrowing of the aortic ostium or removal of one kidney, compensatory hyperfunction of the heart or remaining kidney occurs instantly but does not eliminate the functional insufficiency of these organs. Indeed, heart hyperfunction is complicated in the first few days by signs of cardiac insufficiency, and kidney hyperfunction does not provide excretion of the amounts of water and urea normally eliminated by two kidneys. Only subsequently, after heart and kidney hypertrophy due to activation of nucleic acid and protein synthesis, does compensation become complete and stable.

With administration in the early compensation stage of actinomycin D, an inhibitor of RNA synthesis, in doses not toxic to control animals, hypertrophy of the heart or single kidney proves inhibited. The structural trace does not form in the compensatory system, and cardiac insufficiency, as seen in reduced heart resistance to fatigue, edemas, and ascitis [455], or renal insufficiency, as seen in increased urea content in the blood [346], develop instead of stable compensation. Thus, *the development of a structural trace in the compensatory system constitutes a real foundation, an essential link in compensation.*

As a consequence of the above, the administration of synthesis cofactors and nucleic acid and protein precursors considerably accelerates compensation and renders it more perfect. This was proven for compensatory heart hyperfunction induced in animals by experimental heart disease [455] or by myocardial infarction [299] and confirmed in a cardiological clinic, where a complex involving orotic acid, vitamin B_{12}, and folic acid proved to be a factor positively affecting compensatory hyperfunction and hypertrophy of the intact regions of the heart after myocardial infarction [286]. On the whole, these facts agree with the familiar evidence that nucleic acid and protein synthesis activators promote the establishment of highly variable adaptational responses from adaptation to physical loads to temporary connections (see Chap. 5). This similarity in the results is due to the fact that the systemic structural trace that forms through the activation of nucleic acid and protein synthesis in cells responsible for the system's adaptation forms the common basis for all adaptational responses. This commonness does not make the compensation involved in the injury of various organs and systems less unique. For example, following the formation of a structural trace, specifically with greatly hypertrophied organs achieving compensatory hyperfunction, the growth of key structures responsible for control, ion transport, and energy supply may lag behind the growth of their mass. This process stage leads to reduced functional possibilities of the organ and is best studied in compensatory heart hypertrophy, which first prevents acute cardiac insufficiency, but may subsequently lose its adaptive significance [322].

In conformity with this, Fig. 28 shows that the delay in the growth of key structures in myocytes as compared with the increase in the heart total mass leads, with exceeding heart hypertrophy, to reduced power of adrenergic regulation of the heart, lower transmission of the coronary bed, impaired calcium transport and disturbed glycolysis mobilization, and decrease in the myofibril ATPase activity. Occurring in parallel, these shifts potentiate each other (Fig. 28). For instance, reduced effective concentration of noradrenaline, which in normal condiditions activates calcium absorption by the SR, enhances the disturbance of calcium transport. The increased calcium

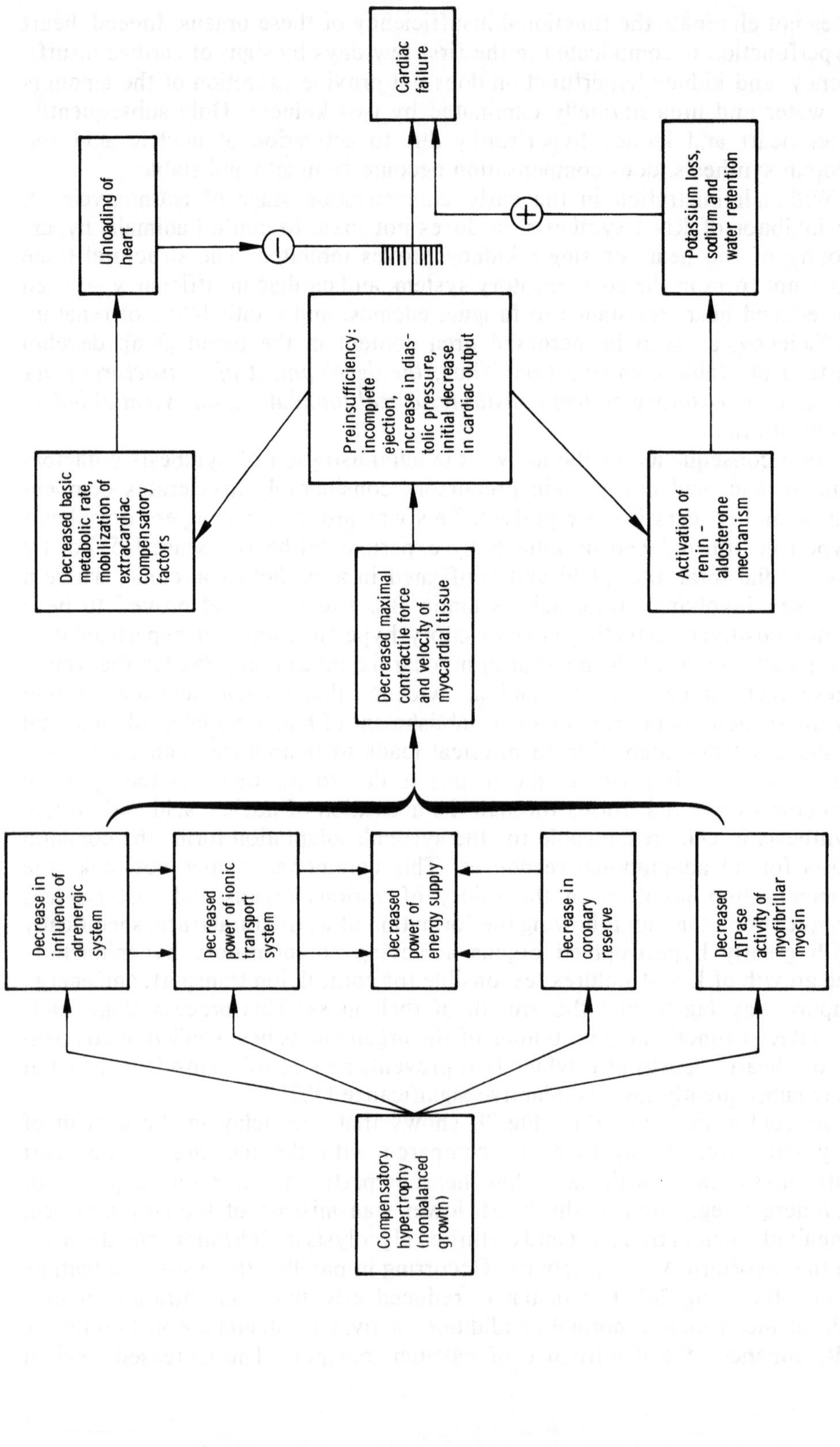

Fig. 28. Transition of compensatory hyperfunction and hypertrophy of the heart to cardiac insufficiency. Explanation in text

load on the mitochondria in conjunction with lower transmission of the coronary bed reduces, as was just seen above, the power of the oxidative phosphorylation system. This decreased power of energy supply is quite definitely combined with a diminished myofibril capacity to utilize the ATP energy, since myosin ATPase activity drops. In effect, this combination signifies a decline in the maximal rate with which the myocyte can perform its specific function, convert the chemical energy of ATP into mechanical energy of contraction. This results in a depressed rate and amplitude of contractions of the hypertrophied heart and, consequently, in incipient hemodynamic disorders characteristic of cardiac insufficiency. Therefore, *in certain conditions, the structural trace constituting the basis of compensation may lose its adaptive significance so that we observe how compensation is transformed into a disease.*

In previously published reviews, we analyzed compensation during injuries to internal organs [318, 322] and to the nervous system [313] to show that differences in injury localization often conceal common regularities of the compensatory process but never eliminate them.

1) The stage of *urgent compensation* was shown to be characterized by the formation and hyperfunction of a compensatory functional system and by a pronounced stress reaction. When other adaptational responses by the injured organism rapidly eliminate the functional defect, compensation may be restricted to this stage. For example, in hemorrhage, restoration of erythrocytes by direct regeneration terminates compensatory hyperfunction of the heart; during unilateral croupous pneumonia, elimination of the inflammation by means of phagocytosis and immunogenesis also makes compensatory hyperventilation of the other lung superfluous. However, in many cases, the injury proves irreversible, and then the stage of urgent compensation restricts, but far from eliminates the functional defect caused by the injury. Disruptions in homeostasis to some extent persist, and compensation continues to develop.

2) *The transitional stage from urgent to long-term compensation is* characterized by – on the basis of the interrelation between function and the genetic apparatus – the hyperfunction of the compensatory system causing nucleic acid and protein synthesis to be activated in the cells of a given system and inducing the formation of a systemic structural trace. As a result, the structural changes that enhance the power of the system, energy shortage in its cells, and the functional defect caused by injury begin to diminish and consequently reduce the stress reaction.

3) *The stage of stable long-term compensation is* characterized either by the elimination or considerable decrease of the functional defect, by the presence of a structural trace in the compensatory functional system, and by minimal activation of protein synthesis in the cells of the system essential for renovating the increased mass of structures. The stress syndrome is absent. With heart diseases, loss of one kidney or lung, and many injuries of the nervous system, this stage may endure in conformity with specific life expectancy.

4) *The stage of exhaustion and functional insufficiency* may develop with a major initial functional defect and excessive hyperfunction and hypertrophy of compensatory systems, and also because of additional injuries. In its general form, this stage is determined by a compensatory hyperfunction essentially differing from the evolutionally envisaged functional growth that periodically occurs in a healthy organism. This functional increase is always temporary and is supplanted by a period of relative quiet. The compensatory hyperfunction of intact nerve centers or a single kidney or heart with an injured valvular apparatus is, conversely, continuous since the irreversible nature of the injuries that have occurred in the organism makes continuous hyperfunction a vital necessity. The stage of exhaustion signifies the transition of compensation to decompensation and may occur on account of the two basic mechanisms discussed above in relation to compensatory hypertrophy of the heart. In the first mechanism, with high hypertrophy, cell growth proves unbalanced and is accompanied by a selective delay of the mass of structures responsible for ion transport, energy supply, and energy utilization. In the second mechanism, after lenthy hyperfunction and hypertrophy, the cells of the compensating system may develop a singular *local wear* complex, which is evident in reduced nucleic acid and protein synthesis, impaired renovation of structures, death of some cells, and development of sclerosis in a given organ (see Chap. 1).

Thus, compensation is one of the adaptive reactions of an injured organism and involves certain differences from the adaptive reactions of a healthy organism. Yet, these differences are not a matter of mechanism, since the forming of a systemic structural trace constitutes the common basis of all long-term adaptational reactions. During compensation, this process passes through the same four principal stages as during adaptation of a healthy organism to the environment.

Chapter 5
Systemic Structural Trace: Foundation of Memory and Higher Adaptational Responses

"Dates, events and countenances
our memories store forever.
The wells of memory are infinite, however."

V. Vysotsky

A significant feature of all the adaptational responses examined so far is that the systemic structural trace, the basis of adaptation, develops within the functional systems and preexists the action of the environmental factor. Accomplished on the basis of the $G \rightleftarrows F$ mechanism, this trace enhances the power of the systems, renders their functioning more efficient, and becomes the basis for long-term adaptation; this is precisely how matters stand upon adaptation to physical loads, hypoxia, toxic substances, etc.

With the development of qualitatively more complex forms of adaptation to biologic and social media, there are no ready congenital systems of this kind. Adaptation may be achieved only at the expense of brain systems that provide new skills and behavioral responses; the systemic structural trace, which secures these new systems, constitutes the basis for new responses and represents an expression of brain memory. In accordance with this, the present chapter examines the relationship between memory and adaptation, the role of $G \rightleftarrows F$ interrelation in the mechanism of brain memory and higher adaptational responses, the role of the stress reaction in forming systemic structural traces in the brain, the principal stages of higher adaptational responses and, finally, the possibility of affecting the memory and behavior by means of factors that cause nucleic acid and protein synthesis in the brain to activate.

Relationship Between Memory and Adaptation

The structural traces forming the basis of adaptation develop as a result of more or less long-term physiologic activity, during which the enhanced cell function activates the genetic apparatus. These changes, which are not injuries and occur due to physiologic activity, may be regarded as utilization changes; this is the shortest definition of memory. A stricter definition may be rendered as follows: The individual memory of living systems represents a change that has occurred due to physiologic activity caused by the effect of environmental factors. This change is stably retained after the activity itself has largely ceased to determine future responses to environmental factors.

These considerations at one time led us to the conclusion that there is always an inseparable relationship between the phenomena of 'memory' and 'adaptation' [317]. Two principal memory levels are distinguished, viz the preneural or elementary level inherent in any living tissue and constituting the basis of the systemic structural traces thus far examined and the higher level of memory constituting the basis for higher adaptational reactions by the brain [245]. The principal difference between the higher and preneural memory levels is that the preneural memory referred to thus far is instrumental in the development and retention of the structural traces of effects that directly influence the organism and in one way or another interfere in homeostasis; conversely, neural memory is instrumental in the development and retention of the traces of effects that in themselves are indifferent to the organism and only become significant as signals of biologically important environmental factors.

In effect, the mnemonic trace secured in the brain, which converts the previously indifferent stimulus into one causing strong responses, may be the relation between the centers excited by action of the previously indifferent stimulus and the centers whose activity underlies unconditioned adaptive activity. In other words, terms like 'memorizing' and 'memory engram,' both widely used in the literature to designate a mnemonic trace in the brain, are synonyms of the familiar concept 'temporary connection.'

It will further be shown that like elementary preneural memory brain memory is essentially based on the activation of nucleic acid and protein synthesis, which occurs in the cells under the action of enhanced function. It should be emphasized that this common mechanism of forming a systemic structural trace in the brain and other organs does not deprive that trace of uniqueness. Indeed, intensified functions of various organs chiefly lead to a greater number and size of the cells therein and to increased activity of definite enzymatic systems, i.e., to structural changes enhancing the power of the preexisting functional system to the needed level. Structural changes occurring with intensive functioning of brain neuronal systems during higher adaptive responses are not exhausted by these quantitative shifts. They result in new connections between the cells of that organ and in new multineuron systems, the very ones that form a systemic structural trace inherent specifically in the brain.

Systemic structural traces, i.e., mnemonic traces, constitute the basis of both simple and higher adaptational responses, and in this sense memory is the basis of adaptation. However, this does not mean that memory is identical to adaptation. The presence in the brain of a memory engram may play its role in adaptation to the environment only with the active functioning of the reproductive apparatus that extracts the mnemonic trace and includes it in the adaptational response. Similarly, the systemic structural trace constituting the foundation of adaptation to physical loads, the basis of the degree of training, may play its decisive role in adaptational response, provided the mobilization apparatus, in which higher control levels play a major part, functions flawlessly. Thus, the systemic structural trace – the mnemonic trace – is an absolutely necessary prerequisite of adaptation, its structural basis; however, it is not identical to adaptation.

Interrelation Between Function and the Genetic Apparatus Is the Basis of Brain Memory and Higher Adaptational Responses

At present, there is a great deal of evidence to suggest that the interrelation between the genetic apparatus and cell function plays the same essential role in the formation of temporary connections as in the mechanisms of the adaptational responses bases on preneural memory mentioned above. The following five groups of facts speak in favor of this concept.

1) The G$\rightleftarrows$F interrelation takes place in the neurons of the cerebral cortex on a large scale and at a high rate. For example, Oppelt [422] studied the effect of light stimulation on the relationship of single ribosomes and polysomes in the cells of the occipital lobe in the cerebral cortex of rats for 3 days in the dark. The total RNA in the polysome fraction was found to have increased by 82% following 15 min stimulation. The polysome is a group of ribosomes, which in the course of protein synthesis are arranged and move along the same messenger RNA molecule. Hence, the obtained result signifies that *excitation of cortical neurons caused intensive programing of messenger RNA ribosomes, this RNA being either synthesized anew or present in the neurons in the form of an active reserve. In any case, this increased ribosomal programing was accompanied in the above experiments by growing intensity of protein synthesis in the cerebral cortex: An enhanced function had urgently received its structural provision.*

2) When the cortex performs its specific function, i.e., closes and fixes temporary connections, its nucleic acid and protein synthesis is naturally activated in its neurons. For example, in the detailed work by Uphouse et al. [554], RNA synthesis in the brain was determined by inclusion of tritium- or carbon-labeled uridine in the polysome fraction. The RNA synthesis in the rat brain was found to have increased in intensity by 40%–70% as early as 30 min after the animals had started to develop a conditioned defense reflex to light or sound. Also, this synthesis activation was precisely connected with the closing of the temporary connection . It was completely absent after the animals had performed ready, preformed conditioned reflexes; it did not occur as a result of sound, light, and unconditioned stimuli (electric stimuli) when under experimental conditions a temporary connection could not be formed between these stimuli. A whole series of similar investigations was performed by other authors [193, 440, 517].

An electron-microscopic study of cortical neurons after the animals had developed a motor-food conditioned reflex showed high activity of the protein synthesis apparatuses, which was evident in a considerably enhanced number of ribosomes and polysomes compared with controls. In addition, the following changes were revealed in cortical neurons. The nucleus and nucleolus markedly increased in size; the synapses grew in length and number of active zones; a large number of synaptic vesicles appeared in the presynaptic region; and spinular apparatuses, differing from the norm in length and formation of intricate membrane complexes, developed on a mass scale [295].

Thus, activated synthesis of summary or specific proteins and specific structural changes in synapses constitute the essential link in the formation and securing of temporary connections.

3) Inhibition of nucleic acid and, particularly, protein synthesis either worsens or prevents consolidation (fixation) of temporary connections. To reveal the effects of inhibition, one must observe certain conditions, which include, first, small 'training intensity' and, second, inhibition duration and extent, which would suppress protein synthesis by more than 90%. The RNA synthesis is inhibited by administration of 8-azaguanine, which causes synthesis of anomalous RNA molecules, or by actinomycin, which blocks DNA-dependent RNA synthesis. Protein synthesis is inhibited by puromycin, cycloheximide, acetocycloheximide, and anisomycin [23, 116, 344]. This indicates the need for protein and RNA synthesis to secure the temporary connection, to consolidate it in long-term memory.

4) Chemical factors cause increased nucleic acid and protein synthesis, promote consolidation, accelerate it and render it more efficient. Precursors of pyrimidine bases incorporated in RNA (orotic acid, uridine monophosphate, etc) or cofactors of nucleic acid synthesis (folic acid and vitamin B_{12}) are usually used in this. For example, preinjection of mice for 7 days with a complex consisting of orotic and folic acids and vitamin B_{12} did not affect the development of conditioned defense reflexes in a T-shaped labyrinth; however, it either eliminated or considerably reduced the 'erasing' effect of an electric shock generated immediately after training.

Similar studies have been published by other authors, too [426]. Thus, the agents that enhance nucleic acid synthesis and, correspondingly, protein synthesis improve the consolidation and preservation of temporary connections.

5) As a rule, interference in nucleic acid and protein metabolism, violating fixation, does not affect the production and reproduction of temporary connections. Hence, protein synthesis is not involved in the production and reproduction of the conditioned reflex, but is necessary to fix the temporary connection to secure it in long-term memory [23, 533]. Thus, changes in nucleic acid and protein synthesis are involved in securing temporary connections. This concept was developed in an interesting manner by Barondes and Cohen [23a]. It appeared that to activate nucleic acid and protein synthesis considerably, i.e., to secure fully the temporary connection, it is not enough for the excitation of certain neurons simply to occur simultaneously and to involve an undisrupted protein synthesis. To activate nucleic acid and protein synthesis considerably, the initial and apparently most intensive excitation, which occurs with conjugated conditioned and unconditioned stimuli, must coincide with the active state of the protein synthesis apparatus.

Hence, to activate the genetic apparatus in the nerve cells and to consolidate the temporary connection, one needs not simply excitation but its high intensity, normally occurring at the very outset, to be further provided solely by additionally applied unconditioned stimuli. From such facts, we arrived at the concept that the $G \rightleftarrows F$ interrelation is achieved in neurons, as in other cells, to constitute a necessary link in the forming of temporary connections. Briefly, this concept may be formulated as follows. In acting in definite sequence on receptors, environmental stimuli engender in the brain a group of intensively functioning neurons, including both neurons responsible for the perception of previously indifferent signals and those responsible for unconditioned adaptive activity. *In the initial, neurodynamic stage of short-term memory, physiologic hyperfunction of each of the excited neurons activates the genetic apparatus through an intraneuronal relationship between function and the genetic apparatus, i.e., enhances nucleic acid and protein synthesis in the neuron. This phenomenon is the essence of consolidation of temporary connection; it leads to increased plasma flow and to the transformation of existing synaptic bonds between neurons and, possibly, to the development of new ones. As a result, the initial functional process is structurally provided to transform the group of excited neurons into a certain integral, structurally secured functional system, preserved long after stimulation has stopped and, hence, comprising the structural foundation of memory* [328].

In effect, this hypothesis is in good agreement with the classic concept of neurophysiology, according to which the transition of functional changes into structural changes constitutes the basis for temporary connections or brain memory [28, 98, 188, 191, 431]. At the same time, it shows the G⇄F interrelation to be a concrete mechanism providing that transition. The principal stages of the transformation of functional changes into structural changes were distinguished by studying the resistance of the temporary connection to the 'erasing' effect of electroshock and to other factors, i.e., in experiments with retrograde amnesia.

The ability of electroshock to cause retrograde amnesia was revealed in the first clinical works involving electroconvulsive treatment. Zubin and Barrera [602] showed that recently studied material was retained less well under electroshock than that memorized long before it. Similar results were obtained later on people and animals by Duncan [92, 93], Croncholm and McLander [70], and some other researchers. On the basis of this evidence, the memory arising directly after the introduction of data into the brain and readily 'erasable' by electroshock has been termed short-term memory. On the other hand, memory that develops a long time after the introduction of data and is stable to electroshock was designated long-term memory. Finally, transformation of short-term memory into long-term memory was called consolidation.

Further study of the effect of amnesic agents on the formation of memory trace has led to the conclusion that numerous amnesic factors do not prevent the fixation of temporary connections but simply impair the possibility of reproducing the developed conditioned reflex. Numerous facts testify in favor of this [215, 245, 446, 456]. These data do not dismiss the existence of short- and long-term memory, but lead to the idea that as early as during reflex development, or in the immediate minutes following reflex development, enhanced neuronal function causes the neurons to develop certain quite stable structural changes that can however be erased by electroshock. These changes subsequently develop due to pronounced protein synthesis activation constituting the essence of consolidation and providing structural fixation involving interneuronal connections and the multineuronol memory engram as a whole. In addition, one has to account for the fact that the occurrence and stability of amnesia depend on the ratio between the initial 'training intensity' and the strength of amnesic effects: With insufficiently intensive training and subsequent intensive amnesic effects, amnesia is generally stable and irreversible; conversely, with excessive training and insufficiently strong amnesic effects, amnesia may never occur or be temporary.

Thus, at least three stages may be distinguished in the development of a temporary connection. The first stage i.e., the stage of short-term memory is characterized by the occurrence of excitation; the second by the existence of excitation and consolidation, which develop by the activation of nucleic acid and protein synthesis; and the third stage, i.e., the stage of long-term memory characterized by the presence of a stable structural mnemonic trace in the absence of excitation.

One may state that these three stages do not basically differ from similar stages in the cells of any adaptation-responsible system in simpler adaptational responses, i.e., from the reaction stages in the genetic apparatus of any cell to long-term enhanced function. Indeed, during adaptation to physical loads, the first stage is characterized by a situation in which motor neurons, myocytes, and other cells of the system responsible for adaptation perform an increased function and have high SFI; in the second transitional stage, high SFI causes the cell's genetic apparatus to activate and bring about the coexistence of hyperfunction and activation of nucleic acid and protein synthesis; finally, in the stage of stable adaptation, structural changes (hypertrophy and enhanced power of mitochondria and systems of ion transport) reduce SFI and eliminate the initially hyperfunction-caused ATP shortage thus terminating activation of nucleic acid and protein synthesis. This common stepwise nature in the forming of systemic structural traces of varying complexity is ostensibly caused by the fact that they are all due to the same intracellular $G \rightleftarrows F$ mechanism, which is apparently similar in cells of very different systems. There is evidence that in brain neurons this correlation occurs in the same way as in other cells, viz due to a shift in energy balance and a relative shortage of energy-rich phosphoric compounds.

It was indeed shown that with intensive neuronal excitation, increased ATP utilization exceeds its resynthesis resulting in a decrease in ATP concentration and glycolysis mobilization [181, 565]. This is followed by increased RNA and protein synthesis and higher activity of respiratory enzymes in neurons with a simultaneous decrease in these indices in glial cells [161, 163, 192, 193]. In addition, the neurons are found to increase not only in size but also in weight [188] and the glial cell nuclei decrease in size [217]. A substantial argument in favor of the notion that the whole set of changes is due to a relative energy shortage is that activation of RNA and protein synthesis in the cortical neurons and also changes in neuroglial relations may be reproduced not only by enhanced function but also by hypoxia, which limits ATP synthesis in the mitochondria (see pp. 224–225, Figs. 29, 30).

The unity of the cell mechanism that forms the basis of systemic structural traces of varying complexity makes it possible to assume that the principle of cell structure ratio changes, in the form of preferential accumulation of the physiologically most important cell membrane structures, occurs – as with other adaptational processes and possibly to an even greater extent – when temporary connections form between the brain centers. Indeed, above we have developed the concept that with increased cell function, energy shortage and the factor-regulator thereby determined activate the transcriptions of the cell's genetic apparatus to a different degree. Under optimal load and corresponding concentration of the factor-regulator, the transcriptons that code for short-lived proteins of membranes responsible for ion transport, perception by the cell of governing control signals, and mitochondrial membrane proteins, are activated to the highest degree. Selective accumulation, with which these structures prevail, sharply enhances the effective functioning and control of a given organ. In relation to heart neural regulation, it was found that adaptation to physical loads or altitude hypoxia results in two major shifts.

The first shift is growing adrenoreactivity of the heart, due to which the need in increased cardiac function may be achieved by a small mediator quantity, i.e., with efficient functioning of adrenergic regulation (see Fig. 10). According to this concept, such phenomena associated with efficient regulation may be provided when the enhanced function preferentially activates certain transcriptons resulting in the selective accumulation of receptor membranous structures, i.e., adrenoceptors, cholinoceptors.

In the second shift, in adapted animals, the decrease in the contractile function of the heart proves to be several times less than in controls after completely desympathizing the heart by pharmacologic shut-off of β- and α-adrenoceptors. In other words, adaptation substantially enhances the resistance of the contractile function of the heart against cAMP shortage. Cyclic AMP does not bring about any previously nonexistent processes in myocardial cells. Its role is reduced to potentiating via the protein kinase system reactions caused by Ca^{2+}, an intracellular regulator, namely, potentiating contraction, relaxation, and activation of glycolysis. It may, therefore, be assumed that the lesser dependence of these reactions upon cAMP caused by adaption may be due to the increased population and power of the protein kinases that achieve vitally important links in cell metabolism despite the cAMP decrease. Thus, in adaptation to physical loads and hypoxia, the relationship between function and the genetic apparatus creates a change in the cellular structure ratio in the organs, this change becoming the basis for economical and effective control of these organs.

As was shown above, the same interrelation occurs in the neurons when brain memory forms; correspondingly, one may assume that in these cells too it provides the selective accumulation of structures that play a decisive role in regulatory interaction, i.e., structures responsible, on the one hand, for synthesis, deposition, and release of mediators and, on the other, for reception of those mediators and for the response of the nerve cell thereto. This selective accumulation of key membrane structures in the neurons that form the systemic structural trace may, in our view, be the basis for reorganization of interneuronal synaptic connections and, hence, the basis for the development economical functioning of multineuronal systems constituting the brain memory engrams. Apparently, this hypothesis still needs to be adequately verified and developed.

Role of Emotional Stress in Forming Higher Adaptational Responses. Unity of Dynamics in Forming Conditioned Reflexes and Other Adaptational Responses

A temporary connection acquires decisive significance in achieving higher adaptational responses not in itself but as a link in the complex architecture of these responses, primarily as a link in the conditioned reflex mechanism, which, besides the temporary connection per se, embraces its formative and reproductive apparatus. In this case, the emotional stress that occurs naturally under substantial environmental changes plays an important role in forming conditioned reflexes. In accordance with Simonov's [522] informational theory of emotion, emotions occur as a result of a gap between real requirements (self-preservation, food, and sex) and the absence of sufficient information on whether or not a given need may be satisfied in given environmental conditions. This disagreement between the need and the information essential to achieve it is perceived at the level of a new cerebral cortex to result in ascending cortical effects. These effects activate a specific neural apparatus of emotions, a system of nerve centers located in the region of the hypothalamus, the amygdaloid complex, and the old cortex, which according to numerous data plays the role of a connective link between the new cortex – in which shortage of information is primarily determined – and the very neural apparatus of emotions in the hypothalamus. Subjectively, these centers are excited as emotions of fear, anxiety, fury, pleasure, etc. and, objectively, as two interrelated phenomena. The first is the occurrence of multiform emotional behavioral responses, with which the organism attempts to fill the 'informational gap' between a given need and the possibility to satisfy it. The defense dominant that forms in children or adults in new complex situations is a relatively simple instance of an emotional attempt to compensate or fill this informational breach. This type of dominant is seen in the occurrence of multiform and often hyperbolyzed defense responses to highly different environmental factors, in many cases lacking biologic significance. By change, one such response may be exact, in which case it is reinforced to form a new conditioned reflex corresponding to the environment, i.e., adaptation.

Other expression of emotional behavior are imitative responses, which are often dictated by fear due to insufficient information about the environment. With repeated, multiform, and frequently biologically unjustified versions of this response one may find a successful version appropriate to the environment, which subsequently becomes established.

No less striking examples exist in the world of laboratory-conditioned reflexes. For example, Popova [444] for a long time developed a conditioned motor reflex to food in a dog by making the animal use one of its forelegs to lift a feed-box by a special lever. After the other foreleg was secured to the lever, the dog persisted in lifting the first leg, but this naturally failed to provide it with food. Ineffective attempts by the very hungry animal to lift the feed box with the 'trained' foreleg evoked generalized emotional excitation, and the dog began to lift all four legs. The animal obtained its food when by chance it lifted the leg to which the lever was secured and thus disclosed the signaling significance of the lever. From that signal, after two to three combinations, the dog formed a new conditioned reflex. Thus, the biologic significance of the outwardly chaotic and energy-wasteful emotional behavior is one of the most creative responses, a quest for new information providing new conditioned reflexes and skills that correspond to the habitant.

In relation to more complex processes constituting the basis of thinking, emotion may provide quick, multiform, and often chance recombinations of previously developed mnemonic traces and, hence, new associations, i.e., in the end, new ideas about the surrounding world, ideas supported by the environment and helping to develop further and change the latter. Hence, like emotional behavior, emotional thinking represents one of the most creative processes of higher nervous activity for providing new ideas and skills.

In relation to all these situation (complex and simple behavioral responses), the criterion for a successful decision correctly found through an emotional quest is the optimally developing organism – habitant relationship, which represent a practical result in the broadest sense of the word. An attained result is relayed via feedback mechanisms, and this sanctioning afferentation becomes a major factor in stabilizing and limiting the functional system 'found' by the organism, a system that acquires decisive significance in adapting the latter to a given environmental situation. In our view, the notion 'atteined result' corresponds to Pavlov's term 'reinforcement,' always regarded as a decisive factor in the formation of a new conditioned-reflex stereotype.

The development of a functional system and particularly of a conditioned reflex adequate to the habitant is insufficient in itself to achieve stable and future-oriented adaptation. In order that the new functional system work infallibly and not be erased by the actions of extraordinary stimulators or simply by inhibition caused by transitory behavioral dominants, it should be fixed by the consolidation mentioned above. Over the past decade, it has been shown that if emotional behavior, the first consequence of emotion just examined, plays a part in the occurrence of functional systems of higher adaptational reactions, the second consequence, examined below, plays a major role in fixing those systems. The second consequence of emotions occurs inside the organism and consists of excitation of the adrenergic and hypophysial-adrenal systems, i.e., in emotional stress itself.

With tense emotions, the main stress components are actuated in definite sequence. For instance, Di Giusto et al. [84] showed that when confronted with a threatening situation rats instantly displayed a catecholamine effect; about 10 s later adrenocorticotropic hormone (ACTH) appeared in the blood and only 15–60 min later glucocorticoids appeared, which immediately began to suppress ACTH secretion.

The mobilization of the organism's energy and structural resources by a set of stress hormones and the transfer of these resources to intensely functioning systems (Fig. 11) play an important role in ensuring emotional behavior. Catecholamines, corticosteroids and, as has now been shown, ACTH simultaneously and directly affect the new functional system forming in the brain by playing a role in its formation and, especially, fixation.

The role of adrenergic regulation in forming temporary connections is determined by the direct action of catecholamines at the nerve cell level. Catecholamines are known to prolong the action potential of calcium influx into the cells; indirectly, via cAMP, they activate the protein kinases localized in the neuronal key points. This shift may entail at least three outcomes: Phosphorylation of B phosphorylase into A phosphorylase resulting in additioal activation of glycolysis; phosphorylation of a definite component of endoplasplasmic reticulum membranes, which enhances the calcium active binding rate by those structures; and, finally, phosphorylation of histones, which naturally leads to a higher RNA transcription rate in the nuclei and to additional activation of protein synthesis.

This effect of catecholamines on the excitation, energy supply, and formation of structures should undoubtedly activate the development of systemic structural traces, in general, and that of temporary connections, in particular. Indeed, the development of various conditioned reflexes was found to be followed by a substantial decline in the truncate regions of the brain of preliminarily administered labeled noradrenaline, this being apparently indicative of increased synthesis, label dilution, and increased noradrenaline utilization. Correspondingly, pronounced disrupted temporary connections resulted when noradrenaline synthesis was suppressed by diethyl thiocarbomate, an inhibitor of dophamine-β-hydroxylase [462]; when introduced immediately following the development of a conditioned reflex, the inhibitor caused retrograde amnesia. When the inhibitor was given 2 h later, this was not observed, and when administered prior to checking, whether or not the temporary connection was retained, the reflex did not show. The effects of diethyl thiocarbomate are accompanied by a considerable decrease in the noradrenaline and dophamine content in the brain and are not evident if noradrenaline is administered simultaneously with the inhibitor [74]. Hence, there can be no doubt that the catecholamine physiologic content must be retained in the brain both for consolidating and reproducing temporary connection. This is in agreement with the evidence on the role of increased catecholamine secretion in human mental activity. For example, when people competed in solving arithmetic problems and were unjustly criticized, the resultant emotional stress led to a considerably increased catecholamine excretion with urine. At the same time, those persons who showed the best results were found to excrete more noradrenaline than the others [119].

An interesting explanation is offered by Kety's [216] hypothesis, according to which the role of the brain's adrenergic systems in securing temporary connections is when reinforced, the activation of these systems supports the trace activity of neurons, which became activiated after perceiving the conditioned stimulator. As a result, the effectivity of synaptic transmission stably rises to form a connection between the corresponding sensor entrance and the systems achieving the reinforcing unconditioned response. Possibly, the adrenergic ascending system examined by Kety is precisely the one in which the sufficiently known activating influence of the reticular formation takes place to form temporary connections in the brain. Weak subthreshold stimulation of the reticular formation after training was shown to improve the retention of conditioned reflexes and also either to eliminate or decrease retrograde amnesia caused by inhibitors or protein synthesis [35]. It was further revealed that weak stimulation naturally causes a considerable rise in the RNA content of the brain.

Generally speaking, there are grounds for assuming that the *excitation of the adrenergic system observed under stress, specifically the ascending adrenergic influence of the reticular formation, activates the formation and subsequent consolidation of temporary connections.*

Excitation of the hypophysial-adrenal system under emotional stress may affect the formation of a temporary connection due to the effects of glucocorticoids and ACTH. At present, there is quite abundant evidence that glucocorticoids, whose content in the blood increases with stress five- to tenfold, may activate the formation of temporary connections. For instance, Barondes and Cohen [23a] showed that a corticosterone-hydrocortisone mixture eliminates amnesia caused in mice by cycloheximide, an inhibitor of protein synthesis.

With emotional stress developing when new skills and behavioral responses are being formed, the question concerns the effect of large glucocorticoid doses that reduce the serotonin content. Excess serotonin is known to inhibit nucleic acid and protein synthesis and to reduce the activity of the dopaminergic system, which plays a major role in mnemonic processes. Hence, one may assume that rising glucocorticoid concentration during emotional stress reduces the serotonin content to a level at which the intensity of nucleic acid and protein synthesis increases as does the activity of the dopaminergic system. As a result, the forming of the memory engram is activated.

There are also other explanations for the activating effect of corticosteroids on mnemonic processes. For example, Flood et al. [116] believe that like other stimulators that have no direct effect on protein synthesis but enhance the intensity of excitation and short-term memory, corticosterone affects the development of the memory trace. In conformity with this concept, a rise in corticosteroid concentration under stress modulates excitation of the central nervous system so as to provide an accelerated transition of short-term memory to long-term memory. Adrenocorticotropic hormone and its derivatives have also proved to have a direct effect on fixing the memory trace in the brain; when excreted in increased quantity during emotional stress they may play an important part in forming new conditioned reflexes.

Ress et al. [467] discovered that after mice had developed a conditioned avoidance reflex, their brain and liver protein incorporated more amino acids. These facts showed that stress to one measure or another accompanying the initial stage of a new reflex causes generalized activation of protein synthesis in the brain. Dunn [94] reproduced this experiment on adrenalectomized animals and showed that the phenomenon of generalized activation of biosynthesis was reproduced despite an exceedingly low corticosteroid level in the blood and, hence, does not depend on their effect. He suggested that, in the first stage of training, generalized activation of protein biosynthesis in the brain is caused by ACTH. Soon, Rudman et al. [485] discovered that the administration of large doses of ACTH and of melanocyte-stimulating hormones of the hypophysis really does stimulate protein synthesis in the brain.

Finally, Reading [466] established that synthetic $ACTH_{4-10}$ (a peptide fragment, which hardly affects corticosteroid secretion) increases the inclusion of ^{14}C leucine in rat brain proteins. It was further revealed that long-term administration of $ACTH_{4-10}$ restores the decreased RNA content and the reduced protein biosynthesis in the brain of hypophysectomized animals and eliminates the behavioral disorders observed in them [577]. From these and other facts, Dunn formulated a highly interesting hypothesis on the role of stress in the mechanism of brain memory. According to his theory, stress caused by any sufficiently strong stimulator and, therefore, observed at the outset of teaching some skill causes an increased release of ACTH and ACTH-like hormones, which in addition to affecting the adrenals directly affect the brain, where they cause activation of nucleic acid and protein synthesis. This generalized activation of biosynthesis is not simply a background but a significant prerequisite for local structural changes, which, as was previously shown, constitute the essence of the memory engram or the temporary connection.

Generally speaking, the above shows that using brain memory and higher adaptational responses as examples, one can essentially concretize the previously formed concept that the role of stress as a nonspecific component of adaptation signifies that it potentiates a systemic structural trace constituting the basis for specific adaptation. Indeed, available facts indicate that, during emotional stress, ACTH-like hormones and glucocorticoids cause a generalized increase in RNA and protein synthesis in the brain, and also generalized changes in mediator metabolism. Excitation of the adrenergic system also promotes this activation. However, unlike the role of ACTH and glucocorticoids, that of catecholamines is not restricted to a generalized effect, for, besides a humoral effect on the whole brain, they act via ascending adrenergic fiber systems directly on the developing multineuronal system of specific adaptation.

When combined, the generalized regulatory activation of nucleic acid and protein biosynthesis and the directed effect of adrenergic control facilitate the occurrence and fixation of temporary connections between previously indifferent and now considerable stimulators and specific unconditioned reflex activity. One of the mechanisms of the facilitating effect of emotional stress on the development of brain memory is apparently the preliminary increase in the RNA and protein reserve in the neurons, which may accelerate the formation of structures that predetermine the reorganization of interneuronal synaptic connections during intensive excitation of the dominant multineuronal constellation responsible for a given response. In these conditions, the $G \rightleftarrows F$ interrelations in each excited neuron are achieved more effectively and quicken the formation of interneuronal connections and multineuronal systemic structural traces constituting the basis for specific adaptation to a given real situation.

Thus, emotions, which inevitably arise when solving any really new task, on the one hand engender diversified and often logically unfounded responses of emotional behavior, which essentially represent an expensive search for an optimal adaptational response and, on the other hand, involve a standard stress reaction mechanism. The latter plays the role of an essential link, first, in energy supply for emotional behavioral activity, i.e., for seeking a solution and, second, in the securing by the brain of a correct solution to an adaptational problem after the solution has been found.

The role of the stress reaction in forming and securing the systemic structural trace of higher adaptational responses is in agreement with the fact that with the further consolidation and perfection of these responses, the stress reaction naturally diminishes. For instance, Marchbanks [297] and Haley [cited according to 149] showed that young bomber crew members naturally urinated more 17-oxyketosteroids during and after flights than their senior colleagues. Similarly, according to Gubachev et al. [149], construction workers working at a height of 24 m instead of 4 m above ground initially excrete two to three times more catecholamines in their urine; however, this phenomenon declines with age.

Thus, having performed its role in forming a systemic structural trace and having potentiated the formation of stable phenotypic adaptation, the stress reaction gradually abates.

The above allows one to chart approximately the basic stages of the development of the conditioned reflex and, apparently, of the more complex adaptational responses of the organism.

Main Stages of the Development of Conditioned Reflex. Differences and Common Features of Higher and Simple Adaptational Responses

The first stage was defined by Pavlov as that of generalization since it is in this period that the organism responds to numerous proximate or even remote stimulators, not to a single specific stimulator, and its response is seen in a generalized motor reaction accompanied by major shifts in circulation, respiration, and other 'vegetative' functions. Osipov [424], an outstanding Soviet psychiatrist, and Simonov [522] justifiably assess this period as an emotional one. Hence, it is a period of search involving the formation of a multineuronal functional system, within the framework of which a temporary connection forms in the future. In this stage, the system is still not sufficiently developed to embrace an obviously excess number of afferent and efferent elements. This is expressed in generalized responses, a pronounced stress reaction, and the generalized activation of nucleic acid and protein synthesis in the brain. In effect, this period combines a search for an adequate functional system and a readiness of the brain to fix that system.

The second transitional stage is characterized by the presence of a temporary connection in the brain, i.e., by a definite, sufficiently restricted multineuronal functional system embracing representations of both conditioned and unconditioned stimulators. An intensive and sufficiently concentrated excitation takes place in this system, which via the $G \rightleftarrows F$ mechanism considerably activates nucleic acid and protein synthesis. In other words, excitation causes consolidation and coexists with it, and this results in the formation of a systemic structural trace constituting the basis of specific adaptation. In this case, generalized emotional behavior and the stress reaction prove superfluous and rapidly diminish.

The third stage of the developed systemic structural trace is characterized by the fact that the new conditioned-reflex stereotype is stable because the new multineuronal temporary connection, i.e., a memory engram system in the brain has become structurally secured. The nucleic acid and protein synthesis in the neurons of that system is not increased. Acquired response is highly economical and is not accompanied by emotional behavior and the stress reaction.

The fourth stage is characterized by destruction of the developed structural trace and loss of adaptation; in physiologic conditions, this stage usually does not develop. In specific conditions involving radical change in the organism's adaptive program, and even under serious pathology, the question usually concerns a situation in which temporary connections are either extinct or their reproductive apparatus disrupted, not true disappearance of the systemic structural trace.

Hence, it may be stated that the development of higher adaptational responses based on brain memory differs from that of simpler adaptational responses based on preneural memory chiefly by the fact that the first stage of the formation of the conditioned reflex and more complicated adaptational responses is always a stage of searching for a new functional system, subsequently fixed in the resultant systemic structural trace. With relatively simple adaptational responses in adult humans and animals, e.g., in adaptation to physical loads or oxygen shortage, the first stage of the process involves intensive hyperfunction and relative functional insufficiency of a ready functional system that has been genetically programed and had formed in early ontogenesis. The subsequently developing systemic structural trace is designed to enhance the power of this preexisting system and to expand the link that had limited its function, but not to secure a new system. Fully realizing the significance of this difference, one may nonetheless indicate a number of common features in simple and more complex adaptation responses. (1) As in all other long-term adaptational responses, the $G \rightleftarrows F$ interrelation is a necessary link in developing higher adaptational responses. This very mechanism inside neurons reorganizes their synaptic connections and forms multineuronal systems constituting the basis of brain memory. (2) The fact that adaptation involves changes in cell structure correlation, proven with relatively simple long-term adaptive responses, i.e., selective increase in the mass and power of the structures responsible for the perception of control signals, generation of excitation, can very probably plan a role in the forming of systemic structural higher adaptational responses. (3) The stress reaction plays the role of a nonspecific but necessary link in both simple and higher adaptive responses. In all cases, this role, by the direct action of hormones and mediators and by the directed redistribution of resources, potentiates the formation of a systemic structural trace constituting the basis for specific adaptation. (4) Four stages, nonidentical but largely similar to those that occur in simpler adaptive responses, may be distinguished in the development of higher adaptive responses.

These common features, primarily the role of the G⇄F interrelation in developing the structural basis for all long-term adaptive responses, are characterized by the important consequence that the activation of nucleic acid and protein synthesis in the cells of a formed functional system constitutes a necessary link both in higher and simplest long-term adaptive responses. This means that the accumulation of certain excess amounts of RNA and proteins after sustained emotional stress, or after gradual adaptation to hypoxia, can potentiate the fixation of temporary connections and higher adaptive responses in general. Thus, an interesting possibility of using relatively simple adaptive responses actively to influence higher adaptive responses begins to appear. In this connection, we shall now discuss the effect of adaptation to hypoxia on animal memory and behavior.

Effect of Adaptation to Hypoxia on the Formation of Temporary Connections. Behavior and Brain Resistance to Harmful Factors

The role of timely activation of nucleic acid and protein synthesis in forming the structural basis of adaptation to environmental factors is seen especially distinctly in adaptation to hypoxia, since in this case the functional system that provides adaptation encompasses a wide scope of organs and activates nucleic acid and protein synthesis in all of them (see Chap. 3). During the initial stage of hypoxic action, restricted oxidative resynthesis of ATP in the brain leads to alternative changes in the neurons and glia [16, 251, 282, 516]; subsequently, these changes largely develop inversely [273, 601] and result in a gradual pronounced activation of RNA and protein synthesis. Correspondingly, at this juncture the following three conclusions have been made in studying the effect of lengthy or periodic hypoxia on higher nervous activity: (a) the effect on humans and animals of acute hypoxia caused by a long stay in mountains or in an altitude chamber at 5,000–8,000 m disrupts conditioned-reflex and intellectual activity [3, 231, 414, 497, 568]; (b) preadaptation to continuous or periodic effects of hypoxia largely eliminates and prevents these disorders [223, 273, 555]; (c) as was proven mainly in the author's laboratory, gradual adaptation to the effects of periodic altitude-chamber and continuous mountain hypoxia is not restricted to simply eliminating the disorders of higher nervous activity caused by the shortage of oxygen in the environment. Activation of RNA and protein synthesis is accompanied by definite positive changes in conditioned-reflex activity. These changes are evident in quicker development and better retainment of conditioned reflexes [333, 334, 350, 555] and also in the increased resistance of temporary connections to the effects of extraordinary stimulators [352] and in significant behavioral changes. Correspondingly, we shall consistently examine the dynamics of nucleic acid and protein synthesis during adaptation to hypoxia, the effect of adaptation to hypoxia in the forming of a temporary connection and in animal behavior in conflicting situations, and the effect of adaptation on brain resistance against harmful factors.

Activation of nucleic acid and protein synthesis in the cortex and lower

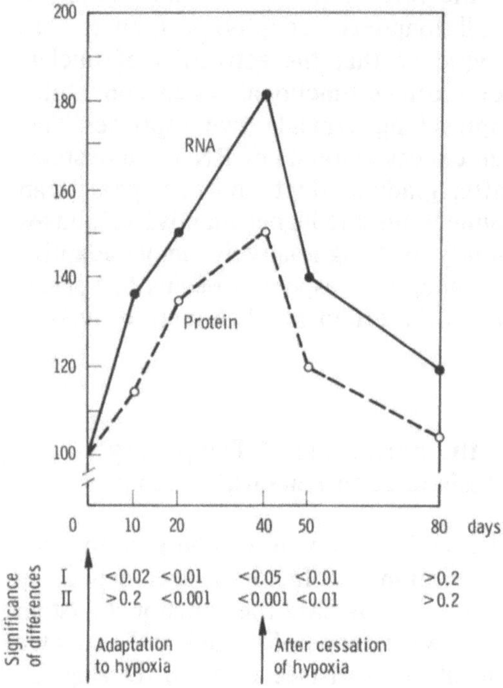

Fig. 29. Dynamics of DNA concentration and protein synthesis in brain cortex, percentage of initial values, during adaptation to periodic effect of altitude hypoxia and after termination of altitude hypoxia

regions of the brain was for the first time discovered with gradual adaptation of animals to periodic hypoxic actions in an altitude chamber [349] and then with adaptation to high-altitude mountain hypoxia [357]. Activation developed gradually; it was more pronounced in the cortex than in the subcortex and was evident both in the neurons and glia.

The curves in Fig. 29 show the dynamics of RNA concentration and protein synthesis in the cerebral cortex of rats 'lifted' in an altitude chamber to 6,000–7,000 m and kept there for 6 h. Again, in Fig. 29, one can see that activation of biosynthesis in the brain cortex develops gradually. Ten days after the beginning of the hypoxic action, the RNA concentration in the cortex proved to be only 15% higher and gradually increased to attain 50% by day 40 of the experiment. The intensity of protein synthesis in the cortex, measured by the incorporation of labeled methionine, grows with the increase in RNA concentration therein to reach 80% by the end of adaptation.

The curves in Fig. 30 show the results of an autoradiographic study of protein synthesis in the cerebral cortex. The upper curves show that during adaptation to hypoxia the inclusion of labeled thyrosine in the giant pyramidal cells of the cortex increases to attain 180% of control values by day 40. Incorporation of thyrosine in glial cell nuclei, the satellites surrounding these neurons, increases in these experiments by 300% compared with controls. The lower curves characterize the dynamics of the cell sizes in the given cortex zone and show that in the course of adaptation to hypoxia the sizes of pyramidal neurons increase by an average of 35%, while those of the glial cell nuclei conversely

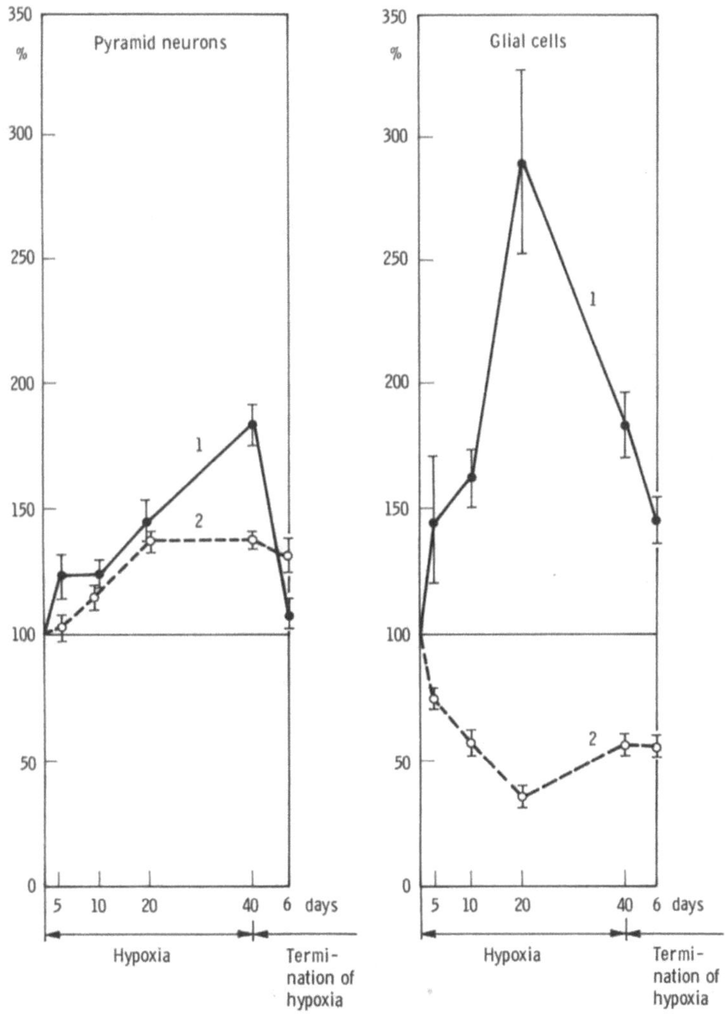

Fig. 30. Dynamics of autographically determined protein synthesis in pyramid neurons and in neighbouring glial cells of the brain cortex *(1)* and glial cell sizes *(2)*, given in percentage of initial values during adaptation to hypoxia

decrease by 50%–60%. Possibly, the neuroglial relationships observed in this case are associated with the glial donor function. At any rate, this picture is distinctly apparent both with adaptation to intermittent altitude chamber hypoxia and with adaptation to environmental conditions at an altitude of 3,200 m.

The physiologic significance of activated nucleic acid and protein synthesis during adaptation to hypoxia requires further study. Quite probably, this adaptation plays a role in providing vessel growth and changes in the glycolysis and oxidative phosphorylation systems, i.e., in developing systemic changes that provide sufficient ATP resynthesis in the brain and its continued function despite hypoxemia. This approach is in line with the fact that protein synthesis is activated in the brain not only in response to hypoxemia caused by hypoxia but also to other disorders of the homeostasis. For instance, during starvation followed by gradual depletion of carbohydrate reserves and hypoglycemia, the brain shows activated nucleic acid and protein synthesis, which in rats results in a ninefold increase in β-hydroxybutyrate hydrogenase activity as early as after 2 days. This provides a possibility for the brain to utilise increasingly the β-hydroxybutyrate substrate retained in the blood. Oxidation of this substrate becomes the basis for ATP resynthesis in the brain of starving people and animals, the foundation for continued consciousness and vitally important behavioral activity [310]. As with hypoxia, in this case adaptive activation of nucleic acid and protein synthesis forms an elementary memory trace in the brain, a trace that becomes the basis for the brain's adaptation to disorders of homeostasis to help retain the higher brain functions and adapt the whole organism to the environment. This conclusion leads to the idea that Claude Bernard's concept that constancy of the internal medium is a necessary condition for brain activity and free behavior in the environment does not fully embrace the situation arising in real conditions. Indeed, the preceding material provides grounds for assuming that in addition to responses that restore homeostasis *a more or less long-term disruption of the latter entails structural changes in the brain, which allow it to adapt to disorders of homeostasis, i.e., to function with a knowingly changed internal medium of the organism.*

In terms of our presentation, it is significant that the adaptive structural trace forming in the brain as the basis of adaptation to hypoxia or hypoglycemia is initially not associated with the specific function of the brain as an organ of behavior and, with regard to its significance for the organism, may be regarded as an example of elementary, primary, 'prenervous memory.' However, by localizing in the brain, this elementary memory trace may come to influence specific memory inherent in the brain and influence the forming of a temporary connection since in the end the same phenomenon, namely activation of nucleic acid and protein synthesis, underlies both forms of memory. One of the concrete ways in which the generalized activation of nucleic acid and protein synthesis during adaptation can influence the development of temporary connections involves the considerably increased RNA concentration observed in the cortex, which unambiguously shows that the number of ribosomes in the neurons had increased. Ribosomes, which constitute over 90% of the total RNA, are known to be nonspecific, i.e., they can be programed by any messenger RNA and used as a site for any protein synthesis. In this connection, excitation of neurons with a ribosomal reserve would quicken the synthesis of proteins necessary at a given moment, e.g., proteins needed to reorganize interneuronal synaptic connections and to fix the occurring memory trace.

From these consideration, we decided to intersect two memory levels, i.e., to study the effect of activation of nucleic acid and protein synthesis caused by adaptation to hypoxia on the forming of temporary connections.

In this case, adaptation was achieved either at a low altitude or by gradually increasing the 'altitude' during 'ascents' in an altitude chamber, since we sought to diminish the first stage of response to hypoxia coupled with alternative changes in the brain and disturbances in higher nervous activity.

In our joint experiments with Isabaeva et al [351], we studied the effect of adaptation to moderate hypoxia at an altitude of 1,600 m (Lake Issyk-Kul) and found that in these conditions animals of various genetic strains develop defensive conditioned reflexes of active and passive avoidance much more quickly and retain them much longer. From Table 21, which illustrates this for BALB/c mice, it follows that the trend showed by the animals on day 6 to develop a conditioned reflex of active avoidance less quickly and retain for less time proved nonsignificant; neither did the results differ from controls on day 10 of adaptation. Later, on day 18, the conditioned reflex was found to develop much more quickly and persist much longer than in controls. On day 37, when adaptation had formed, the result became firmly secured and was evident as, with concentrated and prolonged training, the reflex developed 2.5 times more quickly and lasted 30% and 20% longer, respectively.

In assessing these results, one should keep in mind that all the tests on adapted animals were performed at altitude; hence, at least two factors could affect their higher nervous activity, namely adaptation to hypoxia and hypoxia per se. In line with our problem, we were interested in the effect of adaptation to hypoxia, since adaptation is accompanied by the activation of nucleic acid and protein synthesis. Hence, all further studies were performed in the laboratory; the animals were adapted to periodic ascents in an altitude chamber and their higher nervous activity was studied when the oxygen content was

Table 21. Effect of altitude hypoxia on development and retainment of conditioned reflex of active avoidance and release in BALB/c mice

Duration of stay at 1,600 m	Number of animals	Number of combinations needed to develop conditioned reflex		Retention of conditioned reflex, %	
		Concentrated training[b]	Prolonged training[c]	Concentrated training[b]	Prolonged training[c]
1. Control (prior to ascent)	14	7.2 ± 1.4	5.7 ± 1.1	68.5 ± 9.1	72.2 ± 7.5
2. 6 days	14	8.4 ± 2.2	7.5 ± 1.6	70.1 ± 13.0	59.9 ± 13.0
3. 10 days	13	8.5 ± 2.8	4.7 ± 1.2	79.6 ± 8.5	70.4 ± 10.6
4. 18 days	14	4.8 ± 1.2	1.8 ± 0.5	93.8 ± 6.7	100.0 ± 0.0
5. 37 days	15	2.8 ± 0.9	2.2 ± 0.8	100.0 ± 0.0	91.0 ± 4.1
Significance of differences[a]					
p_{1-2}		n/s	n/s	n/s	n/s
p_{1-3}		n/s	n/s	n/s	n/s
p_{1-4}		n/s	< 0.01	< 0.05	< 0.01
p_{1-5}		< 0.02	< 0.02	< 0.01	< 0.05

[a] n/s: nonsignificant. [b] Interval between combinations, 20 s. [c] Interval between combinations, 90 s.

Table 22. Retention of conditioned reflex of active avoidance during adaptation to altitude hypoxia and in the course of deadaptation

Days after termination of hypoxic actions	Number of animals	Number of combinations		Reflex retention index, %
		During reflex development	During reflex reproduction	
1. Control	20	31.0 ± 1.6	14.8 ± 1.1	40.0 ± 5.3
2. 1 day	10	26.6 ± 0.4	7.3 ± 0.5	71.7 ± 2.8
3. 5 days	10	27.3 ± 1.7	8.1 ± 1.2	69.2 ± 4.7
4. 10 days	10	25.2 ± 1.9	9.9 ± 1.1	58.8 ± 7.4
5. 27 days	10	27 ± 1.5	12.6 ± 2.0	52.7 ± 4.4
Significance of differences[a]				
p_{1-2}		< 0.05	< 0.01	< 0.01
p_{1-3}		n/s	< 0.01	< 0.01
p_{1-4}		< 0.05	< 0.01	n/s[a]
p_{1-5}		n/s	n/s	n/s

[a] n/s, nonsignificant.

normal. In experiments performed jointly with Solomatina [362], mongrel male rats were lifted from an 'altitude' of 1,000–5,500 m for 1 week and were then lifted every day to that 'height' for 40 days. Adapted and control animals were studied for the defensive conditioned reflex of active avoidance in a U-shaped labyrinth. Light served as a conditioned stimulator, and electrodermal irritations as the unconditioned stimulator; the reflex was developed until the criterion of four faultless responses in five consecutive tests was met, and reflex retention was checked 24 h after it had been developed.

In some of the animals, conditioned reflex activity indices were determined 24 h after termination of adaptation, and in others 5, 10, and 27 days after termination of hypoxic actions (Table 22). We could thus assess the effect of adaptation and the dynamics of its disappearance during deadaptation. From Table 22 it follows that adaptation had no noticeable effect on the number of combinations needed to develop the conditioned reflex and decreased by twofold the number of combinations required to reproduce the reflex when checking its retention; the reflex proved essentially prolonged. This was significantly determined in the first 5 days after terminating hypoxic actions; after 10 or more days, the phenomenon substantially decreased to become nonsignificant. *Thus, adaptation to hypoxia seemed to have no effect on the development of a temporary connection, but substantially increased its retention time; the reflex gradually disappears during deadaptation.*

To assess additionally the effect of adaptation on the development of a conditioned reflex of active avoidance (CRAA), we compared specially selected animals including those adapted to hypoxia and control specimens with the same (71%) reflex retention index. Adapted animals were found to need 1.5 times fewer combinations than controls to obtain an equal retention index during reflex development. Hence, adaptation to altitude hypoxia makes reflex development more effective. However, the principal fact gathered from this experiment is that longer reflex retention probably shows that, like previously performed mountain tests, these experiments were indicative of earlier involvement in adapted animals of a temporary connection responsible for maintaining the fixation, rather than a primary effect on reflex development. In turn, this results in adapted animals requiring less training power than controls to retain the reflex for an equally long time. *In other words, activation of nucleic acid and protein synthesis caused by adaptation to hypoxia apparently potentiates consolidation of the temporary connection, and with multiple combinations this has a possitive effect on reflex development.* This explanation is in agreement with the dynamics of the disappearance of positive changes in the formation and retention of conditioned reflexes after the termination of hypoxic action, i.e. during deadaptation. Indeed, as shown in Fig. 31, the process develops within approximately the same time as nucleic acid and protein synthesis in the brain ceases to be activated after the termination of hypoxic actions.

The rate of increasing resistance of the temporary connection to the erasing effect of electroshock is known to be the accepted criterion characterizing the consolidation rate of temporary connections. Consequently, a possible way of checking the assumption that adaptation particularly affects consolidation is to compare the temporary connection resistance to the erasing effect of electroshock in adapted and control animals at various times after reflex formation.

The results of relevant comparative experiments (Table 23) indicate that an electroshock applied 1 min after the reflex had formed completely excluded all conditioned-reflex manifestations in nonadapted animals, whereas adapted animals were found to retain it for a considerably long time. When electroshock was applied 1 h after the reflex was formed, the conditioned reflex proved also to be retained better in adapted animals than in controls. This was most probably due to the fact that adaptation to altitude hypoxia enhances the rate and perfection of the consolidation of temporary connections.

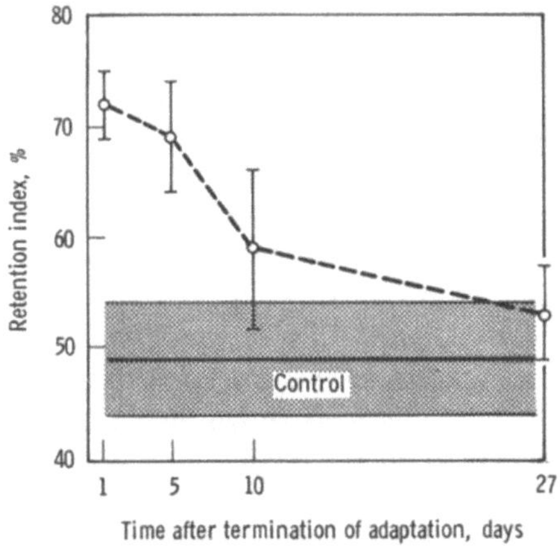

Fig. 31. Retention of conditioned reflex of active avoidance in γ-like labyrinth after determination of adaptation to altitude hypoxia. *Continuous line* and *shaded space,* average value and variation in controls; *dashed line* adaptation

Table 23. Effect of adaptation to hypoxia on consolidation rate of temporary connections in rats

Series	Stay on illuminated platform, s	
	Control	Adaptation to hypoxia
Prior to developing conditioned reflex of passive avoidance	7.73 ± 1.84	13.60 ± 1.73
1 day after developing conditioned reflex of passive avoidance	52.70 ± 3.70	60.0[b]
Under electroshock		
1 min after developing conditioned reflex of passive avoidance	6.00 ± 1.14	34.80 ± 5.39
1 h after developing conditioned reflex of passive avoidance	8.70 ± 4.90	47.40 ± 4.30

[a] Significance of differences between series, $p < 0.05$.
[b] Maximal stay on illuminated platform was 60 s.

Yet, this conclusion could be questioned because in all previously performed investigations conditioned reflexes were not only developed but verified for retention against a background of adaptation. Correspondingly, the following issue remains open: Do adaptation and the related activation of nucleic acid and protein synthesis really selectively affect the consolidation of the temporary connection, or they also affect processes responsible for reproducing temporary connections and occurring when reflex retention is verified? To solve this problem, we subsequently studied the effect of adaptation to altitude hypoxia upon retention of active defensive conditioned reflexes that had developed in a T-shaped labyrinth and the effect of passive conditioned reflexes of avoidance for long periods of time after reflex development [362]. As a result, we found that when defensive conditioned reflexes of active and passive avoidance developed against a background of adaptation to altitude hypoxia, their degree and duration of retention proved significantly increased. This effect was equally pronounced, irrespective of whether the reflex was reproduced against a background of continuing adaptation to hypoxia or after hypoxic actions had ceased and adaptation had undergone inverse development.

Thus, the 'fulcrum' of adaptation is the process of temporary connection fixation itself not reproduction, which takes place similarly both in the presence and absence of adaptation. A temporary connection that forms with adaptation involving generally activated nucleic acid and protein synthesis in cerebral neurons, a phenomenon characteristic of adaptation, proves more long-lived and more resistant to extreme stimulators. Hence, the stability and duration of the temporary connection are largely predetermined by the state of the brain when it is being developed. In this case, the level of nucleic acid and protein synthesis in the brain is one of the decisive parameters characterizing the 'state of the brain.'

The concept that activation of nucleic acid and protein synthesis in the brain enhances the strength and duration of the temporary connection during adaptation agrees with the previously established fact that the administration to animals of chemical substances activating nucleic acid and protein synthesis in the brain, specifically of a combination of orotic and folic acids and vitamin B_{12}, also enhances the degree of conditioned reflex retention [347].

The fact that by adapting the animals to hypoxia one can activate fixation and compel them to memorize certain events more quickly and to remember them longer led us to suggest that activation of nucleic acid and protein synthesis and other metabolic changes in the brain during adaptation to hypoxia may influence not only the performance of specific conditioned reflexes but also animal behavior in 'difficult' situations of the environment. From this suggestion, the author together with Solomatina, Klygul, and Vikhlyaev [358] studied the effect of adaptation to hypoxia on rat behavior in conflicting situations caused by the conflict between two vitally important motivations. The experiments were performed on white female rats, weight 200–250 g. Adaptation to altitude hypoxia was developed in an altitude chamber under daily 5 h exposures. On day 1, the animals were lifted to a 'height' of 1,000 m, which in the subsequent 10 days was gradually increased to 7,000 m. After that, the rats were exposed to 30 hypoxic actions of equal intensity. Control animals kept under normal atmospheric pressure were subjected to the same set of visual, acoustic, and tactile actions as adapted animals.

The course of hypoxic actions was followed by an experiment that reproduced a conflicting situation in adapted and unadapted animals. The conflicting situation model was simulated in an installation designed by Klygul and Krivopalov [222]. The experimental chamber comprised one large (48 × 38 × 33 cm) and one small (16 × 13 × 33 cm) compartment. The large compartment floor was made of metal plates. Rat movements from plate to plate were recorded by a counter and served as an indicator of motor activity. Animal passage from the large compartment to the small one was recorded by a photoelement to determine the number of approaches to a drinking bowl in the small compartment. One of the stimulating electrodes was connected to the drinking bowl and the second electrode was the metal floor plate in front of the drinking bowl. This permitted the passage of an electric current through the rat during drinking. In this way, we created a conflict between the drinking and defensive motivations.

The first series of experiments lasted 3 days. After having been deprived of water for 24 h, during the first 2 days the animals were placed in the installation for 10 min. During that time, the rats got used to the chamber, found the drinking bowl, and satisfied their thirst. The conflict situation was created on day 3: When the rat attempted to drink water from the bowl, it was painfully stimulated with electricity (90 V). During a 20 min stay in the chamber, electronic counters automatically recorded the number of times the rat tried to drink water, the number of approaches to the bowl, and its motor activity in the large compartment. The experimental results were statistically treated by the Student's t-test.

The results of this series of experiments on 40 control and 40 adapted animals are shown in Table 24, from which it follows that adaptation to

Table 24. Behavior of control rats and those adapted to altitude hypoxia in conflict situations

Index	Control	Adaptation	p
Number of drinking times	3.4 ± 0.3	11.1 ± 1.9	< 0.001
Number of approaches to bowl[a]	12.2 ± 1.9	16.0 ± 2.0	n/s
Motor activity (number of movements)	19.5 ± 3.6	9.3 ± 1.3	< 0.01

[a] n/s, non-significant.

altitude hypoxia causes considerable behavioral changes in thirsty rats experiencing conflicting drinking and defensive motivations. These changes essentially reflect the following three facts. First, having been subjected to strong painful stimulation when trying to satisfy their thirst, the animals adapted to hypoxia nonetheless made 11 more such attempts, while control rats made only three (Table 24). In other words, the conflict between drinking and defensive motivations revealed that prior adaptation to hypoxia increases the drinking reflex threefold, i.e., inhibits its suppression under the action of the 'conflictive' pain stimulator.

Second, adapted animals approached the drinking bowl virtually the same number of times as control animals, 16 and 12 times, respectively (see Table 24). However, only 28% of the approaches by the latter ended in drinking, 72% not being consummated by that reflex. Conversely, 69% of approaches by adapted animals ended in the drinking response. Thus, adaptation to hypoxia considerably decreased the number of approaches that failed to end in drinking and made animal behavior in a conflictive situation more effective from the viewpoint of realizing the drinking reflex.

Third, the motor activity of hypoxia-adapted animals in a conflict situation was less than that of controls. For instance, adapted animals were recorded to move nine times in the large compartment, and controls 19 times (see Table 24). Hence, more frequent drinking from the bowl by adapted animals cannot be explained by their generally increased motor activity.

These facts indicate in adaptation to hypoxia the drinking reflex, vitally important to thirsty animals, can be achieved by more effective inhibition of the defense reflex. To verify this assumption in the second series of experiments on control and adapted rats, animals were denied water for 24 h and then placed in the experimental chamber. They approached the drinking bowl and began to drink in the absence of painful stimulation. They were then given minimal stimulation, which was gradually increased until they ceased to drink and jumped away from the bowl. The electric circuit voltage under which the animals stopped showing the drinking reflex in given conditions amounted to 21.9 ± 1.9 V for controls and 48.4 ± 2.2 V for hypoxia-adapted rats. In other words, to discontinue further achievement of the drinking response once started, adapted animals require a painful stimulation twice as much as controls. Thus, hypoxia-adapted rats differ from controls in that they possess much greater possibilities for realizing the drinking motivation despite the pain stimulator action.

The above facts may have several different explanations. First, the development of passive avoidance is violated in hypoxia-adapted animals. Indeed, as in the conflict between the feeding and defense motivations [429], during the conflict we simulated the animals should have developed passive avoidance. One may assume that gradual training to altitude hypoxia weakens or eliminates the animal's ability to inhibit 'punished' responses and sharply worsens the development and occurrence of passive avoidance. However, the following facts directly contradict such an interpretation of the behavioral features of adapted animals. Earlier, it was revealed [320] that in hypoxia-adapted rats the development of passive avoidance by the method of Bures and Buresova [50] was the same as in controls: In the starting test, the duration of stay in the 'safe' compartment was the same for both groups of animals. On the other hand, after 1 day and more, the adapted animals retained the developed reflex considerably better than those in the control group. Thus, when using the traditional and most adequate methods of studying passive avoidance, no disturbances of the latter were revealed in adapted animals. This is in agreement with the results of previously published [357] cytologic investigations of the brains of hypoxia-adapted animals. These studies, performed in the course of a histoautoradiographic inquiry into protein synthesis in nerve and glial cells, revealed no destructive changes in cerebral structures. Hence, one cannot accept the assumption that in conflict situations changes in the behavior of adapted animals are due to hypoxic injury of the brain structures responsible for passive avoidance.

Second, it may be that adaptation to altitude hypoxia enhances the pain sensitivity threshold to make the animals less sensitive to electroshock. Third, adaptation to altitude hypoxia in some way possibly enhances the intensity of the drinking motivation, which is consequently dominant and cannot be suppressed by a pain stimulator. Finally, it may be that adaptation to altitude hypoxia in some way enhances motor activity to result in numerous pointless approaches to the drinking bowl.

The last three possibilities were studied in a final series of experiments, in which animals not involved in conflict situations were measured for the pain sensitivity threshold, drinking motivation intensity, and motor activity.

Table 25 shows that control animals attained the pain sensitivity threshold at 36.9 V, and hypoxia-adapted animals at 27.5 V. After having been deprived

Table 25. Indices of pain sensitivity threshold, values of drinking motivation, and spontaneous motor activity in control and altitude hypoxia-adapted rats outside conflict situations

Index	Control	Adaptation	p
Pain sensitivity threshold, V	36.9 ± 0.7	27.5 ± 1.10	< 0.01
Motivation intensity (by amount of water, ml, consumed in 10 min after 24 h-long thirst)[a]	6.3 ± 0.3	6.7 ± 0.4	n/s
Spontaneous motor activity (number of displacements in 1 h)[a]	24.9 ± 4.6	23.0 ± 5.5	n/s

[a] n/s, non-significant

of water for 24 h, control and adapted animals consumed within 10 min 6.3 and 6.7 ml of water, respectively. Outside a conflict situation, control animals in the narrow chamber moved 24 times during 1 h, and adapted animals 23 times (see Table 25). These data show that outside a conflict situation the pain sensitivity threshold in hypoxia-adapted rats was mainly unchanged, though sometimes it was decreased; the intensity of drinking motivation was unchanged and motor activity was the same as in controls. Thus, increased drinking motivation, motor activity, and pain sensitivity threshold, and also impaired development of passive avoidance, played no role in the mechanism of the given response.

The evidence obtained suggests the conclusion that *adaptation to hypoxia outside a conflict situation changes neither drinking nor defense motivations; in a conflict, adapted animals acquire the ability to suppress their defense motivation and, despite pain, to realize their drinking motivation in this case, the vitally important one. Using what, in our view, is conventional but sufficiently expressive terminology of ethologists, it may be said that adaptation enhanced 'self-control' to allow the animals to achieve their drinking instinct under such intensity of pain stimulation that made it impossible to achieve the instinct in unadapted animals.* Thus, the 'crossing' of two memory levels that we engineered proved highly effective and adaptation to altitude hypoxia could be used as a factor for controlling higher nervous activity.

The activating effect of adaptation to hypoxia on the fixation of temporary connections and its positive influence on animal behavior in conflict situations served as a foundation for using adaptation to enhance brain resistance to harmful factors. As a result of many years of investigations reviewed in a previously published monograph [320], it was shown that *preliminary adaptation of adult animals to altitude hypoxia may prevent or diminish the harmful effect on the brain of extraordinary stimulators such as loud sound and electroshock and toxic factors such as alcohol, corasol, and the hallucinogen LSD. On aggregate, this evidence makes advisable a clinicophysiologic study of the effect of adaptation to intermittent hypoxia upon human resistance to factors that impair higher nervous activity.*

Such investigations are all the more essential since adaptation to altitude hypoxia is only an individual example of how by adapting the organism to a certain physical environmental factor behavioral responses can be developed. Recent studies by Ustinova and myself [337] revealed that prior animal adaptation to cold substantially affected their behavior as assessed by the criterion of aggressiveness. To examine aggression, a pair of mice are usually placed in a chamber, through the floor of which a current of growing intensity is transmitted. In this case, the animals develop aggressive responses in the form of so-called fights. The number of these aggressive responses characteristically changes depending on the intensity of the electro-painful stimulator [545]. The graphs in Fig. 32 show that with current intensity close to threshold level (0.035 mA), the number of such responses in control animals was 1.1 during 3 min. The largest number of aggressive responses (15.7) was observed under 0.5 mA; subsequently, with higher current intensity, the number of aggressive responses sharply declined. Figure 32 shows how

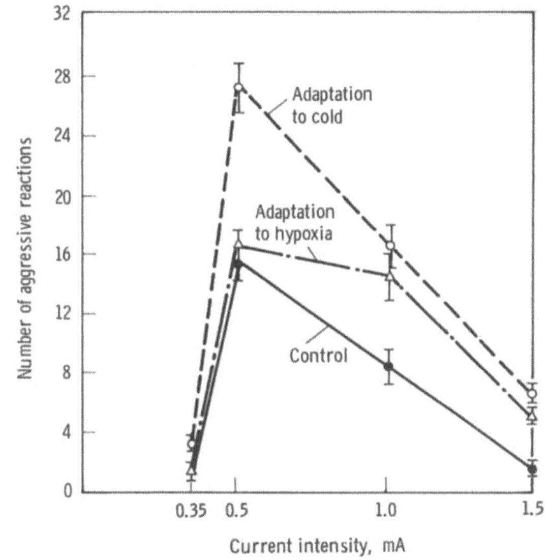

Fig. 32. Effect of adaptation to cold and hypoxia on number of aggressive reactions in mice at varying current intensities

in animals adapted to cold, the number of aggressive responses also depended on the stimulator intensity; however, with all current parameters used it was two to three times higher than in controls. Thus, adaptation to cold caused the number of aggressive responses to increase under painful action [337]. Naturally, this fact requires further analysis. And, yet, together with the above evidence on the effect of adaptation to hypoxia on animal behavior in conflictive situations, it permits us to conclude that *adaptation to physical environmental factors may be used as a means for predetermining behavioral responses in 'difficult' situations, providing possibilities for controlling behavior.*

Chapter 6
Stress and Stress Damage

> "And are there hunters on that planet?"
> "No."
> "Oh, how interesting! And are there hens?"
> "No."
> "There's no perfection in the universe," sighed the Reynard.
>
> Antoine de Saint-Exupéry

The previous chapters showed that a general adaptive stress reaction arises in situations where the environment poses new tasks for the organism. This reaction not only precedes stable adaptation but plays an important role in developing it. This is provided, by mobilizing energy and structural resources, by directionally redistributing them toward preferentially supplying the systems responsible for adaptation to a given factor, and developing a systemic structural trace and, second, through the direct influence of stress hormones and mediators on metabolism and cell function in the system in which the systemic structural trace forms.

As the systemic structural trace and stable adaptation develop, disruptions in homeostasis that stimulate the stress reaction gradually disappear and, having played its role in establishing adaptation, the stress reaction itself is gradually eliminated (see Fig. 17). This concept determines the relationship between stress and adaptation and emphasizes that the stress reaction developed in the course of evolution as an essential nonspecific link of a more complex integral adaptation mechanism. However, the significance of the stress reaction is not limited to this in the life of the organism. In reality, there are the so-called hopeless or dead-end situations, when environmental requirements cannot be satisfied by means of specific behavioral or other responses: The forming of a functional system and systemic structural trace adequate to the environment proves impossible, and adaptation does not occur.

Under the action of an unusually strong and unavoidable unconditioned stimulator, e.g., excessive load, cold, hunger, or pain, these situations are assessed as those of 'insurmountable difficulty' or 'forced patience.' Under the action of stereotypes of conditioned stimulators that come out as signals of approaching harmful factors, and also where there is a long delay between a strong requirement and information about the possibility to satisfy it, these situations are assessed as 'traps,' 'anxieties,' etc. In all situations of this kind, the absence of a possibility to achieve adaptation leads to a situation in which disruptions of homeostasis, constituting stimuli of stress, are retained for a long time. Excitation of the adrenergic and hypophysial-adrenal systems, which is the essence of stress, proves protracted.

The unusually long and intensive actions of high catecholamine and glucocorticoid concentrations may result in highly varied injuries constituting the so-called stress diseases, which play a major role in modern medicine. Stress has now been proven to be a principal or secondary etiologic factor in ulcerous lesions of the mucous coat of the stomach, hypertension, atherosclerosis, impaired function and structure of a previously healthy, and even more a sick, heart, immunodeficient states, and even malignant tumors. In effect, these facts signify that under certain conditions the stress reaction is transformed from a general nonspecific link of adaptation to various environmental factors into a general nonspecific link of pathogenesis in diseases that shorten human life.

The concept that stress plays a role in pathology is so popular that it quite often distracts the attention of investigators from its principal, evolutionally determined role as a link in adaptation and prevents them from focusing their attention on the very important fact that most people and animals placed in so-called hopeless situations do not die, but acquire a certain degree of resistance to those stress situations. Such stress situations in the form of long periods of hunger, cold, natural calamities, and interspecific and intraspecific conflicts are always widespread in any given natural habitat to which animals successfully adapt. In human habitats, qualitatively more complex socially determined stress situations are equally numerous. Indeed, during the recent, relatively short period of recorded history alone, mankind has passed through times of slavery, serfdom, and world wars with no signs of degradation whatsoever, and has thus showed highly effective adaptation to so-called hopeless stress situations. *This means that the temporary transformation of the stress reaction from a link of adaptation to a link of pathogenesis is not the end of the vital process but its intermediate stage. Things do not end with this transformation, and most animals and people do not die of long-term and repeated stress situations; hence, the organism must possess mechanisms that ensure adaptation to stress itself.*

It has been experimentally shown that with long and especially repeated actions of such serious stress factors as immobilization or pain, the adrenergic and hypophysial-adrenal systems show increasingly diminished activation [213, 395]. For example, the first injection of formalin into rat dorsal skin increased the corticosterone concentration in the blood by 4.5 times; but its concentration did not change at all after the 22nd–27th injections. This termination of the stress reaction does not depend on exhaustion of the adrenals: Adrenocorticotropic hormone (ACTH) administered to pain-adapted animals causes higher corticosterone concentration in the blood than in controls.

This evidence implies that activation of the central control mechanisms, which under pain and similar stimulators inhibit the secretion of releasing factors, ACTH, and, hence, corticosterone and catecholamines, is one of the factors of adaptation to environmental stress situations that cannot be escaped through behavioral responses. In the brain, inhibitory neuronal systems synthesize and release inhibitory mediators, viz GABA, glycin, dopamine, serotonine, and, possibly, encephalines. This may suggest that it is precisely these inhibitory systems that restrict the stress reaction and play a role in adapting the organism to seemingly hopeless stress situations. Similarly, control systems of adenine nucleotides and prostaglandins, and antioxidant systems acting in the periphery and playing the role of modulators, may restrict the excessive effects of catecholamines and other factors and thus prevent stress injuries. In effect, the presumable adaptive mechanisms to protracted stress situations are those of natural prevention of stress injuries. Hence, their study is interesting not only theoretically but may reveal certain metabolite-regulators for subsequent use in the active prevention of stress injuries.

To sum up, it may be said that *at present the problem of stress has at least three interrelated aspects, namely:* (a) *stress as a link in the adaptation mechanism;* (b) *stress as a link in disease production and development; and* (c) *adaptation to stress situations and natural prevention of stress diseases.*

The first aspect, the role of stress in the adaptation mechanism, was examined in the previous chapters. The material that follows will discuss the transformation of stress into a link in pathogenesis; the mechanism of stress injury to certain internal organs; natural mechanisms of preventing stress injuries; and possibilities of active chemical prevention of stress injuries on the basis of those mechanisms.

Damaging Stress Situation and Pathogenesis of Stress Stomach Ulcers

Multiform situations causing serious and long-term stress fraught with injury to internal organs are in the end reduced to a conflict between the imperative need for instant defense, feeding, and sex reaction, on the one hand, and an insurmountable prohibition of these reactions, on the other. In its most elementary and exposed form, this conflict is reproduced when animals are subjected to painful stimulation and simultaneously deprived of the possibility to avoid pain. In a qualitively more complex form, this conflict occurs when humans are subjected to socially determined actions endangering their existence or dignity and when the response is banned by other (likewise socially determined) conditions that require self-control in order to avoid still greater dangers.

Indeed, self-control proves to be provided through critical tension in cortical inhibitory mechanisms; however, at the same time, only the external behavioral component in the response proves inhibited or changed in form. Its internal vegetative component, i.e., the stress reaction and mobilization of circulation, respiration, and other functions are retained and may even prove more intensive and protracted than when the behavioral response itself occurs. This situation, which was examined under different angles by Pavlov [432], Lang [253, 254], Anokhin [5], and other researchers, should among other things be characterized by the fact that an enhanced stress reaction is evident in long and significant rises in the catecholamine and glucocorticoid concentrations of blood.

Without dwelling on the quite complex issue of the degree to which stress situations causing human diseases are reproducible in animals, one should regard as quite justified the present position in science, which is that the principal conflict mentioned above between the 'need' to achieve a defense response and the 'prevention' of this response occurring is chiefly used to simulate approximately such situations.

This simulation very soon showed that the extent of the stress reaction and the resultant injuries essentially increase if additional actions that activate the emotional apparatus are superimposed on the basic elementary conflict. These additional actions include the introduction of signals that forewarn of inevitable pain shock and of shocks via chance as well as in frequent shocks that cause the anticipation of pain and concomitant anxiety and fear. The introduction of secondary conflicts in addition to the principal conflict has similar significance, e.g., the infliction of pain in response to the developed feeding response or the infliction of pain in response to a reflex by means of which the animal had previously managed to avoid danger.

To assess the relationship between the principal and secondary conflicts in stimulating stress systems and developing stress injury, it is advisable to turn our attention first to the most thoroughly studied of stress injuries, viz stress stomach ulcers and then to examine stress injuries of the heart and other organs.

Ulcerous lesions of the stomach caused by very different stress actions occur so regularly that they have been entered into the list of essential signs of the first emergency stage of the stress reaction. But in reality they are not. A stress reaction that manifests itself as a nonspecific adaptation link may not necessarily be accompanied by ulcerous lesions of the mucous coat of the stomach; however, there is no doubt that these lesions represent one of the first and quite distinct signs of the transformation of stress into a link in pathogenesis, this sign being subject to exact quantitative evaluation. Hence, specifically ulcerous lesions play the most important role in clarifying the mechanism of stress injuries.

Ulcerous lesions of the mucous coat of the stomach have been revealed virtually with all the actions that lead to serious stress. For instance, they have been proven in humans affected by sepsis [89], burns [420], traumas, and surgical operations [115]. In humans, ulcers of the stomach and the duodenum often start and are aggravated after serious emotional shocks. In animals, chiefly in rats, ulcers naturally develop after long immobilization [48] under the influence of inevitable painful actions, which however do not damage the tissues [580], and still more under the influence of conflict situations that cause fear and anxious anticipation. They are discussed below.

From on a comparison of histologic, electron-microscopic, and histochemical evidence, Goldman and Rosoff [137] arrived at the conclusion that in all cases stress stomach ulcers in rats represent 'linear infarctions of the mucous coat.' Kristt and Freimark [243] produced a stress situation in rats. The rats obtained food by pressing on a pedal to receive a painful electroshock either through the floor of the test chamber or via an electrode implanted in the brain. This conflict between the feeding and defense unconditioned reflexes naturally resulted in ulcerous lesions and other pathologic changes in the mucous coat of the stomach; these changes were subjected to subsequent detailed morphological investigations. When studying these lesions, the above authors confirmed the evidence of Goldman and Rosoff and arrived at the conclusion that serious stress causes two kinds of changes in the stomach, namely, (a) a peculiar metaplasia of the mucous coat and (b) ulcerous-

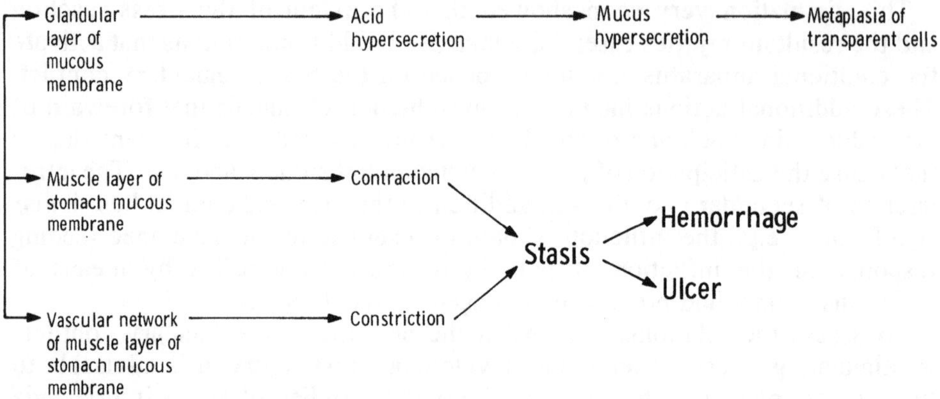

Fig. 33. Scheme of EPS-induced pathogenesis of stomach ulcer. (After Kristt and Freimark 1973)

hemorrhagic lesions, which are normally the subject of attention of investigators. Basing their work on a study of the dynamics of the latter lesions, Kristt and Freimark [243] assumed that stress involves spasms of the arterioles of the muscular layer of the stomach, which entails stasis and subsequent hemorrhage into the mucous coat of the submucous layer. Ischemic lesion of the mucous coat and hemorrhage result in focal necrosis of the glandular epithelium with subsequent ulcerations. These ischemic necroses and the ulcers that form in their place are quite often either linear or narrow and coiled in form. The outcome is a typical clinical picture of quite narrow longitudinal or wound stress ulcers with hemorrhagic edges that show iron-containing blood pigments. The Kristt and Freimark [243] diagram in Fig. 33 provides a more detailed understanding of the dynamics of the events in the mucous coat of the stomach during stress.

In our presentation, the control factors leading to arteriolar spasms and subsequent necrosis of the mucous coat are important. To assess these factors, one must undestand that ulcerous lesions of the mucous coat of the stomach chiefly develop some time after, not during, the stress action. In order that various stress situations produce an ulcerogenic effect, it is highly important that the stomach be empty. In other words, ulcers were found to develop after stress actions on hungry animals.

In studying the significance of these important factors, Desiderato and Testa [79] showed in experiments on rats with stomach fistulas that excitation of the adrenergic system during the stress action itself (when kept in test chambers, the animals were subjected to electric painful shocks through the floor) caused complete inhibition of gastric juice secretion; after the stress action had terminated, the vagus nerve tonus and gastric juice secretion rapidly reached their normal, but not increased, levels. The fact that in these experiments, the animals developed poststress ulcers when the gastric juice showed normal acidity and enzymic activity means that during stress the mucous coat of the stomach developed changes that deprived it of resistance to the digestive action of the gastric juice. In other words, the mucous coat sites injured during stress through adrenergic vasoconstriction and ischemia are digested after stress, when the activity of the parasympathetic part of the nervous system and the secretion of gastric juice determined by the vagus nerve are restored. This explains the fact that stress ulcers may be prevented both by adrenolytics, which block adrenergic spasm of arterioles, and by cholinolytics, which block the poststress restoration of gastric juice.

Indeed, diethyl thiocarbomylpicolinic acid, which inhibits dopamine-β-hydroxylase and the conversion of dopamine into noradrenaline in synaptic nerve endings as shown not to penetrate through the hematoencephalic barrier and to produce no effect in the brain; correspondingly, it does not prevent stress ulcers in rats. However, β-butyril picolinic acid, the butyril derivative of picolinic acid, which permeates the hematoencephalic barrier and suppresses dopamine-β-hydroxylase directly in the brain, naturally prevents ulcerous lesions from developing during stress [173]. Thus, central suppression of the excitation of adrenergic control may prevent stress lesions of the stomach.

In their experiments, Houser et al. [182] placed the animals in special cages, where they were compelled to move constantly, the movement being especially after day 3. These rats received food once a day and died by day 7, with pronounced ulcerous lesions of the mucous coat of the stomach. Scopolomine bromide, an anticholinergic drug suppressing gastric secretion, did not affect the mortality, but substantially reduced the size and number of ulcers. Thus, intense excitation of adrenergic control is the most probable factor that injures the mucous coat of the stomach during stress, and subsequent restoration of the tonus of parasympathetic control causes this injury to materialize as ulcers.

The obvious role of excited adrenergic control in achieving the stress reaction itself and in developing stress injuries has generated a whole series of investigations, in which the extent of the stress action (effectivity of the harmful action of various stress situations) was assessed by the utilization of catecholamines and by the degree to which the stomach was injured by ulcerous lesions.

The noradrenaline content in the brain was found naturally to decrease under intense painful stimulation [547], cold [309], and electroconvulsive action [175]. This gave rise to the idea that this shift is an essential component of stress reactions. However, it was subsequently shown that with less intense stimulation, a higher catecholamine content in the blood and increased catecholamine excretion in the urine are not accompanied by a decline in the catecholamine content in the brain, since increased catecholamine utilization is counterbalanced by higher catecholamine resynthesis [527].

Similarly, a strong but protracted stress action first involves the increased excretion of catecholamines and their decreased content in the brain, and then the developing activation of catecholamine resynthesis restores this content despite increased excretion. Chronic stress situations were noted for higher catecholamine content and considerably increased turnover. In our experiments, the most interesting result was the evidence that suggested the role of the emotional factor in the effect of stress situations on the organism based on changes in the catecholamine content and on the extent to which the mucous coat of the stomach was affected by ulcerous lesions.

Pare and Livingston [429] showed that a 2 h electric painful action (electroshock, 3 mA every 30 s) leads to a relatively low (13%) decline in the noradrenaline content in the cerebral hemispheres. If every shock is preceded by a specific sound lasting 10 s, then the noradrenaline content drops by 35%. In other words, the danger-warning conditioned stimulator, acting in a hopeless situation when the menace could not be avoided, sharply increased the organism's stress reaction.

In the next stage of the research, rats were trained to procure food in a special chamber by pressing the door. They were subjected three times to long-term (22 h) stress, which was that every pressure on the door resulted in electroshock together with food; in other words, a conflict was repeatedly created between the feeding and defense unconditioned reflexes. In controls, only food was given when the animals pressed the door. Electroshocks of the same intensity were inflicted irrespective of food reinforcement; in other words, the same two unconditioned reflexes were realized, but they did not conflict.

A comparison of these two series of experiments revealed that 2 h after they were started, the noradrenaline content in controls declined by 10%, and in conflict-exposed animals by 20%. Subsequently, after three 22-h stress sessions, the brain developed considerably activated catecholamine synthesis and the noradrenaline content in the brains of all the animals increased as compared with the starting content by 10% in controls and 35% in conflict-exposed animals. Further, as a result of the chronic stress actions, 21% of the control rats were found to have developed ulcers, the average number of ulcers per stomach being two. In conflict-exposed animals, ulcerous lesions of the mucous coat of the stomach developed in 61% of the cases, the average number of ulcers per stomach being four. Thus, the presence of a stress conflict situation between two vitally important unconditioned reflexes caused considerably more intense excitation of the adrenergic system and, correspondingly, two to three times larger ulcerous lesions than simple physical painful action alone.

In his research, Weiss [578, 579] created the conflict between a stable predeveloped skill (pain-avoiding conditioned reflex) and unconditioned painful stimulation applied in response to a successful avoidance response. This conflict in the form of 'punishment for a developed skill' likewise led to considerably larger ulcerous lesions of the stomach than unavoidable simple painful stimulation. On aggregate, this evidence signifies that at least two emotional factors, namely anxious anticipation caused by forewarning of pain and a conflict between unconditioned reflexes or between previous experience and reality, sharply enhance the stress reaction in experimentally created hopeless situations and aggravate the harmful effect of stress. One can readily see that similar factors may play a role in complex social conflict caused by stress reaction and stress diseases in humans. Precisely for this reason, in studying the pathogenesis of stress damage to the heart and the organism's natural antistress systems, we used emotional painful stress

(EPS), the so-called anxious neurosis after Desiderato et al. [80], which involves both elements, conflict and anxiety of anticipation, and is examined below.

In assessing the evidence on the harmful effect of stress, one should keep in mind that in addition to a stress situation genetically determined or ontogenically acquired resistance to stress plays a major role in the occurrence or nonoccurrence of ulcers and other lesions. One of the most elementary factors determining this resistance is age. For example, Anokin [6] showed that the starting concentration of noradrenaline in the brain of juvenile sexually immature rats was lower than in adults, and that under a stress situation (immobilization plus pain) noradrenaline concentration decreased less because the catecholamine resynthesis apparatus possessed greater power. Correspondingly, ulcerous lesions of the mucous coat of the stomach in juvenile rats proved less pronounced than in adults. This was further developed and confirmed in the experiments of Ritter and Pelzer [471], who showed that under cold and painful stress actions, the catecholamine concentrations in the hypothalamus and forebrain of old rats decreased more and were restored less than in adult animals. This seems to testify that with age, in addition to accumulating individual experience that helps rats avoid stress situations, their direct resistance to these situations declines.

An interesting investigation by Dirks et al. [86] showed the effect of resistance to stress on feeding in early ontogenesis. Adult rats taken away from their mothers on the 15th day of life proved to develop ulcers under immobilization stress in 90% of all cases; on the other hand, animals that were separated from their mothers on the 22nd day, i.e., within the physiologic period, developed ulcers only in 10% of the cases. Thus, the extent to which the stress reaction and stress injuries are pronounced is in no way proportional to the intensity of the stress action and is strongly dependent on the genetically determined or acquired resistance to a stress situation. As will further be seen, this resistance may be actively changed by guided external actions.

Pathogenesis of Stress and Ischemic Injuries of the Heart

Adaptive changes in heart function constitute an essential link in the defensive, aggressive, and sexual responses of animals and in the major socially determined responses of humans. One can imagine that in the case of the above-mentioned basic conflict between these vitally important reactions and the prevention of their implementation, i.e., under an intensive and long stress reaction, the cardiovascular component in these responses is not simply retained despite the absence of the responses themselves, but may be accompanied by injury to the myocardium. The latter may arise for at least two reasons: First, under the influence of high catecholamine concentrations for an unusually long time and, second, because of overloading of the heart due to the fact that in the absence of the behavioral response itself, the vessels of the skeletal muscles do not dilate and the resistance to expulsion proves unusually high. Yet, there is a significant gap between this assumption and real proof of stress-induced injury of the myocardium. This is determined by the fact that unlike the stomach where ulcers are readily perceptible, the heart signals much less definitely the occurrence of structural damage. Even with serious stress actions, changes in heart activity are often manifested in such quite normal events as tachycardia or bradycardia. Numerous studies, in which such changes were revealed, are highly essential in assessing the cardiac component in the organism's behavioral responses; however, these changes themselves do not decide the presence or absence of stress-induced injuries of the heart.

Even in the presence of stress-induced metabolic and structural disorders of the heart, electrocardiographic shifts may be relatively slight because injuries spread diffusively. Finally, changes in contractile function are often restricted to a set of events indicative of the mobilization of that function. For example, Randall et al. [461] used the catheterization method to study the contractile function of the left ventricle of the hearts of nonanesthetised monkeys placed in a classic stress situation: A high stimulus was created for the defense response, while the response itself was cut off by fixation. The stimulus involved the animals being subjected to sound stimulation followed by a electric stimulus. They responded to the former by increased systolic pressure, higher contraction rate, and greater tachycardia. Experiments with denervation showed the cardiac component in the organism's response to be chiefly determined by excitation of the sympathetic innervation of the heart and to a much lesser extent by the effect of catecholamines circulating in the blood.

Evidently, these and numerous similar data characterize the contractile function of the heart during stress conditions sufficiently deeply. However, they do not answer the question as to whether or not injuries of the myocardium occur during stress. There is abundant evidence of alternative changes in the myocardium under excessive loads during swimming, long stay in a vertical posture, traumas, burns, and/or administration of large doses of biologically active substances. However, in all these cases, it is difficult to decide whether or not the changes were caused by excess catecholamines and glucocorticoids, i.e., by the activation of stress-realizing systems in the exact meaning of the word, or by the action of the load itself, by chemical factors, and so on.

Until recently, there was very little direct evidence on structural and metabolic disturbances in the heart during stress. In the experiments of Miminoshvili et al. [397], stress induced in monkeys by violating flock–sex relationships in a few cases resulted in a negative T wave in the ECG and in one animal in a morphologically proven necrotic focus in the myocardium. However, subsequently performed repeated experiments failed to reproduce that result [97]. In the experiments of Repin and Startsev [468], fixation of monkeys during passive-defense response, i.e., during flight, in several cases resulted in a negative T wave in the ECG and in higher transaminase and creatine kinase activities in the blood, and in one case ischemic necrosis of the myocardium was seen. This experiment was performed on five monkeys and it seems difficult to quantify its results or to reproduce them accurately. Selye [507] observed in rats that had been exposed to immobilization stress, transitory histologic changes evident as fuchsinophilic foci.

Few, but quantitatively estimable, facts on the harmful effect of the stress reaction on the myocardium were obtained in experiments with emotional painful actions; they proved largely similar to those described above in examining stress ulcers. For example, it was shown [24, 55] that rats subjected to painful electric stimulation through the floor of a rather spacious cage made ineffectual attempts to escape. Each electroshock was preceded by a forewarning-conditioned sound stimulator. In the series where electroshocks were inflicted after equal time intervals, the glucocorticoid concentration in the blood was 50–60 μg/100 ml, and with those inflicted at occasional intervals the animals expected them continuously and the glucocorticoid concentration reached 85–95 μg/100 mg. The latter was the principal form used. Single stress actions of this kind lowered the noradrenaline content in the auricles by 15%, less in the ventricles. This was found to have been caused not only by higher catecholamine utilization during stress, but by impaired reevacuation of the released noradrenaline back into the sympathetic endings. This impairment in reevacuation depended on the effect of high glucocorticoid concentrations. Ten days later, these disorders of catecholamine metabolism in the myocardium gradually disappeared despite daily stress actions.

An important study representing the first attempt to assess qualitatively heart structural injuries under stress was performed by Miller and Malov [396]. They generated a stress situation in rats by electroshock stimulation of the paws via the chamber floor, the myocardial injury being assessed by enzyme release from the cells of isolated hearts of stress-exposed animals and by incorporation in the heart of technetium-labeled radioactive pyrophosphate, normally used to diagnose myocardial infarction.

In response to stress actions lasting from 4 to 16 h, the activity in the blood of lactate dehydrogenase and of glutamatoxalate and glutamate pyruvate transaminases was found to have increased by two to three times. This fermentemia is a reliable but integral criterion of injury since enzymes may leave the injured cells of any organ, including the heart. To clarify how much fermentemia is connected with heart injury, Miller and Malov [396] measured in vitro the enzyme release into the perfusate from the cells of an isolated heart to show that release was substantially increased. They subsequently established that stress abruptly increases technetium labeling of the myocardium. In a healthy myocardium, the technetium is incorporated in the form of traces and increases together with increased duration of the stress action and continue to be augmented for a certain period after the stress action has terminated. Maximal labeling was observed 8 h after a 12-h stress. On the whole, the results of this work signify that two classic manifestations of ischemic infarction, viz fermentemia and abruptly increased technetium labeling of the myocardium, are fully realized with stress injury of the heart.

At present, the best explanation of this fact is that in protracted stress situations, the heart is injured by the prolonged action of high catecholamine concentrations. Indeed, it was shown that even 8 h after the termination of the stress induced by painful electric stimulation without the possibility of rescue, adrenaline concentration in rat blood had increased eight times and that of noradrenaline seven times [294]. One can readily imagine that if the stress situation does not culminate in sufficiently rapid adaptation and continues to occur, such high catecholamine concentrations may produce a harmful effect on the heart.

Accordingly, acting indirectly via a adenyl cyclase system, high doses of noradrenaline, and even more of isopropyl noradrenaline (isoproterenol), its synthetic analog, cause higher influx into the myocardial cells of Ca^{2+}, whose excess, combined with excess free fatty acids, results in the uncoupling of oxidation and phosphorylation in the mitochondria and in the decrease of ATP and CP in the myocardial cells. At the same time, the excess Ca^{2+} determined by the adrenergic effect causes contracture-like contractions of the myofibrils. These energy-deficient situation and contractures are of a focal nature and result in the well-known small-focal necrobiotic changes in the myocardium due to excessively intense and prolonged adrenergic effects on the heart [476].

In recent years, this adrenergic injury has been reproduced on an isolated perfused heart, in which the noradrenaline effect took place directly on the myocardial cells and could not apparently be mediated by excess fatty acids, thrombuses in the coronary system, etc. Waldenström et al. [572] performed these experiments to show that increasing noradrenaline doses without changing the coronary flow damages the heart cells. This is evident in a lower ATP concentration and in the release of creatine kinase and transaminases into the perfusate. These shifts are proportional to the catecholamine concentration and the duration of their action in the perfusate and are accompanied by depressed heart contraction. The myocardial cells develop pathologic changes, which can be observed in myofibrillar contracture-like contractions. This whole complex of alternative changes may be largely prevented by administering into the perfusate β-blockers or veraparmyl (isoptine), which is known to block the Ca^{2+} influx into myocardial cells.

The question to what extent initial catecholamine injuries of the heart may occur in previously healthy people requires further study. Some epidemiologic observations seem to favor such a possibility.

According to different authors, the ischemic heart disease (IHD) is found to be spread among people engaged in mental work with intensive nervous overstrains. For instance, 10.8% of 514 males examined aged 35–45 years were engaged in intensive mental activity (scientists, industrial managers, etc) and were found to have IHD; 14.6% comprised a group of persons with doubtful signs of IHD; and 22.8% were noted for indistinct fits of stenocardia, questionable ECG changes under load, and periodic extrasystole [244]. Bogoyavlensky et al. [43] examined 387 industrial managers whose activities were connected with frequent psychoemotional loads. Examination of their ECGs, cholesterol, β-lipoproteins, and lactate dehydrogenase isozymes in the plasma showed that despite their relatively young age (over 90% of those tested were under 50 years old) only 42.3% proved to be essentially healthy. In the same group, 27.4% were diagnosed for IHD and 26.4% for hypertension. According to Sanov and Satpaeva [496], IHD occurs four times more often among people engaged in mental work and those with intensive nervous strain than in persons chiefly occupied in manual work and not subject to strong psychoemotional stress in their jobs.

The results obtained by Ostrovsky [425], who studied the effect of extreme emotional stress on various indices of the functional state of the human cardiovascular system, present major interest. The study involved a group of 30-year-old seamen who were confronted with complex conditions in their struggle to prolong the ship's life in one of the regions of the Pacific. The seamen were professionally classified into two teams, the deck team (32 men) and the engine-room team (29 men). After 4 months had lapsed since the beginning of the voyage, 62% of the deck team and 87% of the engine-room team complained of pains in the heart. In 90% of the men with the painful syndrome, it was accompanied by ischemic and hypoxic ECG changes. The T wave changed more pronouncedly and five engine-room seamen were found to have infarction-like changes in the ECG. In some men, the pulse rate increased to 160 beats per minute, and the blood pressure to 240/130 mm Hg. The sinus rhythm was retained only in four seamen, and the number of various blockades sharply increased. The catecholamine content in the daily urine grew by eight times compared with the initial data. In the fifth stage of investigations, 1 month after the crews had returned to their port of registry, previous complaints were repeated by 19% of the deck team and 29% of the engine-room team. Pathologic changes in the ECG slightly decreased, but nevertheless persisted in 20% of seamen with the painful syndrome. Thirty-one men had incomplete blockade of the right crus of the His bundle. In terms of our presentation, it should be noted that at all the stages of this study engine-room seamen were found to have greater pathologic shifts than deck seamen. To some extent, this may be explained by the fact that the former experienced a higher psychoemotional stress because they were unaware of the real situation on board and had a feeling of uncertainty. Significantly, the changes that had occurred when sailing, i.e., in a stress situation, did not completely disappear after the stress was removed; they partially remained in the poststress period, this being indicative of the persistence of the stress-induced disorders in the cardiovascular system.

On the whole, the above affords grounds to believe that stress injuries of adrenergic origin of a previously healthy heart represent real events in human pathology. This statement itself does not resolve the problem of stress-induced injury of the heart; however, in our view, it permits the problem to be approached in the proper way, namely by putting forward at least four substantial questions requiring experimental analysis. First, via what links is the effect of high catecholamine concentrations produced in the myocardium during stress? Second, what concrete disturbances in metabolism and the contractile function of the heart remain in the myocardium after the stress situation itself has passed? Third, how is the role of stress realized in the etiology and pathogenesis of IHD, the main disease of our time? And, fourth, what are the possibilities of preventing disturbances of heart metabolism and function during stress by blockading the principal links in the pathogenic chain of stress injury?

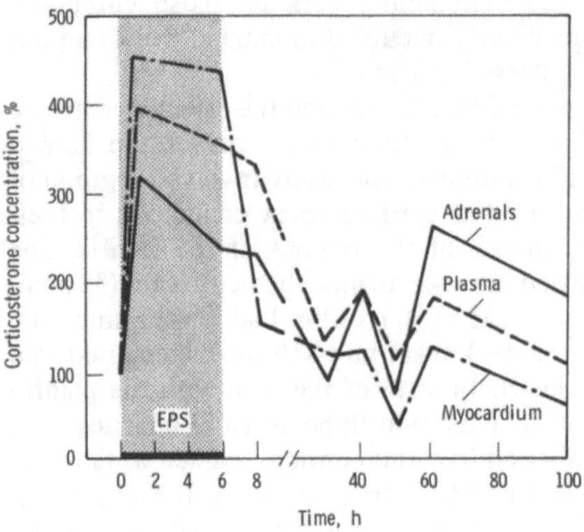

Fig. 34. Variations in corticosterone content of adrenals, plasma, and myocardium during EPS and for 4 days after EPS

To study these questions, our laboratory and integrated institutions[1] performed a system of studies on the already mentioned emotional-painful stress (EPS) after Desiderato et al. [80]. The essence of this model is that male rats (200 g weight) deprived of food for 24 h were placed in a special chamber in which electric current (4 mA) was transmitted through the floor. The rats could avoid the painful electric stimulation only by moving to a platform located in the center of the chamber. As a result, they rapidly developed a conditioned defense reflex of avoidance, and they constantly stayed on the platform. Subsequently, brief strong electroshocks (6 mA for 2 s) began to be inflicted through the platform floor. This was done at occasional intervals for 6 h. The principal features of this EPS model are, first, in the presence of a conflict between the conditioned reflex already developed of avoiding the current by moving to the platform and the unconditioned painful stimulation on that platform and, second, in tense anticipation of the painful electric action due to the fact that on the platform the shocks were inflicted occasionally.

Such EPS is known to develop pronounced activation of the hypophysial-adrenal system and the adrenergic systems. Corticosterone in the plasma, adrenals, and heart was determined by chromatography on columns filled with silicagel, using a modified Botviniev and Veltishev [47] method. The curves in Fig. 34 indicate that with our EPS model, activation of the hypophysial-adrenal system is evident in a three – to fourfold increase in the corticosterone content of the adrenal cortex, blood plasma, and myocardium. To determine the catecholamine content in the myocardium and adrenals, we used the trioxyindol fluorimetric method of determining catecholamines in tissues [105] modified by Matlina and Rakhmanova [304]. The graphs in Fig. 35

1 Laboratory for the Physico-Chemistry of Biomembranes, Department of Biology, Moscow University; Chair of Biochemistry, Chelyabinsk Medical Institute; and Central Research Laboratory, Irkutsk Medical Institute.

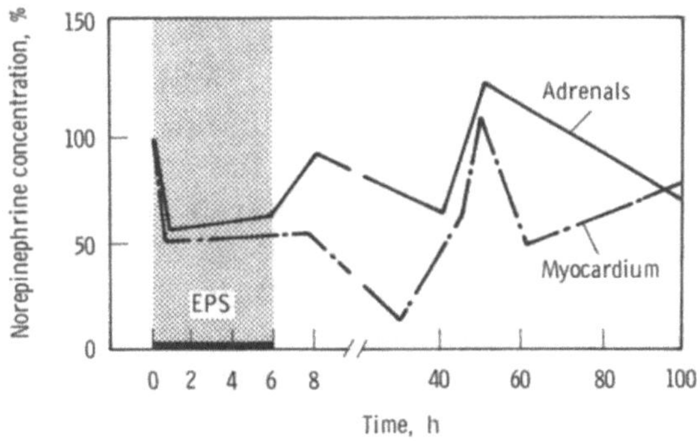

Fig. 35. Variations in norepinephrine concentration of myocardium and adrenals during EPS and for 4 days after EPS

indicate that activation of the adrenergic system is observed, in this case, in an approximately twofold decrease in noradrenaline concentration in the adrenals and myocardium. The curves in Figs. 34 and 35 demonstrate the important fact that stress-induced initial excitation is followed by periods when the stress reaction terminates and resumes after the stress situation has ended.

These delayed periods of activation of the stress system may be interpreted on the basis of the familiar psychological and psychiatrical fact that the state of emotional stress may persist for a long period after the stress-inducing factor has been eliminated. Possibly, this poststress activation of the hypophysial-adrenal and adrenergic systems, determined by emotional stress components, plays its role in developing stress injuries. As in the experiments of other investigators cited above, in our experiments it quite rapidly led to ulcerous lesions of the mucous coat of the stomach that are typical of stress. The average length of ulcers in animals autopsied 2 h after the stress procedure had ceased was 10–15 mm per stomach and no ulcerous lesions were observed in controls. Thus, we reproduced a really serious and hopeless stress situation that naturally injured the animal.

Over 2 h after termination of the stress situation, the animals were decapitated and the heart, brain, liver, and stomach were subjected to the investigations discussed below. In studying the heart, we found that two interrelated sets of changes took place in the myocardium under EPS: First, those that may be assessed as the regulatory effect of high glucocorticoid and catecholamine concentrations, and second, those that express stress damage of the myocardium by these hormones.

Two hours after EPS, a set of regulatory changes is evident in the mobilization of glycogen reserve in the myocardium and in inhibited oxidation of certain substrates of the tricarbonic acid cycle, primarily α-ketogluterate, with simultaneous activation of the pentosophosphate pathway and the GABA shunt in the myocardium. This shift provides increased production of a semialdehyde of succinic acid and of succinic acid itself, whose utilization in the cycle of tricarbonic acid is not decreased but, on the contrary, increased. This compensatory mechanism and possibly other regulatory mechanisms of the myocardial cell lead to a situation in which the ATP concentration in the myocardium after stress does not differ from controls during relaxation nor under maximal load caused by clamping of the aorta. This set of changes in energy metabolism is apparently one way or another due to injury of the myocardium; however, in itself, this is no injury in the exact meaning of the term. It represents the result of catecholamine and glucocorticoid actions on corresponding receptors in the myocardial cells, these effects being the result of strongly activated normal regulatory mechanisms in the cell. Below, we shall examine changes in energy metabolism in greater detail when we compare the states of the antistress GABA system in the brain and heart; in this section, the presentation is focused on the metabolic link, via which the protracted stress reaction harmfully affects the myocardium.

Considering this issue, we turned our attention to a work by Horgan and Philpot [180], who studied the effect of ionizing radiation on the organism and, to that end, measured the intensity of free-radical oxidation in the tissues by such an indirect indication as oxidation in anaerobic conditions of the brilliant cresyl blue. Unexpectedly, they revealed that in control animals manually held by the tail, the estimated activity of peroxidation increased by 20% and in animals held by a neck-gripping device by 75%. This was explained by the fact that anxiety activated peroxidation. Recently, it has become known that when administered in considerable doses, catecholamines, the principal mediators of anxiety, i.e., stress, cause lipid peroxidation (LP) to be activated in the myocardium and, apparently, in other organs as well [248].

Thus, it has been suggested that under stress, high catecholamine concentrations may activate LP and that the resultant excess lipid hydroperoxides becomes the biomembrane-injuring factor. In verifying this suggestion several years ago, the chemiluminescence of lipids extracted from the myocardium of EPS-exposed animals was found to have increased threefold and it was for the first time inferred that LP is activated during stress [364]. However, these measurements do not represent a direct quantitative method of estimating the content of LP products in the tissues. Hence, we [370] subsequently determined the primary molecular LP products (hydroperoxides of polyene lipids) and final LP products (fluorescent Schiff's bases) in the heart, skeletal muscle, and brain of EPS-exposed rats. The method of Folch et al. [117] was used to extract lipids from these organs.

Accumulation of hydroperoxides in polyene lipids was estimated by the ultraviolet (UV) absorption spectrum of the lipid solution in methanol hexane (5 : 1), this spectrum being characteristic of diene conjugates; in this case, the molar extinction coefficient at maximal wave length (232 nm) was taken to equal $2.1 \times 10^4 \, m^{-1} cm^{-1}$ [44]. The spectra were recorded on an MPS-50 spectrophotometer (Shimadzu, Japan). The LP final products, that is, the products of the interaction of short-chain dialdehydes with aminophospholipids, were recorded on a spectrofluorimeter by the fluorescence spectra of lipid solutions in chloroform with a fluorescence excitation maximum at 360 nm and emission maximum at 420–440 nm [71].

Figure 36 shows the typical UV absorption spectra and the fluorescence emission spectra of solutions of lipids extracted from the myocardium of

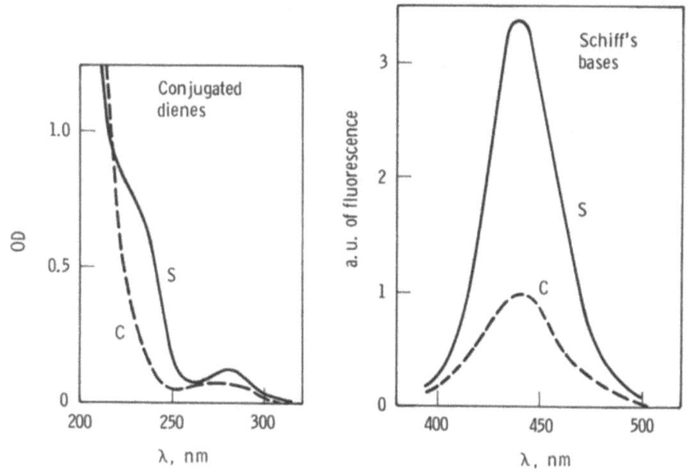

Fig. 36. Effect of stress on UV spectra of absorption and emission of lipids extracted from myocardium. *Dashed line,* control; *continuous line,* stress

controls and EPS-exposed animals. Lipids extracted from the hearts of the latter possess absorption spectra with maximums at 230–235 and 270–280 nm (Fig. 36A) characteristic of polyene lipid hydroperoxides, these maximums being virtually absent in control. It is further obvious that the fluorescence of Schiff's bases, the final LP products, is essentially higher for the lipids of EPS-exposed animals than those of controls (Fig. 36B). In Fig. 36, the shaded sections represent the accumulation of intermediate and final products of lipid peroxidation under EPS.

Table 26 shows that under EPS, the hydroperoxide content in the myocardium increases threefold and the fluorescence intensity of the final LP products fivefold; for skeletal muscles, these values increase 2- and 2.7-fold, and for the brain 1.7- and 2.1-fold, respectively.

Table 27 shows data characterizing the accumulation dynamics of Schiff's bases after EPS and permitting one to assess the LP dynamics after sustained stress. As is apparent from Table 27, as soon as 2 h after EPS the amount

Table 26. Accumulation of lipid peroxidation products under EPS

Tissue	Group	Number of animals	Fluorescence of Schiff's bases, rel. units	Lipid hydroperoxides, optical density units	*p*
Myocardium	Control	10	19.1 ± 2.9	0.35 ± 0.5	< 0.001
	EPS	11	95.5 ± 20.8	0.90 ± 0.05	
Skeletal muscle	Control	10	13.0 ± 1.7	0.40 ± 0.05	< 0.01
	EPS	11	35.3 ± 7.3	0.80 ± 0.10	
Brain	Control	10	9.3 ± 0.4	0.15 ± 0.05	> 0.1
	EPS	11	20.0 ± 1.5	0.25 ± 0.05	

Table 27. Accumulation dynamics of LP products in rat tissues under EPS (fluorescence of Schiff's bases)

Tissue	Time when material was obtained after EPS				Control
	2 h	2 days	5 days	10 days	
Myocardium	29.2 ± 4.6	27.6 ± 4.3	12.9 ± 1.9	9.0 ± 1.1	10.5 ± 3.5
Skeletal muscle	20.6 ± 2.8	27.0 ± 9.7	10.8 ± 1.7	9.0 ± 1.2	10.3 ± 0.9
Brain	20.8 ± 3.4	26.9 ± 3.3	23.7 ± 4.1	10.6 ± 2.2	8.9 ± 1.8
Aorta	31.2 ± 7.5	28.0 ± 2.8	10.5 ± 2.6	–	11.5 ± 1.9

of fluorescing LP products had increased almost threefold in the heart and the aorta and twofold in the skeletal muscle and the brain. During the 2 days that followed, the amount of secondary LP products in the heart and aorta declined insignificantly to reach the control level by day 5. In the brain and skeletal muscle, the amount of Schiff's bases continued to rise for 2 days and reached the control level by day 10 after EPS. Thus, following sustained stress, LP is activated in the organism, being especially pronounced and prolonged in the heart.

The question as to how this situation proven in the above experiments on animals is applicable to humans is obviously quite important for medicine. Recently, it was proved that LP activation produces low-molecular-weight hydrocarbons in the animal organism due to the consecutive decomposition of lipid hydroperoxides and β-splitting of alcoxyl radicals. These gases are released with exhaled air, in which their content is a specific and highly sensitive criterion of LP intensity.

From these data, Prilipko et al. [450] performed a study whose purpose was to clarify, while measuring the content of gaseous LP products in the exhaled air, whether or not LP is activated in humans under stress caused by intense intellectual activity when time is scarce. To solve this problem, we used the method of determining gaseous LP products in the exhaled air. Stress was induced by the operator working out consecutively the solution to seven quite complicated mental problems and knowing that time was limited; the problems lasted a total of 2 h. The content of gaseous LP products in the exhaled air and of catecholamines in the blood was determined 24 h and 15 min prior to the stress, and also 15 min, 1, 2, 3, and 5 days after it. When quantitatively analyzing the hydrocarbons in the exhaled air, we proceeded from the concept that omega-6-polyunsaturated fatty acid residues of phospholipids constituted the preferential LP substrates and that pentane was correspondingly the principal quantitatively predominant hydrocarbon gaseous LP product. Hence, to impart maximal sensitivity to the method, we determined the pentane content in the exhaled air. Twenty healthy people were examined and 13 of them were exposed to stress.

Figure 37 shows the typical chromatograms obtained on examining the exhaled air prior to (1) and after (2) stress. The extreme left peak in both chromatograms corresponds to pentane and indicates that the pentane content

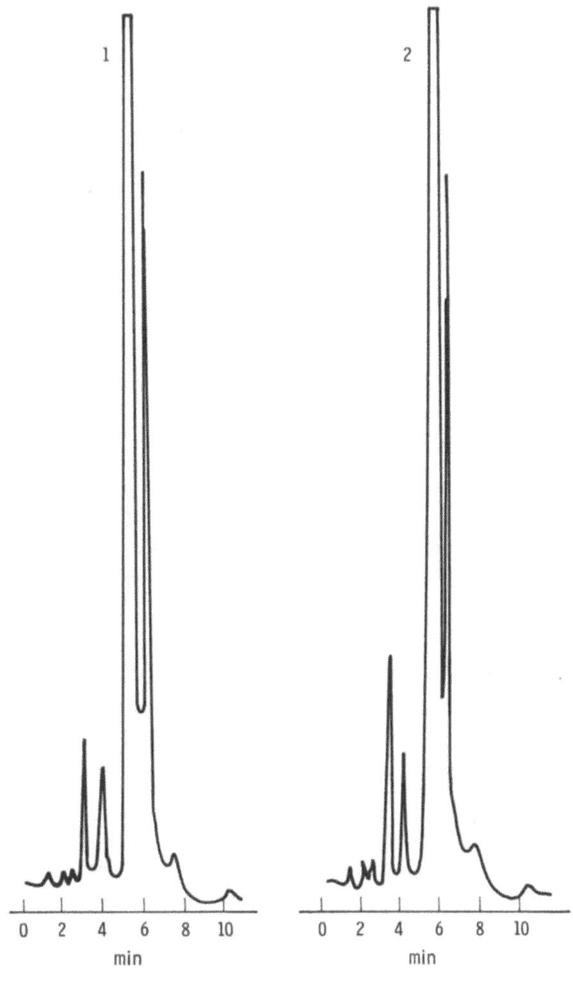

Fig. 37. Chromatograms of exhaled air prior to stress effect *(1)* and after stress *(2)*. *Extreme left peak* showing pentane content had increased after stress

in the exhaled air after stress was substantially increased. Quantitative analysis showed that the pentane content in the exhaled air prior to stress constituted on average 1.71 ± 0.34 nmol/l, and the adrenaline and noradrenaline content in the blood 0.70 ± 0.05 and 0.80 ± 0.05 μg/l, respectively. Instant but relatively short-term increase in the catecholamine content of the blood and slower but prolonged elevation of the pentane content in exhaled air were observed after stress. The curves in Fig. 38 show that 15 min after stress, the catecholamine content in the blood increased by 140%–150% to reach a maximum and returned to the initial level after 24 h. Fifteen minutes after stress, the pentane content in the exhaled air was increased only by 33%, and after 2 days by 72%; only by day 5 did the hydrocarbon content return to the initial level. Thus, the increased catecholamine content reflecting an excitation of the adrenergic system characteristic of stress precedes maximum LP activation, which is evidenced by increased pentane content in the exhaled

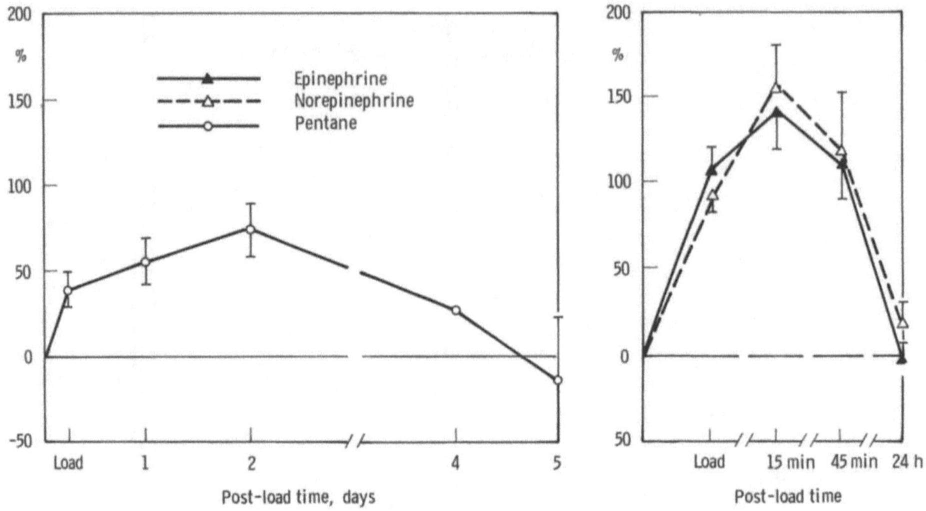

Fig. 38. Variations in catecholamine content of the blood *(right)* and in pentane content of exhaled air *(left)* after stress action; percentage of initial value

air. Significantly, LP activation, which as will be further shown plays a major role in injuring cell membranes, proved to be quite protracted after the stress action applied in the given study. Its dynamics and duration principally coincide with the dynamics and duration of the above LP activation in internal organs during EPS in animals. This provides grounds for assuming that an increase of gaseous LP products in the exhaled air reflects the degree of LP activation in the cells of vitally important organs and, possibly, the extent of stress-induced injuries therein. Thus, the obtained results unambiguously show that *in humans, as in animals, stress actions not connected with direct injuries by exogenic physical or chemical factors cause considerable activation of peroxidation, which is evident in an increased content of gaseous LP products in the exhaled air.*

It has been unambiguously proven that LP free-radical products injure the lipid bilayer in cell membranes and when accumulated in excess amounts may degrade or inactivate membrane-bound proteins, viz enzymes, receptors, permeability channels, and the biomembranes themselves (see Chap. 8). Hence, the fact that LP is activated with human stress may in perspective make possible the use of antioxidants, which are LP inhibitors, to protect the human organism from stress-induced injuries (Chap. 8) and permit utilizing the content of LP products in the exhaled air as an important criterion for assessing the intensity of stress and the efficacy of factors that protect the human organism from stress damage. The nature of LP, its activation mechanism under stress, and the essence of its harmful effects will be examined below (Chap. 7), after we have shown that excessive LP activation is nonspecific to stress and is a common pathogenic link with highly variable injuries and diseases.

It should be pointed out that the effect of catecholamines on the membrane lipid bilayer is not restricted to activating LP; hence, in no way does LP activation therein during stress represent an isolated event. Indeed, it has been proven that LP products cause labilization of the lysosomes in which a considerable number of cell phospholipases are concentrated. The phospholipases released in increased numbers from lysosomes may play a major role in injuring the membrane lipid bilayer and in forming lysophospholipids [209]. As with high concentrations of fatty acids observed in the blood during stress and myocardial infarction, lysophospholipids, on account of their detergent-like action, impair the ordered arrangement of phospholipids in the membranes and this so-called chaotropic effect has now been shown to activate LP [166]. The increased membrane permeability for calcium caused by this chain reaction and the excess of the cation in the cell that occurs as a result in turn activate phospholipases and, possibly, LP [195, 566]. This may result in the closure of the vicious circle which plays an important role in injuring cardiomyocytic cell membranes.

The three interrelated processes that take place in the membrane lipid bilayer, viz LP activation, phospholipase activation, and the detergent action of high fatty acid and lisophospholipid concentrations, we shall henceforward term the lipid triad [384]. Since the effect of excess catecholamines during stress is primarily achieved in the sarcolemma, where the adrenoceptor-adenyl cyclase system is localized, it may a priori be anticipated that the lipid triad too should take place first of all in the sarcolemma. This should, in turn, lead to changed activity in the sarcolemmal lipid-dependent enzymes, e.g., Na^{+}-K^{+}-ATPase, whose activity has been shown to be highly dependent on its lipid surrounding.

From this angle, we investigated the effect of EPS on Na-K-ATPase activity in the myocardium and the possibility of preventing the depressed activity of the given enzyme then discovered by administering Inderal, a β-blocker [385]. In this case, the membranous Na-K-ATPase preparation was separated from rat myocardium by means of NaI according to the method of Matsui and Schwartz [306]; the enzyme activity in the preparation was determined using phosphorus after the method of Rathbun and Betlach [464], and the protein content in the method according to Lowry.

The data shown in Table 28 indicates that after EPS, Na-K-ATPase activity in the myocardium proved to be reduced by more than 25 %, and admini-

Table 28. EPS and Inderal effects on Na-K-ATPase activity of the myocardium, nmol/P_{inorg}/mg protein/h

	Number of animals	Na-K-ATPase activity
1. Control	7	3.86 ± 0.24
2. Stress	6	2.98 ± 0.25
		$p_{1-2} < 0.05$
3. Inderal	10	3.62 ± 0.15
		$p_{1-3} > 0.5$
4. Inderal + stress	6	3.51 ± 0.12
		$p_{1-4} > 0.5$

stration of the β-blocking Inderal prior to stress completely prevented this shift. Na-K-ATPase activity constitutes the basis for the functioning of the Na-K pump of cardiomyocytes and, therefore, plays a decisive role in maintaining the Na^+ and K^+ concentration gradients between the sarcoplasma and the extracellular medium. Correspondingly, a decline in Na-K-ATPase activity may lead to an increase in the Na^+ content and decrease in the K^+ content of the cardiomyocytes. A lower Na^+ concentration gradient may result in a less active Na-Ca exchange mechanism, which functions exclusively on the basis of the sodium concentration gradient. During the action potential, this mechanism provides the influx of calcium into the cardiomyocytes essential for myofibrillar contraction [256] and, in the intervals between the action potentials, plays a major role in bringing calcium out of the myocytes to achieve relaxation. It is, therefore, understandable that a decrease in Na-K-ATPase activity and, consequently, in the Na-Ca exchange mechanism, may play a substantial role in the heart contractile disorders occurring after EPS, and will be examined below.

The fact that Inderal prevents the decrease in Na-K-ATPase activity usually occurring after stress agrees with the concept that LP in the myocardium is activated by catecholamines and that catecholamines damage the sarcolemma, specifically causing its lipid-dependent enzymes to become less active under serious stress.

In terms of our presentation, it is significant that increased LP and other components of the lipid triad may not only damage the sarcolemma but other cell membranous systems as well. From this angle, subsequent research has been aimed at studying two points: First the state of the lysosomes whose labilization may have constituted the next probable link in the pathogenic chain of stress injury, and second, the activity of the lysosomal and cytoplasmic enzymes in the blood plasma, since their release from the cells into the blood may be the direct result of cell membrane injury [373].

The diagram in Fig. 39 shows stress leading to the decline in acid cathepsin activity in the lysosomal fraction and myocardial mitochondria by 25%, and simultaneously increasing the activity of these enzymes in the supernatant by 45%. This shift, to a certain extent expected, indicates that EPS promotes the labilization of lysosomal membranes and release of proteolytic enzymes therefrom. This conception is in agreement with the evidence shown in the same diagram (Fig. 39) that stress results in a twofold increase in cathepsin activity in the blood plasma.

The scheme in Fig. 40 shows that 2 h after stress, the enzymic activity of aspartate transaminase, alanine transaminase, lactate dehydrogenase, and malate dehydrogenase in the blood plasma was about two times higher. Thus, after sustained stress along with LP activation and, apparently, as a result of the lipid triad, lysosomal membranes are found to be labilized and lysosomal proteolytic enzymes to be released into the cytoplasm and blood plasma. At the same time, the cell membranes develop more serious injuries that result in pronounced fermentemia.

In the next investigational stage, we sought to estimate quantitatively the injury to the genetic apparatus of the myocardial cells induced by a seroius

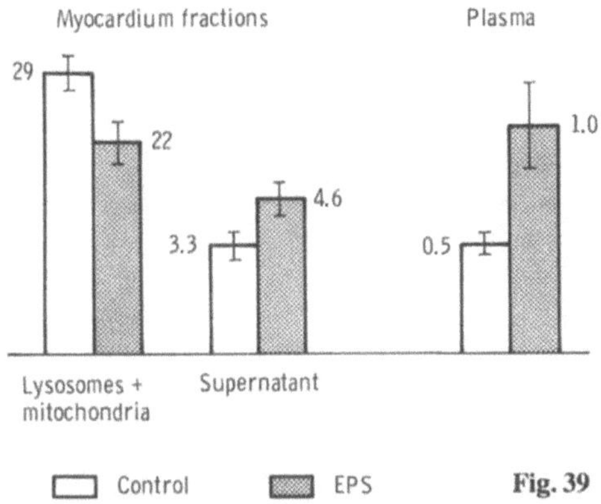

Fig. 39

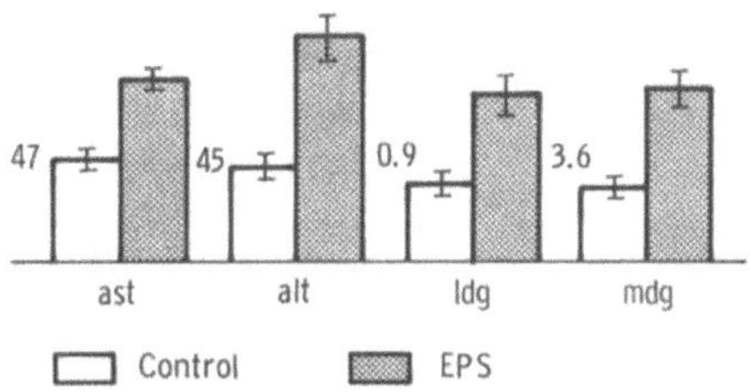

Fig. 40

Fig. 39. Effect of EPS on activity of acid cathepsins (μg thyrosine/mg protein/h) in the heart and plasma

Fig. 40. Effect of EPS on activity of plasma enzymes: AS transaminase and AL transaminase content, μmol/ml/h; activity of lactate dehydrogenase, μmol $NADH_2$/min/ml; malate dehydrogenase content, μmol/min/mg protein

stress reaction. Since in radiation sickness, accumulation in the cells of free-radical LP products is known to be accompanied by discontinuties in the DNA spiral, one cannot exclude that a similar event may take place in stress-induced LP activation. Basing their work on this observation, Meerson and Vasiliev [340] carried out experiments aimed at studying the polymeric nature and reparative synthesis of DNA in myocardial cells after EPS. The polymerism of the DNA of cell nuclei isolated from the heart was assessed by sedimentative analysis in a sucrose gradient. The reparative synthesis of DNA was assessed in vivo and in vitro. To determine it in vivo, oxyurea (500 mg/kg weight), which inhibits semiconservative DNA synthesis, was injected intraperitoneally into the animals with the subsequent administration of ^{3}H-thymidine (100 μCi/100 g weight) [69]. After 1 h, oxyurea was injected again; 1 h after this, the animals were killed to isolate the DNA. To assess DNA reparative synthesis in vitro, a DNA-polymerase system was used to determine ^{3}H-thymidine triphosphate incorporation in the DNA from the myocardial cell nuclei. The acid-insoluble residue obtained in vitro and the DNA alcohol residue obtained in vivo were collected on membrane filters (Millipore, HAWP, 0.45 μ) and the radioactivity was recorded for 10 min using a toluene scintillator in a counter (Intertechnique, France); DNA was determined with diphenyl amine [52].

The curves shown in Fig. 41 reveal that the myocardial DNA is partially depolymerized under the influence of EPS. Directly following EPS (Fig. 41a), this is evident in the sedimentation peak in controls in fraction 8 being displaced to fraction 9, i.e., to a region of lower molecular weights. This decrease in polymerism is most probably due to the occurrence of single-chain DNA ruptures. Figure 41b shows that 2 h after EPS, the DNA sedimentation profile substantially changed. The sedimentogram peak returned to

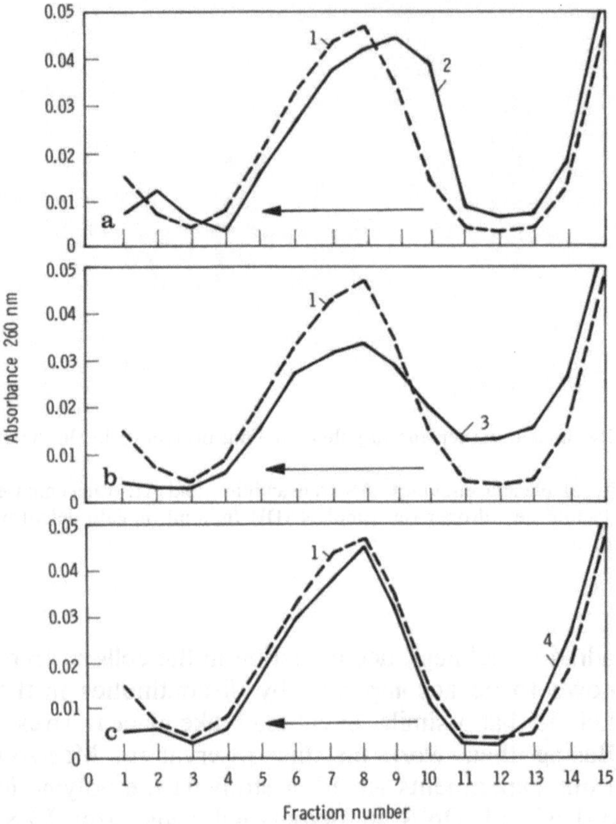

Fig. 41. Sedimentation curves of lysate nuclei isolated from myocardia of stress-exposed rats compared with controls. *1,* control (peak localized in fraction 3 with 95% probability); *2,* nuclei isolated immediately after termination of stress (peak localized in fraction 9 with 70% probability); *3,* nuclei isolated 2 h after termination of stress (peak localized in fraction 8 with 70% probability; total absorption of fractions 5–9 was 15% less than in controls, 95% probability); *4,* nuclei were isolated 24 h after termination of stress (peak localized in fraction 8 with 70% probability). To obtain control sedimentograms, 12 independent nuclei preparations were used. Experimental sedimentograms were plotted using three independent preparations for each version. Two or three sedimentograms were obtained for each nuclei preparation. Direction of sedimentation is shown by *arrows*

the same position as in controls, but at the same time a noticeable number of low-molecular weight DNA fragments appear in fractions 11–14. A possible mechanism for the dynamics of this situation is that single-thread DNA ruptures are rapidly repaired in most myocardial cells that continue to function after stress. Simultaneously, DNA nuclease splitting occurs in cells located in the micronecrotic foci, which as will be shown below develóp under stress; as a result of DNA alkaline denaturation, the above splitting results in low-molecular weight single-chain fragments.

From Fig. 41c it is apparent that 24 h after stress, the DNA sedimentation

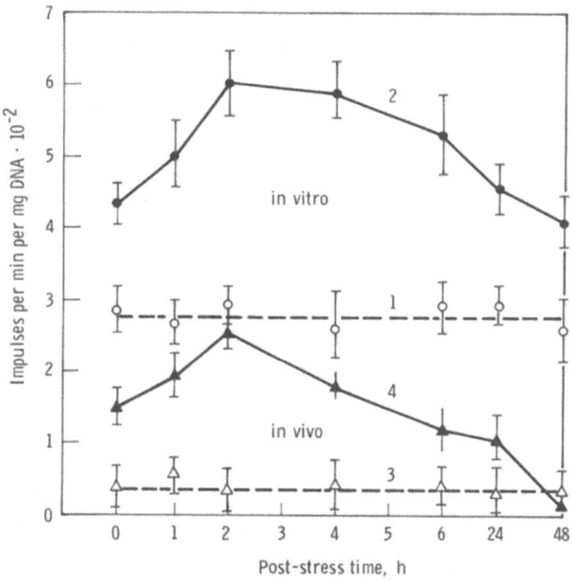

Fig. 42. Stress-induced reparative DNA synthesis in rat myocardium. *1*, control level of ^{3}H-thymidine triphosphate (TTP) incorporation into the system's nuclear DNA; *2*, incorporation of ^{3}H-TTP into the nuclei of animals taken for the experiment at given intervals after termination of stress action. The nuclei were isolated for 3 h at 0 °C and used in a DNA-polymerized system immediately after isolation; *3*, control level of ^{3}H-thymidine incorporation into DNA of rat myocardium; *4*, ^{3}H-thymidine inclusion into DNA of rat myocardium against a background of oxyurea injection at various intervals after termination of stress. Interval from incorporation of ^{3}H-thymidine to decapitation–2 h. Degree of variation is shown in the form of standard deviation

profiles in experimental animals do not differ from those in controls; with some probability, this may be explained by DNA reparation in a live cells and the elimination of disintegration products of dead cell DNA. These assumptions are confirmed by the experimental results for the reparative synthesis of myocardial DNA obtained in vivo and in vitro in nuclear preparations (Fig. 42).

The DNA reparative synthesis in vivo in controls is virtually absent; however, it is observed directly after stress, and 2 h later attains a maximum and stays at a given level for at least 24 h. Basically, the same result was obtained in vitro: DNA synthesis in nuclei isolated from the hearts of stress-exposed animals attained a maximum value after 2 h to remain at an elevated level for at least 48 h.

Labeling in vivo was considerably higher than that in vitro. This difference may be explained by the presence in isolated nuclei of DNA damage caused by the work of nucleases during the isolation of the nuclei. The noticeable level of ^{3}H-thymidinetriphosphate (TTP) incorporation 48 h after the termination of stress in the absence of ^{3}H-thymidine incorporation in vivo was due to the fact that the activity of the DNA-reparative system continues for some time after the reparation of DNA caused by EPS is actually over. On the whole, the obtained facts indicate that injuries and subsequent reparation of the DNA chain, i.e., events quite similar to those that take place under ionizing radiation, carcinogens, and mutagens, occur in the myocardium under EPS simultaneously with the activation of free-radical processes.

Further studies have shown a similar situation involving disintegration and subsequent reparation of DNA to be involved with stress not only in the heart, but in the kidneys, liver, brain, and other organs as well. In this connection, 'DNA injury-reparation' cycles, which repeatedly occur under the influence of stress situations during an individual life, may probably lead to impaired DNA nucleotide sequence, and this accumulation of errors in the genetic code becomes a cause of aging in general and of wear of the heart in particular.

I have already cited data on the effect of stress on RNA and protein disintegration and they show that after stress, the half-life and turnover of ribosomal RNA in the internal organs substantially decreased, while the percentage of RNA substituted for 24 h essentially increased. This set of shifts, directly indicative of higher RNA disintegration, is strongly pronounced in the myocardium: The half-life and turnover times decreased by more than one-fourth, and the substitution percentage grew by more than one-third. In other words, EPS caused considerably increased protein disintegration in the myocardium [388]. Thus, under the influence of stress in the myocardium, DNA ruptures and subsequent reparations become increasingly activated and the RNA and protein disintegration rates become considerably increased simultaneously with the accumulation of lipid hydroperoxides and labilization of lysosomes.

In subsequent investigations, it became important to reveal precisely what structures of the myocardial cell involve protein disintegration and damage under EPS [377]. In morphological studies performed to this end, serial topographic sections of the myocardium (5–7 μ) were stained with pycofuchsin after the method of Van Gieson, with colloidal iron after the method of Hale, and with fuchsin by the Selye method; the procedure also involved Perls' reaction with subsequent staining with hematoxylin-eosin. We also performed the PAS and Brachais reactions. Polarization microscopy was used to assess the state of the contractile apparatus of the cardiomyocytes and the nature of miofibrillar changes. The principal fact established in our morphological studies was that as early as the first hours following stress, the cardiomyocytes developed focal changes of the contractural type; in progressing, these changes attain a maximum, i.e., result in complete contracture of the myocytes, in 39–45 h. In some cases, contracture is accompanied by pronounced necrobiotic changes and leads to death of the myocytes with their subsequent resorption and the formation of fibroblastic granulomas; in other cases, contractures develop inversely to result in the restoration of cell structures.

Figure 43 shows the myocardium 45 h after stress, i.e., in the phase of maximal morphological changes. Figure 43a shows the results of polarization microscopy and demonstrates a deep contracture with the merging of myofibrillar disks A, which form continuous anisotropic conglomerates in individual

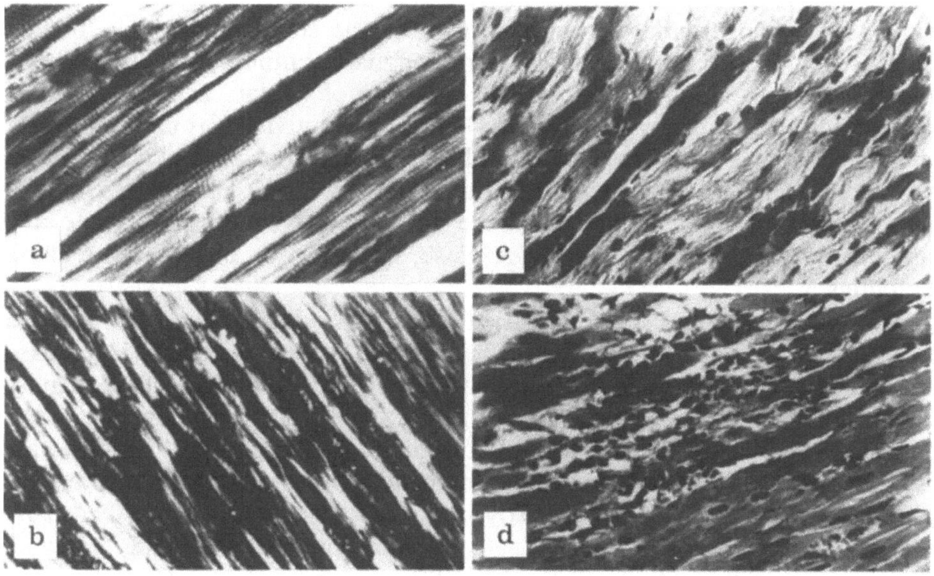

Fig. 43. Injury of rat myocardium 45 h after stress. **a, b** polarized light, × 197; **c** Perls' reaction, × 126; **d** staining with hematoxylin eosin, × 49. Explanations in text

fibers; the microphotograph in Fig. 43b shows with lesser magnification a large number of such contracturally changed myofibrils with large anisotropy. Figure 43c shows the results of the Perls' reaction and groups of necrobiotically changed myofibrils with a positive reaction to iron; finally, Fig. 43d shows the formation of a cell infiltrate around necrotized muscular fibers. Pyknotically changed nuclei, homogenization and fragmentation of the myofibrils are perceptible in the necrotic focus.

In assessing the fact that focal contractural changes develop in the cardiomyocytes under the influence of stress, one should keep in mind that the ATP concentration in the myocardium of stress-exposed animals did not differ from controls both in relaxation and under maximal load on the heart. Hence, focal contractures of myocytes can hardly be explained by hypoxic or other deficiencies in energy-rich phosphorous compounds. The most probable cause of this kind of contracture is impaired membranous transport of Ca^{2+}, for active removal of this cation from the myofibrils in the reservoirs of the sarcolemma and SR constitutes the basis for normal relaxation, and the focal contractures after stress represented nothing other than focal disturbances of relaxation. This caused us to think that under stress, lipid peroxides and proteolytic lysosomal enzymes damage the membranous apparatus of Ca^{2+} transport and cause focal contractural injuries in the myocardium.

To assess the reality of this assumption, we conducted joint studies with Arkhipenko and Rozhitskaya [381] to investigate the effect of EPS on the state of the calcium-transport system. In these studies, the animals were decapitated 2 h after stress, and the microsomal fraction from their hearts was separated by differential centrifugation using the standard method. Judging by enzymic analysis, the resultant membranous fraction consisted chiefly of the SR. Accumulation of Ca^{2+} in the presence of oxalate and Ca^{2+} binding were recorded by ^{45}Ca radioactivity. The ATPase activity was determined by means of the pH using the ATP hydrolysis rate and concomitant acidification of the medium. As an outcome, the following basic parameters of the Ca^{2+} pumps of myocardial cells were calculated – accumulation rate and amount of Ca^{2+} bonded and Mg-ATPase, Ca-ATPase, and Ca-Mg-ATPase activities.

The experimental results shown in Table 29 indicate that EPS reduced Mg-Ca-ATPase activity by 15 %, this happening exclusively due to a decrease in Ca-ATPase activity, which declines by almost 30 %; the Mg-ATPase activity remains the same. In accordance with this, the Ca accumulation rate in the presence of oxalate proved to be reduced by 38 %; the Ca^{2+} bonding rate was reduced by 41 %; and the Ca amount that can be bonded by SR membranes in the absence of oxalate decreased by 46 %. This considerably diminished capacity of SR membranes to accumulate calcium apparently depends not only on a decline in the activity of the Ca-ATPase responsible for Ca^{2+} transport but also on greater permeability of calcium in damaged SR membranes. In terms of this presentation, it is significant that the violated ability of SR membranes to bind and accumulate calcium should with a higher extent of probability lead to increased Ca^{2+} concentration in the sarcoplasma of cells in an intact functioning heart.

To assess the changes in the content and, particularly, in the localization of calcium that develop in cardiomyocytes under stress, we performed a study based on a combination of electron microscopy and histochemistry

Table 29. Effect of EPS on parameters of the enzymic system of Ca^{2+} transport in SR vesicles

	1 Control	2 EPS
Ca, Mg-ATPase	nmol, P_{inorg}/mg protein in 1 min	
	1,267 ± 163 $n = 12$	1,026 ± 249[a] $n = 11$
Mg-ATPase	694 ± 113 $n = 12$	642 ± 153 $n = 10$
Ca-ATPase	502 ± 68 $n = 12$	359 ± 127[b] $n = 10$
Ca^{2+} accumulation rate	nmole Ca^{2+}/mg protein in 1 min	
	37.2 ± 4.0 $n = 10$	23.1 ± 4.2[c] $n = 8$
Ca^{2+} binding rate	3.8 ± 0.42 $n = 10$	2.23 ± 1.14[a] $n = 8$
Amount of bound Ca^{2+}	nmole Ca^{2+}/mg protein in 5 min	
	7.20 ± 2.78 $n = 10$	3.90 ± 1.22[a] $n = 7$

[a] $p_{1-2} > 0.01$. [b] $p_{1-2} < 0.05$. [c] $p_{1-2} < 0.01$. [d] n, number of experiments.

[389]. The animals were taken for investigation 2 and 48 h after the termination of stress. We studied the posterior papillary muscle of the left ventricle. Fixation was performed directly after killing the animals, the fixative being 4% paraform on a phosphate buffer, 0.6 M, pH 7.2. In parallel, calcium localization in the same muscle was studied by the pyroanothymonate method. To reveal Ca^{2+} at electron-microscopic level, we used a 2% solution of potassium pyroantimonate ($K_2H_2Sb_2O_7 \cdot 4H_2O$) in a mixture with 1% OSO_4; the fixing solution pH was reduced with 0.01 M acetic acid to 7.0–7.1 and fixed at 4 °C for 1 h [494].

Dehydration was performed in alcohols of rising concentration and in propylene oxide; the tissue sampler were poured into an epone-araldite mixture. Sections were obtained on an ultratome, type LKB-8 800. Staining was completed with lead hydroxide and uranyl acetate. The specimens were examined in an electron microscope. The use of potassium pyroantimonate to reveal calcium localization in myocytes was largely based on the studies of Legato and Langer [264]. They perfused a dog's papillary muscle with a potassium pyroantimonate solution and in a subsequent electron-microscopic study revealed, first, a residue localized in the exterior side of the sarcolemma, and second, a residue situated in several intracellular cardiomyocytic zones, namely, in the subsarcolemmal space, in SR terminal cisterns, in the region of I disks of the myofibrils, and in the mitochondria. In special experiments using radioactive sodium, Legato and Langer [264] showed that the coarse precipitate on the outer surface of the sarcolemma was formed by sodium pyroantimonate; the precipitate was not observed upon perfusion with a sodium-free solution. On the other hand, the precipitates localized in various myocytic zones proved to have been formed by calcium pyroantimonate.

Despite the well-grounded nature of this conclusion, we performed a special control study in which some of the samples were incubated for 60 min in 5 mM ethylene glycol tetraacetate (EGTA) prior to fixing the papillary muscle; this provided complete disappearance of the pyroantimonate salt precipitate studied and guaranteed that we really studied calcium localization in cardiomyocytes. The electronographs in Fig. 44a, b were obtained when studying the papillary muscles of control animals. From Fig. 44, where EGTA treatment resulted in the absence of a calcium pyroantimonate precipitate, it is apparent that the cardiomyocytic ultrastructure is normal. Figure 44b traces calcium localization in the cardiomyocytes of control animals: A calcium pyroantimonate precipitate is naturally found in the terminal cisterns of the SR and in the H bands of the myofibrillar disk and is absent or proves to be insignificant in the subsarcolemmal space and the mitochondria. The small precipitate granules, occasionally noted in the mitochondria, should apparently be regarded as nonspecific – not representing calcium pyroantimonate. A similar result with regard to mitochondria was obtained in the previously mentioned work by Legato and Langer [264] and corresponds to biochemical evidence that the calcium concentration in the mitochondria of the papillary muscles of normal rats is relatively low [495].

The electronograms in Fig. 44c, d characterize the ultrastructure of the rat myocardium 2 h after stress. In addition, in Fig. 44c, where the calcium pyro-

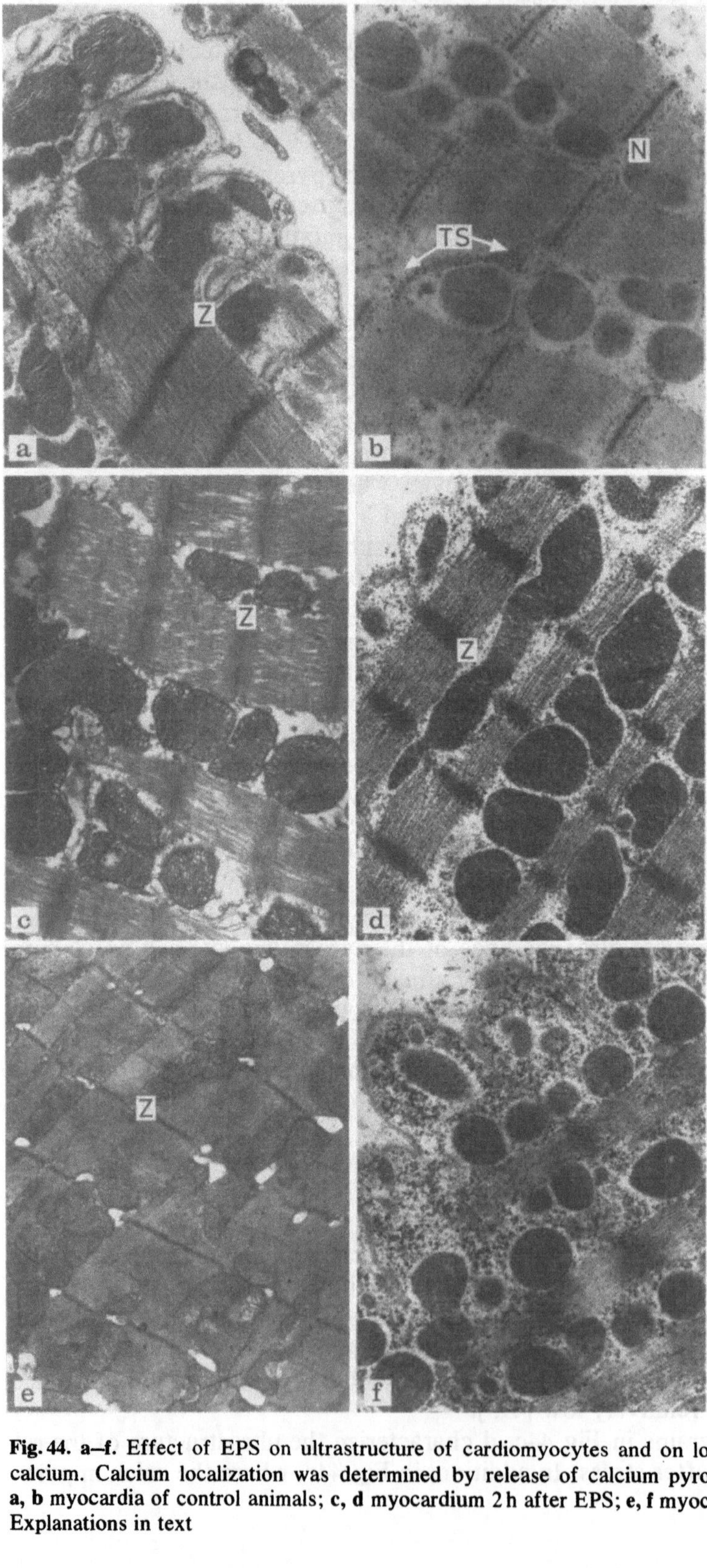

Fig. 44. a–f. Effect of EPS on ultrastructure of cardiomyocytes and on localization therein of calcium. Calcium localization was determined by release of calcium pyroantimonate (20,000). **a, b** myocardia of control animals; **c, d** myocardium 2 h after EPS; **e, f** myocardiá 48 h after EPS. Explanations in text

antimonate precipitate is absent, one can note that the period immediately following stress is characterized by relatively small changes in cardiomyocytic ultrastructure in the form of swelling of the T-system elements and contracted state of the sarcomers, which is evident in the formation of contraction bands in lieu of the Z lines.

Figure 44d indicates a certain change simultaneously taking place in intracellular calcium localization, this being observed in a greater calcium pyroantimonate precipitate in the subsarcolemmal region, in the sarcoplasma between the myofibrils, and around the mitochondria. An increase in the amount of dispersed precipitate may also be noted directly in the myofibrils, especially in the contraction band zone.

The main result was obtained in studying the ultrastructure of cardiomyocytes 48 h after stress (Fig. 44e, f). From Fig. 44e, where the precipitate of calcium pyroantimonate is absent, one can see that the majority of cardiomyocytes show a considerable increase in the number of widened elements of the T systems and the SR. The mitochondria are also noted to be compressed in a peculiar way such that their matrix is condensed and the mitochondrial membranes are more osmiophilic. Figure 44e reveals a simultaneously pronounced increase in the calcium pyroantimonate precipitate in the subsarcolemmal region, sarcoplasma, myofibrils, and especially around the mitochondria. In addition, the sites where calcium is deposited around the mitochondria are separated from these organelles by a characteristic light interval. According to Onishi and Onishi [421], the occurrence of this type of light aureole around the mitochondria may be due to contraction of the mitochondria owing to a strong adrenergic effect characteristic of stress.

According to contemporary evidence, the increased calcium content observed in these experiments in the cardiomyocytic sarcoplasma constitutes a common pathologic link in highly varied forms of injuries of the myocardium and becomes the cause of the so-called calcium triad [336]. The calcium triad involves disturbances in myofibrillar relaxation, activation of phospholipases and especially proteases that disintegrate the myofibrillar Z disks, and impaired oxidation and phosphorylation in calcium-enriched mitochondria. Indeed, the experiments performed by Malyshev et al. [292] showed that after stress, disturbed oxidation and phosphorylation in myocardial mitochondria develop simultaneously with the decline in the Ca-transporting activity of the SR systems. The polarographic investigation results presented in Table 30 show the changes in the principal oxidation and phosphorylation parameters in mitochondria isolated from the myocardium after stress. As is apparent from Table 30, oxidation and phosphorylation disturbances during oxidation by the mitochondria of NAD-dependent substrates, glutamate and malate, appear as early as 2 h after stress to increase twofold and attain a maximum after 1–2 days. Indeed, after 2 days the rate of ADP-activated respiration and the index of respiratory control decline by 40%, and the ADP/O_2 ratio, i.e., the index of phosphorylation efficiency, declines by almost one-third.

This set of shifts unambiguously shows that stress results in impaired coupling of oxidation and phosphorylation in the mitochondria. It is further apparent that after deliberate complete dinitrophenol-induced uncoupling of

Table 30. Oxidative and phosphorylating functions of rat heart mitochondria in EPS dynamics (M ± m)

	V_3	V_4	V_3/V_4	ADP/0	ADP/t	V_{DNP}
Glutamate + malate						
Control	173.2 ± 12.3	47.6 ± 5.9	3.95 ± 0.3	2.9 ± 0.2	469.2 ± 19.5	134.8 ± 5.5
2 h after EPS	123.3 ± 5.2[c]	38.3 ± 1.1	3.2 ± 0.1[b]	2.4 ± 0.15	300.9 ± 30.1[d]	69.0 ± 4.2[d]
24 h after EPS	133.4 ± 9.2[b]	54.6 ± 3.0	2.45 ± 0.1[d]	2.0 ± 0.1[c]	275.2 ± 29.2[d]	87.2 ± 3.3[d]
45 h after EPS	104.3 ± 5.5[d]	41.7 ± 4.2	2.4 ± 0.15[d]	2.0 ± 0.1[c]	233.9 ± 11.3[d]	78.9 ± 8.0[d]
72 h after EPS	154.2 ± 6.1	58.2 ± 4.6	2.7 ± 0.2[c]	2.55 ± 0.2	434.0 ± 43.8	96.8 ± 7.4[c]
96 h after EPS	170.6 ± 7.8	38.9 ± 2.6	4.0 ± 0.3	2.4 ± 0.15	421.7 ± 48.0	98.9 ± 6.2[d]
Succinate						
Control	250.5 ± 17.2	87.9 ± 5.7	2.85 ± 0.2	1.95 ± 0.15	524.2 ± 37.5	199.8 ± 11.2
2 h after EPS	225.7 ± 7.8	75.4 ± 4.8	2.9 ± 0.2	1.7 ± 0.1	412.6 ± 26.7[b]	162.7 ± 14.6
24 h after EPS	215.4 ± 19.6	117.6 ± 9.6[b]	1.85 ± 0.15[c]	1.4 ± 0.15[b]	322.6 ± 26.8[c]	119.8 ± 8.4[c]
45 h after EPS	189.2 ± 12.9[c]	102.3 ± 9.8	1.8 ± 0.1[d]	1.3 ± 0.15[b]	279.7 ± 24.4[d]	127.9 ± 7.7[d]
72 h after EPS	240.5 ± 14.7	106.8 ± 11.1	2.0 ± 0.15[d]	1.85 ± 0.2	468.6 ± 31.7	214.8 ± 13.6
96 h after EPS	271.8 ± 19.2	98.4 ± 5.1	2.6 ± 0.15	1.7 ± 0.1	439.4 ± 32.2	161.5 ± 16.5

$n = 10$. [b] $p < 0.05$. [c] $p < 0.01$. [d] $p < 0.01$.

oxidation from phosphorylation, oxygen utilization by the mitochondria of stress-exposed animals was reduced twofold. This means that the changes developing in mitochondria under stress not only lead to impaired coupling but limit oxidation itself, i.e., the transport of electrons in the disconnected respiratory chain. When mitochondria oxidize the flavin adenine dinucleotide (FAD)-dependent substrate, namely succinate, similar though quantitatively less pronounced disturbances in respiration and phosphorylation were observed in stress-exposed animals.

Thus, there are grounds to believe that under EPS the lipid triad and proteolytic enzymes lead to damage to the membranous apparatus of Ca^{2+} transport and that excess Ca^{2+} causes myofibrillar relaxation to become impaired and, via activation of phospholipases and proteases in combinations with lipid hydroperoxides, may injure the mitochondria. Evidently, in real conditions in a whole organism, this set of changes should inevitably lead to disturbance of the contractile function, whose rhythmic implementation is known to be controlled by Ca^{2+} transport. Indeed, the investigations performed in our laboratory on an isolated papillary muscle by the Sonnenblick [529] method, on an isolated working heart by the method of Neely et al. [413], and on an isolated isovolumic heart by the method of Fallen et al. [107] showed the contraction amplitude, the myocardial contraction and relaxation velocity, and also the developing pressure and the stroke volume of the whole hearts of stress-exposed animals to have been significantly reduced by 30%–50% [323]. In terms of this presentation, the evidence indicative of changed heart responses in stress-exposed animals to fluctuations in Ca^{2+} animals to fluctuations in Ca^{2+} concentrations in the fluid washing the heart is of greatest interest [379].

In experiments devoted to this question, the contractile function of and oxygen utilization by a working heart were studied by the method of Neely et al. [413]. The stroke volume and systolic pressure developed by the hearts of EPS-exposed animals proved to be reduced by 10%–20%; as a result, the external work of such hearts in a steady state was decreased by one-third. Oxygen utilization does not differ from that in controls, and heart efficiency is, correspondingly, also reduced by one-third. The principal result of the experiment was, however, not in this decline in function and efficiency but in an abrupt increase in the response of the hearts of EPS-exposed animals to changed Ca^{2+} concentrations in the perfusive solution.

The lower section in Fig. 45 shows that upon the decrease in Ca^{2+} concentration in the perfusate from 2.5 to 1.25 mM, the maximal corresponding decline in the stroke volume of isolated hearts from control animals constituted less than 40%, while the hearts of EPS-exposed animals lowered their strokes to zero, i.e., by 100%. The upper section of Fig. 45 shows that with a subsequent increase in Ca^{2+} concentration from 1.25 to 7.5 mM, the corresponding increase in stroke volume was 20%–40% in controls and 80% in the hearts of EPS-exposed animals. These shifts were pronounced during the initial 3–5 min following changes in Ca^{2+} concentration; subsequently, the heart function in both series of experiments gradually returned to the initial level due to adaptation of the mechanisms of ion transport.

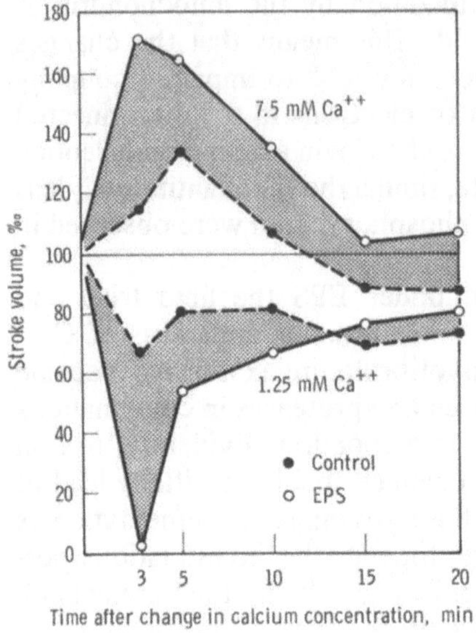

Fig. 45. Response of isolated working heart of control and stress (*EPS*)-exposed animals to changes in calcium concentrations in perfusate. *Below*, response to decrease in calcium concentration from 2.5 to 1.25 mM; *above*, response to increase in calcium concentration from 1.25 to 7.5 mM. *Shaded area* indicates stress action

The curves in Fig. 46 permit one to trace the same EPS effect on heart reactivity to Ca^{2+} concentrational changes by aortal pressure curves. For instance, it was shown that 3 min after the decrease in Ca^{2+} concentration from 2.5 to 1.25 mM, the systolic pressure in control diminishes by about 20 mm, while the heart of an EPS-exposed animal responded to the same decline in calcium concentration by tremendously depressed systolic pressure.

Thus, it is doubtless that after stress, the myocardium develops changes that abruptly enhance the dependence of the heart contractile function upon Ca^{2+} concentration in the washing fluid [379]. Assessing this fact, one should

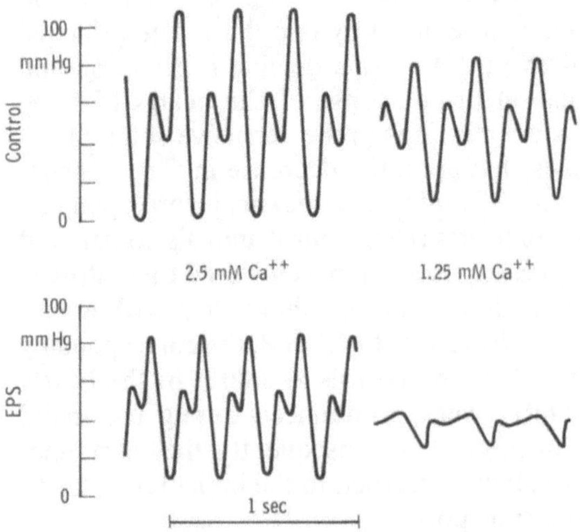

Fig. 46. Effect of reduced calcium concentration in perfusate on pressure developed in aorta by working heart of control and stress (EPS)-exposed animals

keep in mind comparative physiology data concerning the dependence of muscle response to external Ca^{2+} concentration on the power of the cation transport mechanisms. For example, in the skeletal muscle, a developed SR system provides a situation in which almost the entire Ca^{2+} arriving during the action potential in the sarcoplasma and causing myofibrillar contraction subsequently returns to the SR channels. As a result of such 100% Ca^{2+} recirculation, the skeletal muscle may contract in the calcium-free solution for several hours. In the myocardial cells of mammals, the power of the Ca^{2+} transport mechanisms localized in the SR and the sarcolemma is not so high; recirculation is not 100%, and, correspondingly, in a calcium-free solution the heart stops after having made several dozen contractions. Finally, in a frog heart, the SR is developed slightly and the heart stops momentarily after the removal of Ca^{2+} from the solution [403].

The above evidence permits the assumtion that enhanced dependence of the hearts of EPS-exposed animals on external Ca^{2+} concentration results from injury to the membranous mechanisms of Ca^{2+} transport in the myocardial cells, which in turn result from the reduced ability of the sarcolemmal and sarcoplasmic membranes to capture and accumulate Ca^{2+}.

Yet another fact revealed in studying the contractile function of the heart papillary muscle, namely reduced adrenoreactivity, is also apparently due to the EPS influence on cardiomyocytic membranes [368]. The graphs in Fig. 47 show that the papillary muscle in control animals responds to the administration of 10^{-6} M noradrenaline to the perfusate by a considerably increased contraction velocity and amplitude (Fig. 47A, B) and by an even further increased relaxation velocity (Fig. 47C). Correspondingly, the relaxation index, which equals the ratio of the contraction amplitude to the relaxation velocity, increases. For papillary muscles of EPS-exposed animals, the positive inotropic and relaxing effects of noradrenaline proved substantially reduced. Thus, after sustained stress, the myocardial adrenoreactivity was found to decrease considerably.

In assessing the mechanism of this event, one should keep in mind that the noradrenaline effect materializes by means of at least two processes: First, because the noradrenaline acting on an adrenoceptor activates the adenyl cyclase system localized in the sarcolemma and the cyclic monophosphate that forms in increased amounts activates both Ca^{2+} transport and the forming of actomyosin bonds in the myofibrils, and second, due to an increased influx of Ca^{2+} to the myocardial cells. This means that the EPS-induced decrease in adrenoreactivity is caused either by impaired Ca^{2+} influx to the cell and by disturbed myofibrillar response to Ca^{2+} or by malfunctioning of the adrenoceptor – adenyl cyclase system in the sarcolemma. The fact that the positively inotropic effect of excess Ca^{2+} on isolated hearts of EPS-exposed animals is not reduced but, conversely, considerably enhanced, permits the exclusion of the first of these possibilities and we arrive at the conclusion that the post-EPS decrease in heart adrenoreactivity is due to malfunctioning of the adrenoceptor – adenyl cyclase system located in the sarcolemmal membrane.

On the whole, the disorders discovered in contractile function, in reactivity

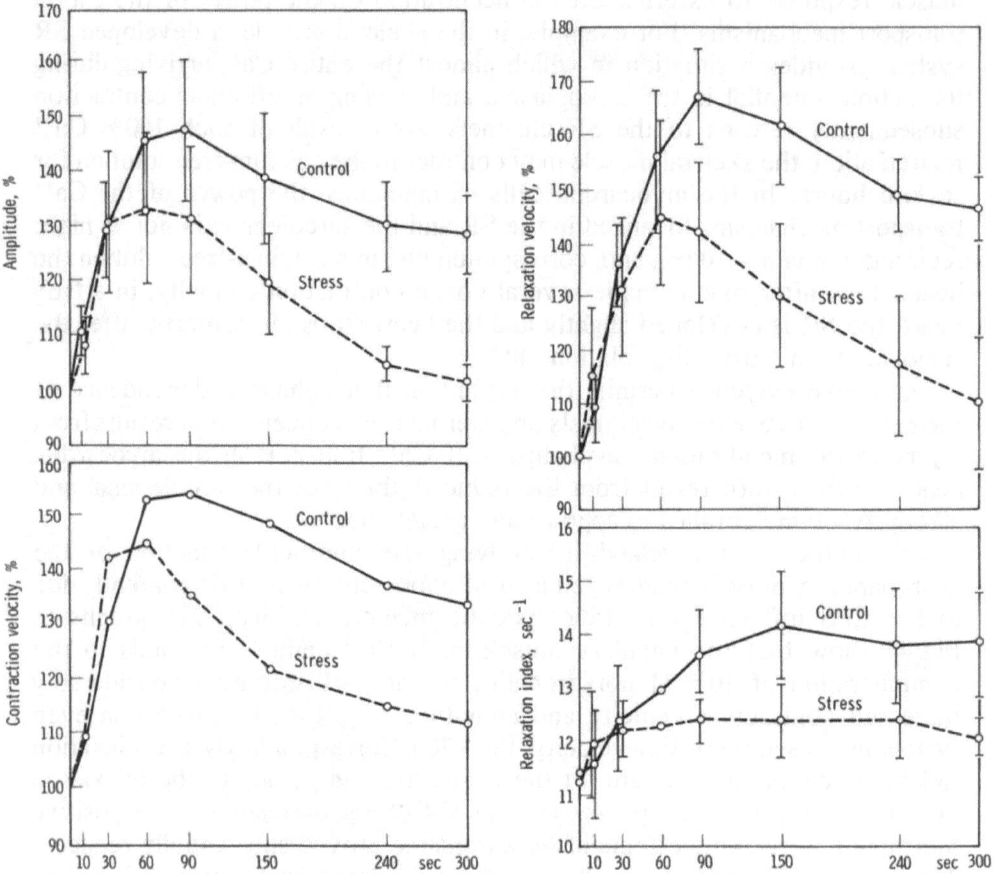

Fig. 47. Effect of sustained EPS on adrenoreactivity of papillary muscle

to Ca^{2+}, and in adrenoreactivity in the myocardium under stress directly agree with the above evidence on damage to the sarcolemma and cardiomyocytic membrane cation pumps and the increase of the Ca^{2+} content therein. Further studies of intact papillary muscles have permitted the comparison of the disturbances in contraction and relaxation observed under stress and have shown that the factor that prevents LP activation under stress is preadaptation to hypoxia, which also prevents all poststress disturbances in the contractile function [390].

The evidence presented in Table 31 shows the comparison of the poststress contractile function of the papillary muscle in adapted and control animals. This defense effect is apparently caused by the fact that preliminary adaptation prevents stress-induced LP activation and, hence, membrane injuries caused by LP products. This adaptation effect agrees with the fact that antioxidants, LP inhibitors, prevent stress-induced disturbances in cardiac function and

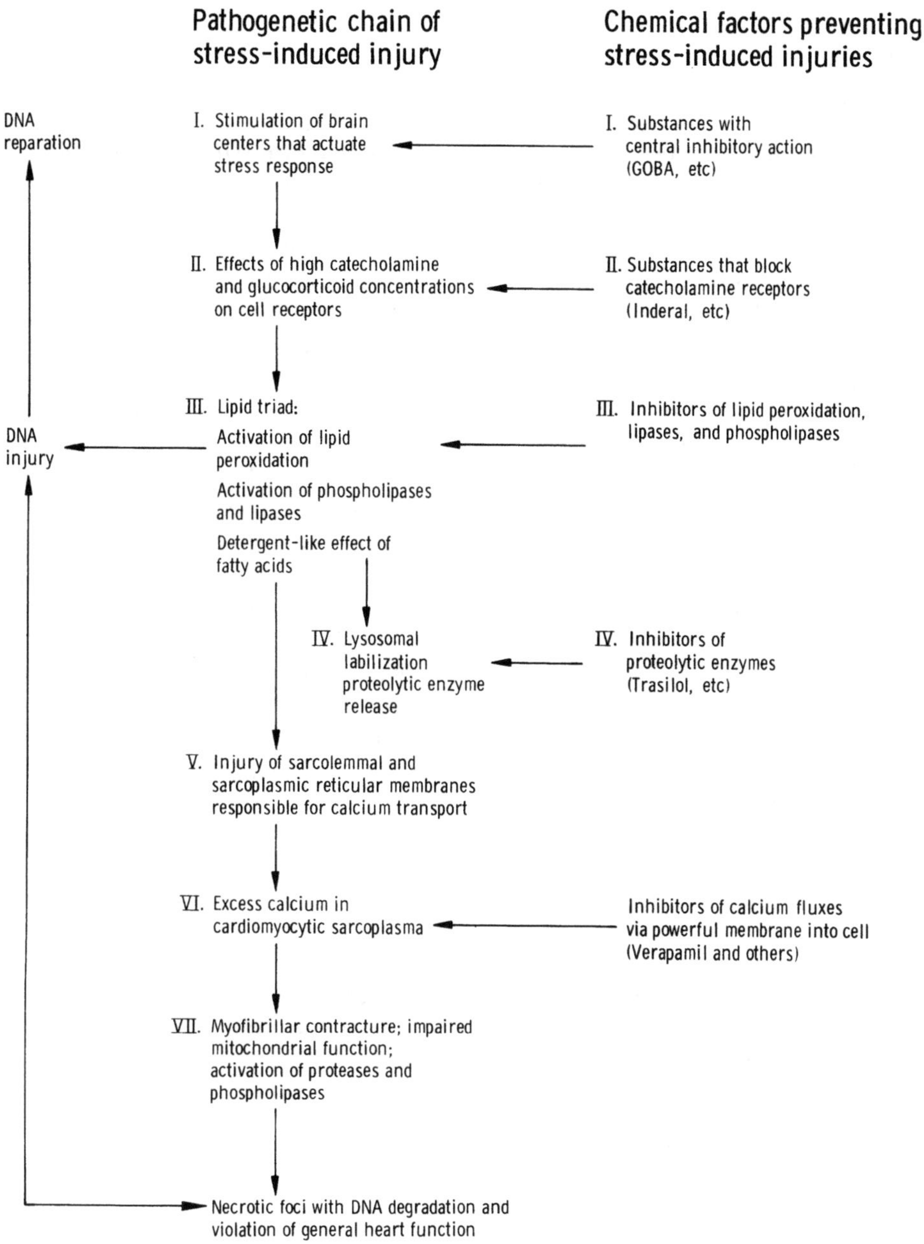

Fig. 48. Pathogenesis of stress-induced heart injury. Explanation in text

heart injuries, estimated by CPK release, as radically as adaptation to hypoxia (see Chap. 8).

On the whole, the presented experimental data lead to the hypothesis on the pathogenesis of stress-induced injuries of the myocardium in Fig. 48. The diagram in Fig. 48 shows that under stress situations, excitation of the higher vegetative centers that determine stress reaction (first link) leads to a manifold increase in the catecholamine concentration affecting the heart (second link). In addition to the usual activation of the adenyl cyclase system, this leads to the lipid triad cited above, i.e., first of all there is LP activation and then phospolipase and lipase activation and also the detergent action of high fatty acid concentrations (third link). The lipid triad apparently forms the basis of lysosomal membrane injury, as a result of which proteolytic enzymes and nucleases are released from the lysosomes (fourth link). The effects of the lipid triad and lysosomal enzymes result in damage to the sarcolemmal and SR membranes responsible for calcium transport and in impaired functioning of the cation pumps and increased permeability of the sarcolemma for calcium. At the same time, ruptures occur in the nuclear DNA molecules; in most nuclei, these ruptures are eliminated by repair and in some cases lead to disintegration of DNA and the death of cells (fifth link). Subsequently, increased permeability of the sarcolemma for calcium and malfunction of the calcium pump result in excess calcium in the cardiomyocytic sarcoplasma (sixth link). This shift has two consequences. First, it may activate the complex of processes comprising the lipid triad and thus close the vicious circle that enhances myocardium injury. Second, the excess Ca^{2+} has its own harmful effect: It causes a set of changes inside the cells previously denoted as the calcium triad [336]. This triad consists of myofibrillar contracture, disturbed function of calcium-overloaded mitochondria, and the activation of myofibrillar proteases and mitochondrial phospholipases, which aggravates injury (seventh link). This results in the irreversible contracture and necrobiosis of individual cell groups and in pronounced disturbances of the entire contractile function.

This concept of the role of excess calcium in the final link of the pathogenic chain of stress-induced heart injury implies that the lipid triad, induced by excess catecholamines, leads to injury to the membranous mechanisms responsible for active bonding of Ca^{2+} and its transport against the concentrational gradient into the environment. As a result, excess Ca^{2+} appears in the sarcoplasma, which itself becomes the cause of further damage, namely, myofibrillar contractures and impaired ATP resynthesis in the mitochondria.

In effect, this signifies that the end links in the pathogenic chain of stress damage are not the original attributes of stress, for quite similar events take place when the myocardium is injured by highly diverse factors. Indeed, comparisons of data on membranous Ca^{2+} transport in heart hyperfunction and hypertrophy (see Chap. 3), ischemia (Chap. 8), hypoxia [83, 261], myocardial reoxygenation [442], toxic injuries [549], and hereditary myocardiopathies [9, 82] have in the past led us to assume that acting via various pathogenic mechanisms, the principal etiologic factors of heart diseases in the end lead to one and the same result. This phenomenon involves decreased power, shut-off,

Table 31. Effect of prior adaptation to hypoxia and of EPS on the contractile function of isolated papillary muscles of the left ventricle of rat hearts

Index	Control $n = 10$	Stress $n = 11$	Changes, %	Adaptation to hypoxia, $n = 11$	Adaptation to hypoxia + stress, $n = 10$	Changes, %
Cross-sectional area, mm^2	0.82 ± 0.05	0.98 ± 0.06	+19	0.78 ± 0.04	0.78 ± 0.07	0
Contraction amplitude, % of initial length	8.41 ± 0.42	4.91 ± 0.25[a]	−42	9.71 ± 0.49	7.80 ± 0.39[b]	−20
Maximal contraction velocity, muscle units/s	0.93 ± 0.04	0.52 ± 0.03[a]	−44	0.96 ± 0.05	0.87 ± 0.04	−9
Maximal relaxation velocity, muscle units/s	0.80 ± 0.04	0.54 ± 0.03[a]	−33	0.93 ± 0.05	0.81 ± 0.04	−13
Relaxation index, s^{-1}	8.97 ± 0.45	10.72 ± 0.54[a]	+20	9.34 ± 0.47	10.48 ± 0.52	+12
Time un til contraction peak, ms	177 ± 9	149 ± 7	−16	180 ± 9	196 ± 9	+3

[a] $p < 0.05$ in relation to control.
[b] $p < 0.05$ in relation to adaptation to hypoxia.

or damage of membranous mechanisms responsible for Ca^{2+} bonding and its ative removal from the myocardial cell against the concentration gradient into the extracellular space. This results in excess Ca^{2+} in the sarcoplasma and causes a standard set of shifts that does not depend on what had previously caused higher Ca^{2+} content. This unity of the end links in the pathogenic chain of stress-induced heart injury and in its injury by such traditional etiologic factors of cardiac insufficiency as long-term overload or hypoxia is a strong argument favoring the view that recurring stress may become an independent cause of chronic heart insufficiency. Indeed, the metabolic, structural, and functional disturbances that we discovered were observed after the stress situation was over; they represented not simply responses to stress factors, but relatively stable consequences of stress-induced damage. This fact in conjunction with vast clinical data on the role of emotional stress in the etiology of circulatory diseases affords grounds to assume that the relatively stable disturbances of metabolism and function that continued to persist after stress was over, and those that accumulate from one stress episode to another, may play a role in the gradual development of the forms of the so-called primary noncoronarogenic cardiosclerosis and chronic cardiac insufficiency, which often occur in persons previously unaffected by diseases of the circulatory system. Quite probably, however, the described complex of stress-induced injuries may overlap with the changes that had taken place in the heart as a result of basic circulatory diseases and thus potentiate the disruption of compensation and the development of cardiac injury and insufficiency during the ischemic disease; this question will be examined below.

It should be emphasized here that the given scheme of pathogenesis of stress injury in Fig. 48 approximately characterizes not only the basic pathogenic links of stress-induced heart injury but also the basic possibilities for preventing the latter. The right-hand side of the scheme lists the factors by means of which we selectively inhibited each of the links in the pathogenic chain of stress-induced heart injury. One can see that inhibited excitation of the brain centers determining the stress reaction by means of an inhibitory metabolite, namely by gamma-oxybutyric acid, Inderal-blockaded heart adrenoceptors via which the catecholamine effect comes about, blockade of the lipid triad by inhibitors of LP, phospholipases, and lipases, Transilol-inhibited lysosomal enzymes, and, finally, verapamyl-blockaded calcium influx in the cells are the five groups of factors that all differently, but with invariably conclusive efficiency, prevent stress-induced injuries.

In this context, this is significant chiefly because it confirms the above idea on the pathogenesis of stress injury. The factual evidence on the prevention of stress damage is discussed in the next chapter.

Stress: A Basic Etiologic Factor of Noninfectious Human Diseases. Its Role in the Origin of Ischemic Heart Disease and Tumors

Similar to the way the adaptational effect of stress (see Chap. 2) encompasses the organism as a whole, the harmful effect of protracted and excessively intensive stress during unrealized adaptation is not restricted to the heart, but is addressed to the whole organism. The stress activation of peroxidation in numerous organs and the pronounced stress fermentemia cited above, indicative of direct cell injury, permit a more specific estimate of the role of stress in the pathogenesis of certain major pathologic processes. Among other things, this relates to contemporary concepts on the role of stress in the origin of major endogenous or noninfectious diseases of the century, i.e., ischemic heart disease and tumors.

The fact that ischemia is now widespread and even marked by an epidemidogy [58] in industrially developed countries has made the elucidation of its etiology a major issue in modern medicine. Also, among the etiologic factors causing ischemic heart disease, stress situations in reality occupy an important place in addition to excess cholesterol in food, excess weight, limited motor activity, and hereditary disorders in lipid metabolism. Hence, major importance is assumed by the largely vague question as to what specific mechanisms responsible for long and protracted stress reactions may actuate or potentiate the pathogenic chain of the ischemic disease. At present, one can distinguish several such mechanisms, which may be actuated *primarily,* i.e., exogenously as components of stress reactions caused by emotional environmental situations, or *secondarily,* i.e., endogenously as components of EPS caused by stenocardia or myocardial infarction themselves. Irrespective of whether or not the stress reaction is caused exogenously or endogenously, one may distinguish at least seven such mechanisms.

The first pathogenic mechanism of this kind is stress hypercholesterolemia, hyperlipidemia, LP activation, and other disturbances in lipid metabolism engendered by the excessively intensified primarily adaptive, lipotropic effect of stress. Experimental data showing that neurogenic, essentially stress-induced animal atherosclerosis may be obtained by frequently changing the stereotypes of conditioned-reflex activity and motor activity, and also by emotional excitation caused by intermittent hunger and protracted electrostimulation of the ventromedial hypothalamus nucleus appeared relatively long ago [218–220]. This evidence seems to correspond to the results of epidemiologic studies, in which it was noted that in humans whose profession requires constant emotional stress, the incidence of atherosclerosis of coronary vessels, ischemic heart disease, and myocardial infarction caused as a result is especially high [18, 136]. For example, the death rate from ischemic heart disease among surgeons was found to be ten times higher than in gardeners [489].

Stress-induced lipemia, specifically hypercholesterolemia, which promotes cholesterol deposits in vessel walls, is usually regarded as the principal mechanism through which stress can potentiate atherosclerosis [37, 130]. Yet, the considerable stress-induced activation of lipid peroxidation in the aortic wall, a fact we have established, provides grounds for assuming that in addition to hypercholesterolemia, direct injury by LP activation of the vessel wall plays a major role in developing stress-induced injuries in the vascular bed. Indeed, we presently know of an experimental model of peroxide injury of the vascular bed, usually denoted peroxide arteriolosclerosis and caused by the so-called antioxidant-free diet enriched with lipid peroxides. In this case, the developing injury of the vascular wall is essentially based on the degradation by lipid peroxides of the elastic framework of the vessel with subsequent calcinosis and lipoidosis; this process may be accelerated by animal immobilization and inhibited by the antioxidant ionol [81, 569]. At present, it is difficult to decide how disturbances in lipid metabolism and injury to the vessel wall by lipid peroxides are combined in the pathogenesis of human atherosclerosis. However, it should be recognized now that stress potentiates both factors and may, therefore, play a role in the development of atherosclerosis of the coronary vessels and of the ischemic disease.

The second pathogenic mechanism, as a result of which stress may potentiate the ischemic disease, is primary stress injury of the myocardium, i.e., it involves a pathogenic chain to which the entire previous section of this book is devoted (see scheme in Fig. 48). Presumably, this type of noncoronarogenic stress injury of the myocardium may play a definite role, first in the occurrence of cases of the so-called ischemic disease, in which coronary vessels are not found to be stenotic and, second, in the progression of myocardial infarction, the principal disease of ischemic origin, in which the EPS induced by infarction may be a major factor causing injury of the nonischemic regions of the heart. This last situation deserves attention since the latest studies of Hearse [167] show the demarcation line between the necrotic zone and the ischemic noninjured myocardium to be highly distinct. There are a number of similar works; as a result, the very existence of the so-called perinfarction, a partially ischemic zone, which was a major object of remedial measures for a whole generation of cardiologists, has suddenly become a debatable issue [221]. At the same time, there is no doubt that heart injury in myocardial infarction is not restricted to the ischemic zone. Hence, one must reckon with the possibilitiy that injuries of the nonischemic zone may essentially be not only ischemic, but even stress-induced. To verify this, Dosmagambetova et al. (unpublished data) ligated the descending branch of the left coronary artery to reproduce experimental infarction of the left ventricle in rats and to study the region of the heart that with this operation could not a priori become an object of ischemic injury, namely the right auricle. It became clear that 24 h after infarction was induced, a set of stress-induced changes described in detail above developed in the auricular myocardium and the isometric pressure developed by that auricle functioning in isolation proved to have decreased by about 2.5-fold. The principal fact obtained from this work was that these structural and functional disorders may have largely been prevented by administering Inderal, and also inhibitors of LP, lipases, and phospolipases (see Chap. 8), prior to producing infarction. This permits one to assume that the EPS arising with infarction may really, among other factors, injure the nonischemic regions of the myocardium; correspondingly, such stress-induced injuries and their prevention play a role in the fate of patients with myocardial infarctions.

The third pathogenic mechanism is apparently that the strong adrenergic component of the stress reaction may lead to spasms of smooth muscles in anatomically unchanged coronary arteries and these quite stable spasms cause secondary ischemic injury to the myocardium. This phenomenon is in many respects essentially vague for, in anesthetized animals, a coronary vessel spasm caused by the stimulation of adrenergic nerves is naturally replaced by metabolic vasodilation, which, according to modern concepts, is caused by the fact that the adrenergic, positive inotropic effect is accompanied by increased ATP utilization and the release from cardiomyocytes of adenosine, a powerful vasodilator.

With the intercoronary administration of catecholamines to conscious animals, one may observe vasoconstrictory, vasodilatory, and phasic responses [439, 562]. Accordingly, it is known that coronary vessels have two types of adrenoceptors: (a) α-receptors, whose stimulation leads to vasoconstriction, and (b) β-receptors, whose activation leads to the dilation of coronary vessels. All coronary arteries have the β-receptor apparatus, irrespective of order, and only the large arterial trunks possess considerable α-receptor activity [477].

The impression is that adrenergic spasm or, conversely, vasodilation depend on the ratio of various catecholamines in the blood, on the ratio of alpha- and beta-receptor activity, and finally, on the rate and intensity of the vasodilative effects of adenosine and prostaglandins, whose formation also increases under the catecholamine action on the target organs (Chap. 7). In terms of our presentation, it is significant that in patients with ischemic disease, the spasm of the coronary vessels is a real, unambiguously proven fact. The role of this link in the pathogenesis of heart ischemia has become obvious due to the extensive introduction of coronarography. Investigations involving this technique showed that in a considerable number of patients, symptoms of ischemic disease develop in the absence of stenotic atherosclerosis, thrombosis, or embolism of the coronary arteries. According to different authors, the number of these patients may reach 10%–15% [142] and even 25% [459] of the total number of persons affected by the ischemic disease. Among the probable causes of the ischemic disease in these patients the most frequently encountered is smoking [512], which is known to activate adrenergic effects on the heart.

From these preliminary clinicophysiologic data, some authors have carried out investigations in which they visually observed a stable spasm of human coronary arteries. For example, Petrosyan et al. [438] showed that, in 4 of 36 patients with typical clinical symptoms of myocardial infarction, the administration of nitroglycerin during emergency coronaroangiography resulted in the observed complete occlusion of one of the large coronary arteries. Similar results were obtained with heart ischemic disease in a number of other studies, too [303]. The role of coronary spasm in the origin of Prinzmetal-type stenocardia [451], in which the attacks tend to occur at night and in the morning, may be provoked by stress actions [124, 591], and are apparently due chiefly to insufficiently effective vasodilative effects of adenosin and prostaglandins. In assessing the role of stress-induced spasms of the coronary vessels, it should be emphasized that only persistent spasms may have pathogenically injurious significance since, unlike normal regulator increase in vascular tone, it is not eliminated by physiologic dilators, i.e., by adenosin, group E prostaglandins. In effect, this should not be simply a transitory spasm, but a phenomenon approaching contracture of the smooth muscles of coronary arteries, i.e., a phenomenon based on disturbed relaxation. Hence the idea that the chain of events that we have examined in detail above for stress-exposed heart striated muscle also takes place in vascular smooth muscles. Contractural spasms of these muscles result in focal injuries of the heart and other very different organs typical of stress. At present, this course of events has been proven for the stomach ulcers mentioned above resulting from ischemic necroses of the mucous coat owing to adrenergic spasms of vessels and subsequent digestion of the ischemic sites by that coat. In full conformity with this, the chemical factors that we used to prevent stress damage not only prevent heart injuries but stress-induced stomach ulcers as well (see Fig. 48). One can readily imagine that a similar event may take place under serious stress in the smooth muscles of coronary arteries, where the pathogenic chain cited above (see scheme in Fig. 48) leads to persistent contracture-type spasms and, consequently, to ischemic lesions of the myocardium.

The fourth pathogenic mechanism, through which stress can potentiate or cause ischemic injuries involves catecholamines and various forms of stress [156]; these activate blood coagulation and, under certain conditions, may promote the development of thrombi. The resultant aggregation of thrombocytes leads to the release of powerful vasoactive substances from the blood platelets, viz prostaglandin G_2 and thromboxane A_2, which a priori enhance the spasm and prolong its duration [101, 418]. This is indicative not only of the relationship between spasms and thrombi in large coronary vessels but also explains the violations arising under the influence of strong adrenergic effects in the myocardial microcirculatory system. For instance, it was shown that the administration of large doses of isoproterenol, a synthetic analog of noradrenaline possessing chiefly β-adrenergic effect, sharply increased the functional intercapillary distance, i.e., the distance between the neighboring perfused capillaries [280]. Restriction of the α-receptor apparatus in the myocardial microcirculatory system and the preferentially β-receptor effect of isoproterenol make it difficult to explain this fact on the basis of the concept of a primary capillary spasm. More probably, violated perfusion in capillaries is in this case due to thrombotic changes therein. In this connection, it should be noted that the adrenal cortex hormones that in addition to catecholamines take part in developing a stress reaction, also affect the rheological properties of the blood: They reduce the coagulation time and fibrinolytic activity [513] and this may sharply worsen the microcirculation in the myocardium and intensify thrombogenesis in all the regions of the coronary system.

The fifth mechanism through which stress is realized in the pathogenesis of ischemic and generally hypoxic injuries was revealed in recent experiments and it involves a protracted stress reaction reducing the resistance of the heart to hypoxia. The evidence presented in Table 32 reflects experimental results obtained by Vorontsova, Pshennikova, and ourselves [390a], which permit the effect of sustained EPS on the contractile function of the myocardium of isolated auricles to be assessed. In the beginning, these auricles

Table 32. Effect of sustained EPS on the resistance to hypoxia of the parameters of the contractile function of rat auricle

Contractile function parameters	Version (number of animals)	Experimental stages						
		Prior to hypoxia	Hypoxia			Reoxygenation		
			1 min	5 min	20 min	1 min	5 min	20 min
Developing pressure, mg	Control, $n = 20$	354 ± 14.5	123 ± 10	14.5 ± 5	0	210 ± 13	253 ± 11.9	252 ± 11.4
	Stress, $n = 16$	201 ± 8.7	69.3 ± 7.5	8.4 ± 3.1	0	126 ± 19.1	131 ± 16	139 ± 9.3
Contraction frequency per minute	Control	250 ± 4	181 ± 4	113 ± 15	0	188 ± 6	238 ± 10	243 ± 4
	Stress	255 ± 5	188 ± 5	65 ± 11	0	166 ± 16	217 ± 15.5	244 ± 4
ISF[a], g/mg × min	Control	3.6 ± 0.20	0.90 ± 0.08		0	2.1 ± 0.13	2.42 ± 0.14	2.52 ± 0.14
	Stress	2.01 ± 0.16	0.51 ± 0.05	0.04 ± 0.001	0	1.01 ± 0.2	1.15 ± 0.13	1.38 ± 0.1
Rest tension increment, mg	Control	–	13 ± 3	219 ± 27.50	510 ± 35	148.5 ± 20	97 ± 12.5	35.5 ± 9.0
	Stress	–	32.2 ± 6.5	244 ± 33	602 ± 20	224 ± 24	152 ± 17	51.4 ± 10
Contracture index, %	Control	–	4 ± 0.7	63 ± 8	139 ± 9	43 ± 5	23 ± 3	10 ± 3
	Stress	–	18 ± 4	130 ± 16	312 ± 15	119 ± 10	78 ± 10	28 ± 8

[a] ISF, intensity of structure functioning equal to ratio of product of tension by frequency to auricular mass.

contracted isometrically in aerobic conditions, then under anoxia, and finally under reoxygenation. From Table 32 it can be seen that in aerobic conditions, the developing tension or auricular contraction intensity of stress-exposed animals was reduced almost twofold; spontaneous contraction frequency was substantially unchanged and the integral index characterizing the intensity of structure functioning (ISF) of the myocardium was correspondingly decreased by more than 40% (see Table 32). The auricular myocardium responded to hypoxic action by depressed contractile force and elevated tension of rest, i.e., by hypoxic contracture. For auricles of stress-exposed animals, both these phenomena were much more pronounced than in controls. For instance, hypoxia-induced tension of rest in controls increased from 13 mg during the 1st min of hypoxic action to 510 mg in the 20th min. For the auricles of stress-exposed animals, the increase in the tension of rest in the course of hypoxic contracture was approximately 1.2–2 times higher than in controls. During reoxygenation, hypoxic contracture disappeared and contractile functions became restored in the auricles of poststress animals much slower than in controls. Further experiments showed that this poststress decline in myocardial resistance to hypoxia could be reduced by increasing the glucose concentration in the perfusive medium or by adding to that medium uridine, a cofactor of glycogen resynthesis. Hence, the stress-induced disturbances in glycolysis played an important role in the poststress disturbance of the heart contractile function and in the decrease in its resistance to hypoxia [390a]. In terms of our presentation, it is significant that the poststress decline in myocardial resistance to hypoxia is a real phenomenon and may, consequently, be one of the factors that potentiate ischemic injury of the heart.

The sixth mechanism or component of the stress reaction that may potentiate ischemic injury of the heart is the adrenergic mobilization of its contractile function. In combination with the regulatory-determined increase in resistance of the vascular bed, this creates a considerable load on the heart and may substantially potentiate ischemic injury not only by spasms or thrombosis but even by simple stenosing of coronary vessels in an atherosclerotic process.

Finally, the seventh and last pathogenic mechanism of stress-potentiated ischemic injury involves a component of the stress reaction – hyperventilation, which naturally entails higher pO_2 in the blood and hypocapnic alkalosis. In turn, these shifts decrease the coronary blood flow [479]. In addition, a decrease in H-ion concentration in the blood during alkalosis causes the coronary arterial tone to rise since this increases the accessibility of Ca^{2+} to troponine binding sites in the myofibrillary apparatus of the myocytes of vessels and leads to their subsequent contracture [114]. Indeed, clinicophysiologic studies have shown that hyperventilation and concomitant hypocapnic alkalosis lead in IHD patients to a decline in coronary blood flow of 15% [410] and to attacks of stenocardia [145]. It is characteristic that these attacks were arrested in 100% of the cases by administering dilthiasem, a Ca^{2+} blocker, which in this case did not affect the blood gas composition in the coronary bed: pCO_2 was reduced and pO_2 remained increased by 20 mm Hg. During hyperventilation, coronary vasoconstriction is also observed in virtually healthy people [479].

Thus, the above evidence is not only indicative of the role of stress hyperventilation in the occurrence of stenocardia attacks but reemphasizes that this pathogenic mechanism, like all others, is realized via the same final link, viz through the occurrence of excess Ca^{2+} in the myocytic sarcoplasma of the smooth muscles or the myocardium and through multiform harmful effects of the given shift.

This brief survey of the mechanisms at the expense of which an excessively intense and protracted stress reaction may play a role in the pathogenesis of heart ischemic disease is summarized in Fig. 49. The scheme shows that the harmful factors examined – components of the stress reaction – may equally occur both with stress exogenically induced by difficult environmental situations as well as by endogenically caused EPS recurring because of the appearance of an ischemic painful focus in the heart and fear of death.

On the whole, the assumed concept denotes that excess stress reaction is an inevitable pathogenic link in ischemic disease. Correspondingly, restric-

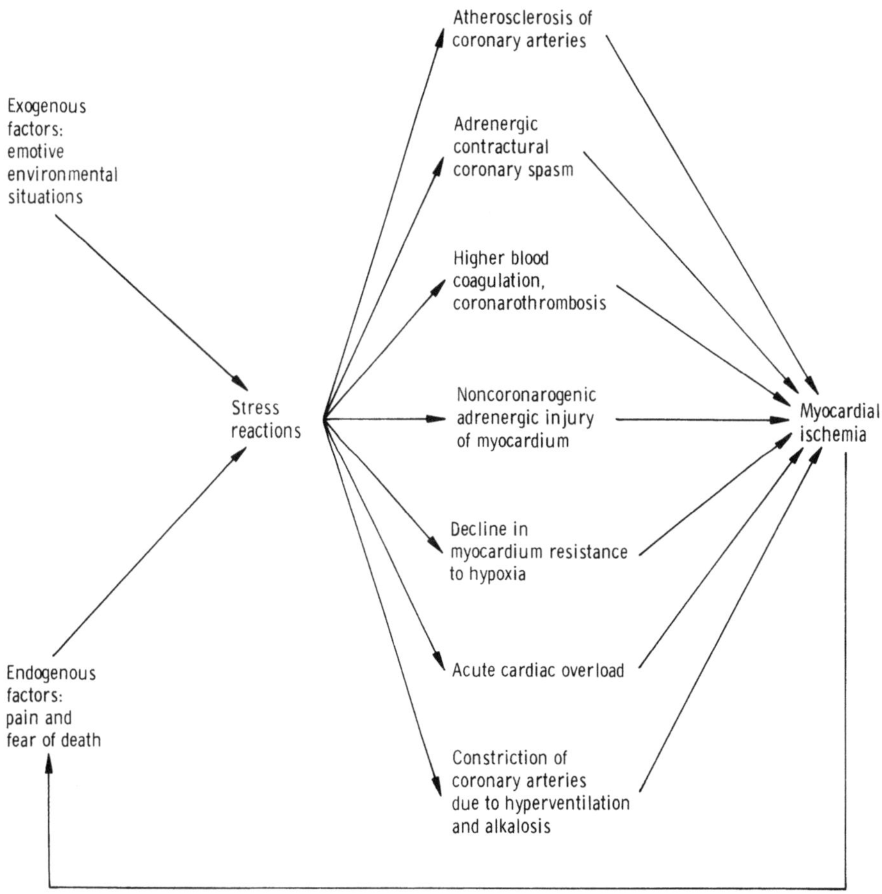

Fig. 49. Role of excessively intensive and long stress reaction in the origin of ischemic heart disease. Explanations in text

tion of the stress reaction by modulatory systems of the organism itself or by the active interference of man plays an important role in preventing not only stress-induced, but also ischemic injuries. This close relationship between stress-induced and ischemic injuries and also the resultant similarity in the factors that protect the organism, specifically the heart, from those injuries, becomes increasingly obvious when comparing the cell link in the pathogenesis of heart injuries under serious stress and myocardial infarction. The evidence on the basic links in the pathogenesis of stress-induced cell injuries was presented above (see scheme in Fig. 48). The existing concept on the pathogenesis of myocardial infarction is far from clear and is rapidly developing. The scheme shown below is our working hypothesis which combines facts obtained in recent years[2]; in accordance with the given concept, the development of a pathogenic chain of myocardial infarction involves four stages (see Fig. 50).

The first stage is designated as that of actuation of the basic pathogenic factors. It is apparent from the scheme (Fig. 50) that two etiologic factors – environmental stress situations on the one hand and hereditary and alimentary disturbances in lipid metabolism and regulation on the other – form at least three primary shifts in the organism. These are stenosing atherosclerosis of coronary vessels, the stress reaction with its components cited above, and changes in the rheological properties of blood and more or less pronounced thrombosis of the coronary vessels. The stress reaction mediated by coronary vessel spasm as well as other components and the two other shifts directly lead to myocardial ischemia, the principal phenomenon constituting the essence of the disease. In addition to its principal result, i.e., shutoff of the mitochondrial respiratory chain, myocardial ischemia naturally causes pain to intensify or reactuate the stress reaction and to enhance or engender considerable excitation of adrenergic regulation and excess catecholamines in the organism. In this sense, one should regard as well grounded the concept developed by the clinic of Chazov [58] that infarction always involves stress, and we can only add that this stress is essentially emotional-painful, i.e., formed by pain and fear of death. The scheme in Fig. 50 shows that the outcome of the examined stage consists of two events: Shutoff of the mitochondrial respiratory chain in the ischemic site of the myocardium and catecholamine excess determined by the stress reaction.

These two shifts engender the second stage of the process, namely that of primary ATP shortage, inhibition of the tricarbonic acid cycle, glycolysis, ATP-dependent cation pumps, and the accumulation of fatty acids. The scheme in Fig. 50 shows that shutoff of the respiratory chain in this stage resulted in inhibition of the tricarbonic acid cycle and the shortage of energy-rich phosphoric compounds. Due to the now known chain of events,

---

2 See: Abstracts of the Meeting of the American Section of the International Society for Heart Research, Pasadena, California, 13–14 May, 1977; Abstracts for the Ninth Congress of the International Society for Heart Research, New Delhi, India 28 Sept.–2 Oct., 1978; Abstracts of the Second Meeting of the Japanese Section of the International Society for Heart Research, Osaka, Japan, 3–4 Nov. 1979; Abstracts for the Tenth Congress of the International Society for Heart Research, Moscow, USSR, 23–28 September, 1980.

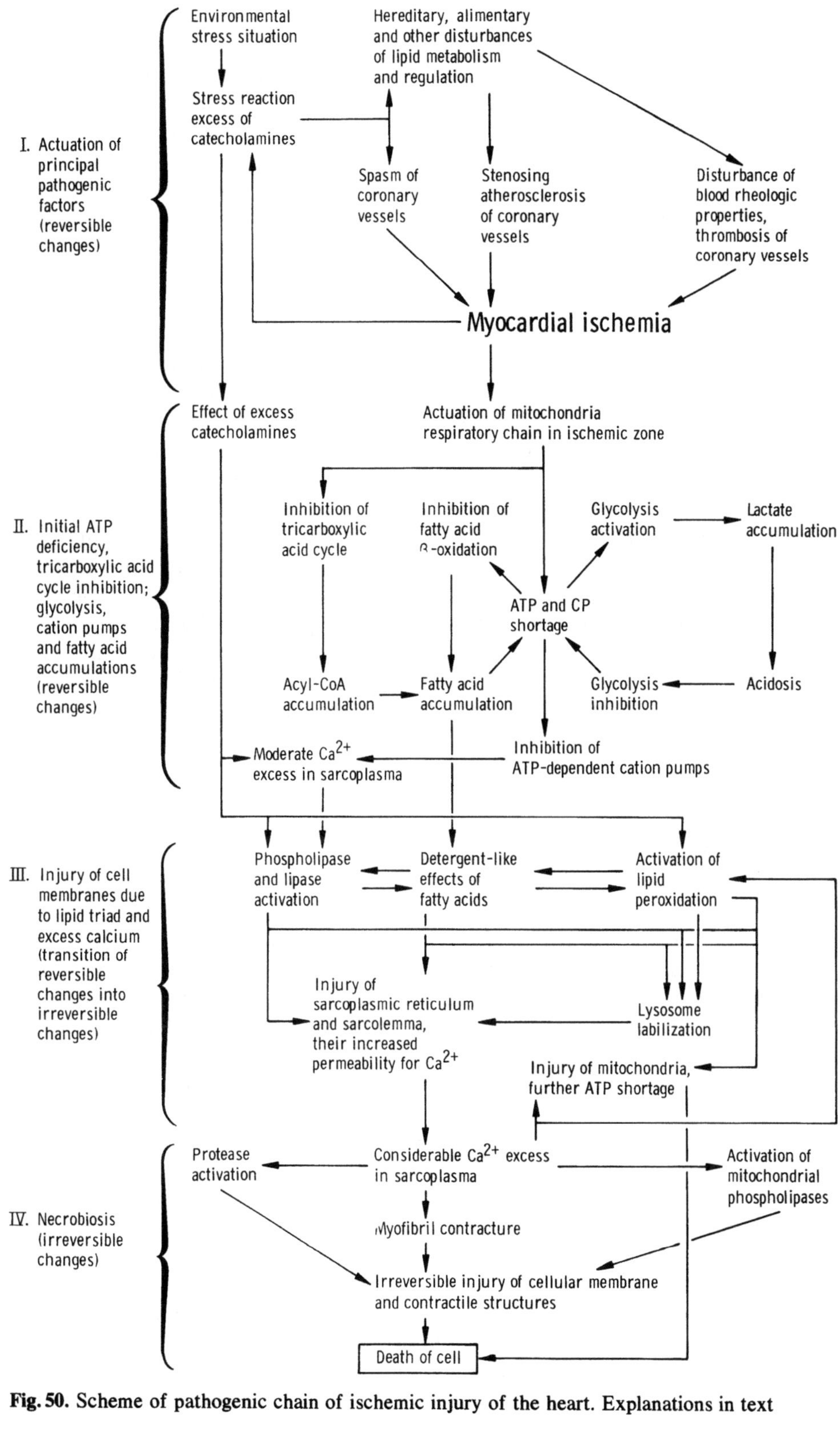

Fig. 50. Scheme of pathogenic chain of ischemic injury of the heart. Explanations in text

these two shifts cause activation and subsequently inhibition of glycolysis and fatty acid accumulation. At the same time, the slow channel for electrogenic Ca^{2+} influx into the cardiomyocytes and the instant reversible depression of the heart contractile function are inhibited; later, this is followed by inhibition of ATP-dependent cation pumps and the resultant initially moderate excess Ca^{2+} in the sarcoplasma.

All the above changes are basically reversible; however, they in turn cause major changes in the lipid bilayer of the myocardial cell membranes. These changes represent the previously examined 'lipid triad' of membrane injury and constitute the essence of the third stage of the process shown in the scheme (Fig. 50), namely, the stage in which reversible injury of the myocardium becomes irreversible. It is clear that the lipid triad is essentially potentiated by excess catecholamines and involves activation of lipases and phospholipases, detergent action of excess fatty acids and lisophospholipids on membranes, and activation of lipid peroxidation. Assessing the components of this triad, one should keep in mind two circumstances. First, LP and phospholipases are activated within cardiomyocytes and the third component of the triad acts due to the activation of lipases by catecholamines in fatty and other extracardial tissues, the concentration of free fatty acids that affect heart membranous structures increasing only as a result of this activation in the blood in infarction and concomitant stress. Second, unlike lipid hydrolysis, lipid peroxidation as a component of the membrane-injuring triad is irreversible since the cell does not possess any mechanisms capable of utilizing the LP final products. The scheme (Fig. 50) further shows that development of the lipid triad leads to labilization of the lysosomes, to injury to the sarcolemma, SR, and mitochondria, and to increased permeability of these membranous structures to calcium. Evidently, the reversibility or irreversibility of these injuries depends on their extent.

With protracted ischemia, injuries to cell membranous structures progress to predetermine the fourth stage of the process, the stage of necrobiosis. The scheme shows that the excess Ca^{2+} resulting from injury of the sarcolemma and other cell membranous structures leads to myofibrillar contracture and degradation by myofibrillar proteases; it induces additional activation of the lipid triad and thereby stimulates membrane deterioration. As an outcome, all the principal cardiomyocytic organelles become irreversibly injured; these cells die, their death being the cause of irreversible termination of the contractile function of an ischemic myocardium and of myocardial infarction.

Assessing this concept on the pathogenesis of myocardial infarction and comparing it with the pathogenesis of stress-induced injury of the heart (see scheme in Fig. 48), one can readily appreciate that these two pathogenic chains in addition to important differences have a number of common links. These are: Presence of the stress reaction and the harmful effect of excess catecholamines; lipid triad of membrane injury and LP activation as its most substantial component; and excess Ca^{2+} in the sarcoplasma and its harmful effects. We shall examine further some of these components in greater detail. In this context, it is significant that the close relationship between the stress-induced and ischemic injurious factors makes it understandable that the organism restricts their action by means of the same or similar modulating systems and that in simulating the organism the physician often seeks to prevent both stress-induced and ischemic injuries by the same factors – central inhibitory metabolites that remove the stress reaction, beta-blockers, LP or lipase inhibitors, Ca^{2+} antagonists, etc. We shall take this essential factor into account in Chap. 7 when we examine antistress systems and the principles of preventing stress-induced and ischemic injuries. Here, it is necessary to emphasize that the role of the stress reaction in IHD, the disease of this century, is undoubtedly large and this must be taken into consideration in its prevention and treatment.

The role of the stress reaction in the etiology of tumors, the second in significance in the of group of noninfectious human diseases, long ago attracted attention because of empirical clinical experience, subsequently reflected in systematic epidemiologic investigations. For example, Le Shan [267] and Le Shan and Worthington [268, 269], basing their work on thorough studies of humans affected by cancer, arrived at the conclusion that in the 2–4 years immediately prior to the disease many patients had sustained serious events in their lives, most frequently such events as the death of a close relative or friend, which had caused a state of considerable depression. These studies were of a retrospective nature, i.e., they were performed by examining people already suffering from cancer. Hagnell [157] performed so-called prospective examinations of 2,550 residents of Sweden who, when first examined, were healthy. He revealed that in the following 10 years, cancer more often affected people with unstable mentality, inclined relatively easily to become depressed on account of everyday stress situations; temperamentally cold people inclined to assess analytically the surrounding world and whom the author calls 'superresistant' were afflicted with cancer less frequently. These and other similar epidemiologic studies do not in themselves decide the role of stress in the etiology of cancer. However, they catalyzed interesting experimental studies in that direction. It has lately been shown that excess glucocorticoids during stress may cause involutionary changes in organs of immunogenesis and immunodeficient states. In turn, this potentiates the development of oncogenic viruses to induce multiform viral tumors [505].

This evidence seems to create a definite premise for explaining the role of a protracted stress reaction in the pathogenesis of viral tumors. The recently established concept that most currently known chemical and physical carcinogens, from cyclic carbohydrates to ionizing radiation and plastics, can like stress activate peroxidation and thereby injure cell membranes [225] provides broader possibilities for understanding the role of stress in the etiology of cancer. The significance of this concept may be evaluated in the light of the theory of oncogenesis developed in recent years by Mekler [391] and some investigators abroad [197, 250].

According to this theory, the starting point of a malignant tumoral cell is the merger of two somatic cells with varying degrees of differentation. This event takes place as a result of injury to the plasmic membrane of cells in contact and leads to the formation of a hybrid cell possessing all the properties of a cancer cell. In terms of this presentation, it is significant that this theory regards episodes connected with injury to the plasma membranes of numerous cells as a factor that increases the probability of the formation of a primary hybrid cancer cell and the beginning of blastomatous growth. As was shown above, stress leads to the accumulation of lipid hydroperoxides, which injure cell membranes, and to the occurrence of pronounced fermentemia, reliable proof that injury really does take place. This affords grounds to assume that the presence of stress-induced cell injuries may prove to be a factor that predetermines cancer hybridization and the appearance of the initial cancer cell. Naturally, this hypothesis requires experimental verification; however, it is nonetheless a possible explanation of the fact that in the clinic and in experiments, serious stress situation often precede the beginning of malignant growth. Malignant tumors rank second among the causes of mortality in the modern world. The possible role of the stress reaction and of the LP activation observed in stress in the occurrence of tumors is in itself noteworthy.

In addition to ischemic disease, hypertension, tumors, and ulcerous disease, stress situations are known to play an important part in the etiology of diabetes and skin diseases and often potentiate major mental diseases. Thus, the transformation of the stress reaction from a link in adaptation into a link in pathogenesis plays a paramount role in pathology, primarily in the pathogenesis of basic noninfectious diseases, whose prevention and treatment constitute the principal unresolved issue of contemporary medicine. Hence, the prevention of stress-induced injuries constitutes an essential and, possibly, major stage in solving the problem of preventing human noninfectious diseases.

In assessing the possibility of preventing stress-induced injuries, one should keep in mind that even a protracted stress situation and a serious stress reaction that may, in principle, cause a disease do not always turn this potential into reality. Indeed, in the course of individual life, all humans and animals are repeatedly subjected to the effects of inevitable, prolonged, and serious stress situations. Yet, most of them are not affected and, even if they were, they recover; as a result, the stress situations continuously present and doubtlessly active in developing the phenotype do not terminate the life of the species; the potential role of stress as an etiologic factor in diseases is either not realized or is restricted by active poststress recovery. In our view, this obvious and simultaneously remarkable resistance to the harmful action of stress situations deserves not less extensive study than the pathogenesis of stress damage.

By analogy with other forms of resistance, one may assume that resistance to stress may be provided by nothing other than the existence in the organism of specialized systems that opportunely restrict the excitation of stress-realizing systems and also prevent the harmful effects of catecholamines and corticosteroids at the level of target organs. The fact that the effects of stress situations on the organism gradually diminish with the persistence of those situations led us to the idea that it is precisely the adaptation to stress that forms antistress resistance and that such adaptation may be based on the increased power of the systems mentioned above of natural prevention of stress injuries.

From this concept, the next chapter examines evidence on adaptation to stress situations and cites certain examples of systems that play a role in the natural prevention of stress damage and the possibilities of preventing such damage by active chemical prophylaxis.

We used the above evidence on the major role of the stress reaction in the etiology of ischemic injuries and on the close similarity in the pathogenic chain of stress-induced and ischemic injuries to assess these systems as systems of natural prevention not only of stress but also ischemic injuries; in addition, in cases where facts were available, we concentrated our attention on utilizing endogenous metabolites of these systems and their chemical analogus to prevent actively both stress-induced and ischemic injuries.

Chapter 7
Adaptation to Stress Situations and Systems for Natural Prevention of Stress-induced and Ischemic Injuries

"When you clean volcanoes accurately, they burn evenly and quietly, without any eruptions."

Antoine de Saint-Exupéry

Adaptation to essentially dangerous environmental stress situations is an obvious trait in animal and human behavior. An ancient Chinese philosophical treatise indicates that adaptation was experimentally studied as early as 2,400 years ago. Changtzi, the author of the treatise, reported those experiments as follows. Chi Hsingtzi undertook to train a gamecock for his sovereign. Ten days later the sovereign asked him:

"Well, is he ready?"

"Not yet," replied Chi Hsingtzi, "he is still full of vanity, conceited to no purpose." Ten days later the sovereign inquired again, and again the answer was: "Not yet. He responds to every sound, rushes at every shadow." Ten days later, the sovereign asked once again and again the reply was: "Still not yet. He glares maliciously, full of fury." Ten days later, the sovereign repeated his query to hear in response: "Now almost ready: if he hears another cock he doesn't even stir; you look at him, he looks as if he's wooden. His will and self-control are immaculate. No gamecock would dare respond to his challenge, it would turn around and flee."[1]

The above behavioral transformation in the course of adaptation to a stress situation should naturally be accompanied by changes in the stress reaction itself, i.e., by changed responses in the adrenergic and hypophysial-adrenal systems. Indeed, correct experiments free of any exotic admixtures have been performed over the past decade to show that these responses decreased when stress situations were repeated. For example, Smookler and Buckley [527] studied the effects of long-term (lasting many weeks) noise, light flares, and mechanical vibrations upon rats placed in small chambers. They also summarized the results of a number of similar investigations. It appears that initially a stress situation causes a considerably increased catecholamine content in the blood and increased catecholamine excretion in the urine; in this case, the noradrenaline content in the brain decreases. Catecholamine excretion is then somewhat reduced; catecholamine resynthesis increases and the content in the brain normalizes to stay at that level of increased renovation intensity. Subsequently (after 4–6 weeks), the level of the catecholamine pool in the brain – the utilization and resynthesis of catecholamine – returns to normal notwithstanding the continuing stress action.

1 Poetry and Prose of the Ancient Orient. Moscow: Khudozhestvennaya Literatura, 1973, p. 326 (Russian translation).

In full conformity with this evidence, Zigmond and Horvey [cited according to 527] showed that with repeated pain stimulation and the lack of avoidance possibilities, in the initial period the rats had a highly decreased catecholamine content in the brain and high mortality; subsequently, despite continuing stress actions, the catecholamine content in the brain was restored to normal and mortality decreased.

Using the criterion of increased corticosterone content in the blood, adaptation to stress actions was observed even more definitely. The curves in Fig. 51 [213] indicate that 30 min after the first immobilization stress action, the corticosterone concentration in the blood increased tenfold; after the second 30 min stress action, the corticosterone concentration was found to increase even more. It sharply decreased under subsequent actions. As a result, in the fifth immobilization action, the corticosterone content in the blood increased only twofold. However, this lower response of the hypophysial-adrenal system to the stress situation was accompanied by its increased background activity: The initial corticosterone level in the blood prior to the fifth action increased by more than twofold compared with the starting background level. This adaptation stage, characterized by a lower response to stress combined with increased background activity of the hypophysial-adrenal system, may also be noted in the material of other authors [564]. Presumably, this energetically and structurally wasteful phase of 'continuous readiness' is transitional to fuller adaptation. The most important studies on adaptation to unavoidable stress were performed by Mikulaj et al. [395],

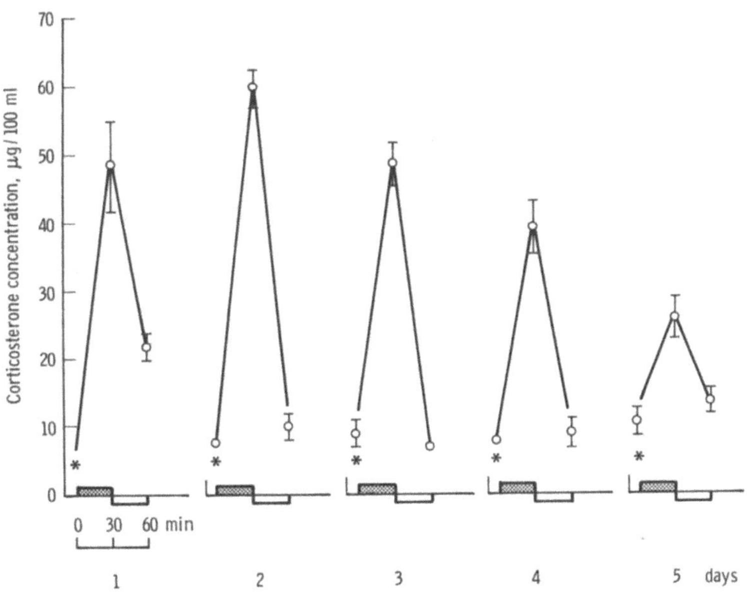

Fig. 51. Effect of 30-min immobilization (conducted for 5 days in succession) on corticosterone (CS) concentration in blood plasma. * CS in control. (After Keim and Sigg 1976)

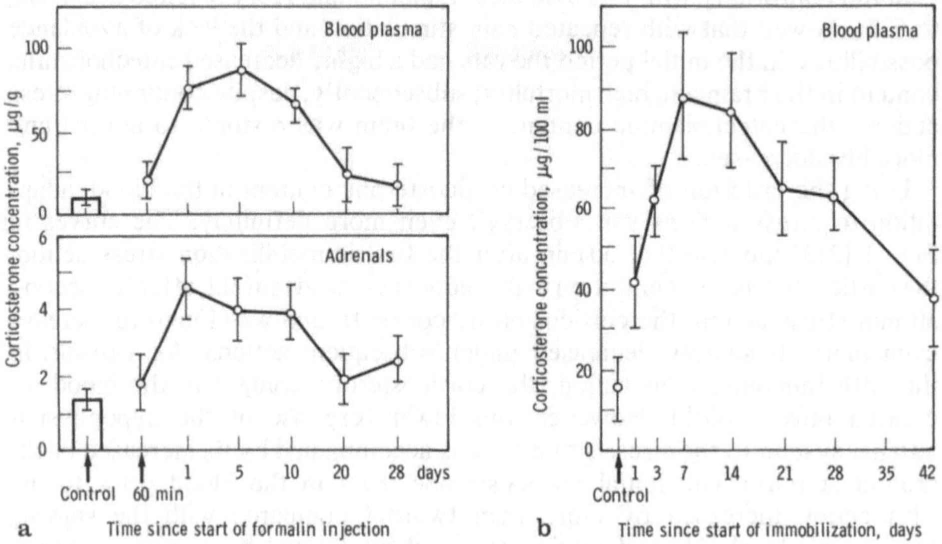

Fig. 52 a, b. Decreased effect of daily stress action on corticosterone concentration in blood plasma and adrenals. **a** injection of formalin; **b** immobilization. (After Mikulaj et al. 1976)

who showed that daily injections of formalin into the dorsal skin of rats and constant immobilization result in multiple increases in the corticosterone concentration of blood. Subsequently, these increases gradually decrease and disappear altogether in response to the 22nd–27th stress actions.

This stop in the activation of the hypophysial-adrenal system is shown by the curves in Fig. 52a, b; it does not depend on exhaustion of the functional possibilities of the adrenals, for administration of ACTH to animals adapted to stress situations causes a higher increase in the corticosterone concentration of blood than in controls. This important fact indicates that the organism's adaptation to the multiple or long-term action of a stress situation is based on the activation of certain central inhibitory mechanisms, which act at the level of the brain despite the pain inhibit excitation of higher adrenergic centers, the efflux of releasing factors, and the ACTH itself to prevent a rise in the catecholamine and corticosterone content in the blood. This concept corresponds to the decrease, well-known from practice, in the circulatory component of the response – the decreased rise in arterial pressure and tachycardia as the animal becomes used to unavoidable dangerous situations or painful actions. In the interesting work of Desiderato et al. [80], the same regularity is shown in relation to another effector organ, namely the stomach. They showed that under the initial stress action – painful stimulation of the foot pads of rats where there was no possibility of avoiding the pain – a deeply depressed secretion of hydrochloric acid in the stomach occurred because of the previously indicated strongly stimulated adrenergic regulation with a simultaneous decrease in the vagus nerve tonus. No such depression was observed when the same stress situation was applied for the fourth time: The general and free acidity of the gastric juice changed little.

In assessing adaptation expressed by a gradually decreasing excitation of the stress system and the target-organ responses, one should note two significant features. First, the process takes place in conditions in which emergence from the stress situation is unfeasible due to external, behavioral adaptation and it represents a peculiar internal adaptation to seemingly helpless situations – a probable physiologic equivalent of what in plain language is defined as *patience.* Second, the termination of excitation of stress-realizing systems in the course of adaptation to stress signifies a decrease in the concentrations of catecholamines and glucocorticoids acting on target organs, i.e., decreases the probability of stress-induced injuries of internal organs. In other words, it appears that the central inhibitory mechanisms, which under repeated or protracted stress situations suppress the excitation of stress-realizing systems, are essentially mechanisms of 'patience' and natural means of preventing stress-induced injuries.

This effect is more probably provided by coordinated GABA-ergic, DOP-aminergic, serotoninergic, and other previously cited inhibitory systems of the brain. As an example, we experimentally analyzed the role of the GABA-ergic inhibitory system, whose activation in response to a stress situation really proved to play a major role in natural prevention of stress injuries.

Activation of the GABA-ergic Inhibitory System Under Stress as a Natural Mechanism for Preventing Stress-induced and Ischemic Injuries

Neurons that synthesize GABA, the inhibitory mediator, are chiefly localized in the black substance, caudal nucleus, pale globe, and to a lesser extent in other regions of the brain. This GABA-ergic system provides the GAMA efflux into the synaptic structures of the cortex and underlying regions, the GABA possessing pre- and postsynaptic inhibitory effects. The function, structure, and metabolism of the GABA-ergic inhibitory system have by now become the subjects of basic reviews and monographs.

Here, it would be sufficient to remind the reader that GABA forms from glutamate due to the action of glutamate decarboxylase (GDC). In achieving its inhibitory effect, GABA is degraded by GABA-transaminase (GABA-T), another enzyme. This last reaction results in a succinic semialdehyde, which may either turn into succinic acid, utilized in the tricarbonic acid cycle, and thereby leave the GABA-ergic inhibitory system or, under the action of the NADPH-dependent dehydrogenation of succinic semialdehyde, a special enzyme, turn into a gamma-oxybutyric acid (GOBA), one of the final inhibitory metabolites of that system. As has now been shown, GOBA acts independently, not via GABA; it possesses a direct and strong inhibitory effect and in large doses induces sleep [305]. Unlike GABA, GOBA permeates the hematoencephalic barrier well and is hence widely used in anesthesiology and psychiatry. In terms of our exposition, what is important is not the pharmacologic

Table 33. Effect of EPS on activity of GABA-ergic system in rat brain hemispheres

	Content, % mg		Activity	
	Glutamate	GABA	GDC, μmol, GABA, h/g	GABA-T, μmol GDC, h/g
Control	104.8 ± 2.1	14.8 ± 0.2	13.08 ± 0.8	13.0 ± 0.7
EPS, 8 h	30.5 ± 1.3[b]	10.4 ± 0.7[c]	24.7 ± 0.8[b]	18.3 ± 0.8[c]
EPS, 24 h	68.1 ± 2.0	10.6 ± 0.6[c]	31.6 ± 1.2[b]	35.4 ± 1.0
EPS, 72 h	72.5 ± 2.5[b]	8.23 ± 0.3[b]	22.7 ± 1.0[b]	34.0 ± 1.18[b]
EPS, 96 h	70.5 ± 2	7.37 ± 0.17[c]	21.25 ± 0.34	31.6 ± 0.5

[a] Ten animals were used in each series. [b] $p < 0.001$. [c] $p < 0.01$.

aspect of the use of GOBA but the possibility that the stress-induced activation of the GABA-ergic inhibitory system may result in increased amounts of the inhibitory metabolite and provide a timely restriction of the stress reaction and perhaps also prevent stress-induced injuries.

Based on this assumption, we studied the state of the GABA-ergic inhibitory system with the above-mentioned EPS [366]. To assess the state of the system in the cerebral hemispheres, accepted methods were used to determine the glutamate and GABA content [514], the activity of the enzyme responsible for GABA formation from the glutamate (GDC), and the enzyme responsible for GABA degradation in the course of its inhibitory effect (GABA-T) [539]. The glutamate and GABA biosynthesis was determined by incorporation of ^{14}C-labeled glucose injected into the animals prior to killing them [453].

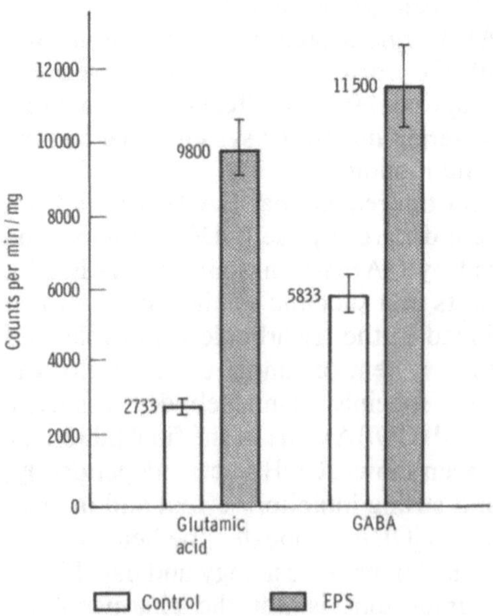

Fig. 53. Effect of stress (EPS) on glutamate and GABA synthesis in the brain

The results of these studies (Table 33) show that as early as 8 h after the beginning of EPS and 2 h following its termination the GABA-ergic system was found to be considerably activated in the brain, this being evident in a twofold increase in GDC activity and a 1.5-fold rise in GABA-T activity; correspondingly, the content of the glutamate that turned into GABA decreased threefold and that of GABA by about one-third. In the following 4 days, the GDC activity remained at the previous elevated level, while the GABA-T activity continued to be increased 2.5-fold. Correspondingly, the glutamate content decreased by one-third and that of GABA almost twofold.

The diagram in Fig. 53 shows that according to the evidence on the incorporation of ^{14}C-labeled glucose, the glutamate synthesis for brain unit mass increased twofold and that of GABA by 80%. Thus, the GABA-ergic system and the synthesis of GABA, an important inhibitory mediator, were substantially activated under stress.

To understand the mechanism and results of activation of the GABA-ergic system of the brain under stress, it is important to take into account the evidence on the effect of EPS on utilization of the main substrates of the tricarbonic acid cycle (TAC) in the large cerebral hemispheres. Oxidation of the main TAC substrates was studied by the incorporation of ^{14}C in the CO_2 that forms during respiration in sections of brain in an incubation medium in which one of the substrates was present [453]. Table 34

Table 34. Effect of EPS on the main substrates of the TAC

Indicator	Animal group		
	Control ($n=8$)	EPS ($n=8$)	p
1,2 ^{13}C-α-ketoglutorate	13,200 ± 1,340	7,900 ± 853	<0.05
1,4 ^{14}C-α-succinic acid	10,760 ± 1,490	4,160 ± 540	<0.05

shows the results of these studies and that EPS substantially decreased the utilization of two major TAC intermediates, viz α-ketoglutorate and succinic acid.

To assess the significance of this event, one should refer to Fig. 54, which characterizes the relationship between the GABA-ergic system and TAC. The shaded section in the scheme (Fig. 54) shows the inhibition and utilization of α-ketoglutorate and succinic acid in EPS.

Inhibited utilization of α-ketoglutorate unambiguously shows that the sharply increased glutamate oxidation to CO_2, which was discussed in connection with EPS, takes place due to decarboxylation of the glutamate in the GABA system, not via α-ketoglutorate. The increased amounts of GABA as a result turn into succinic semialdehyde under the action of activated GABA-T. As Fig. 54 shows, this compound may either be transformed into succinic acid and be utilized in the TAC or, conversely, be converted into

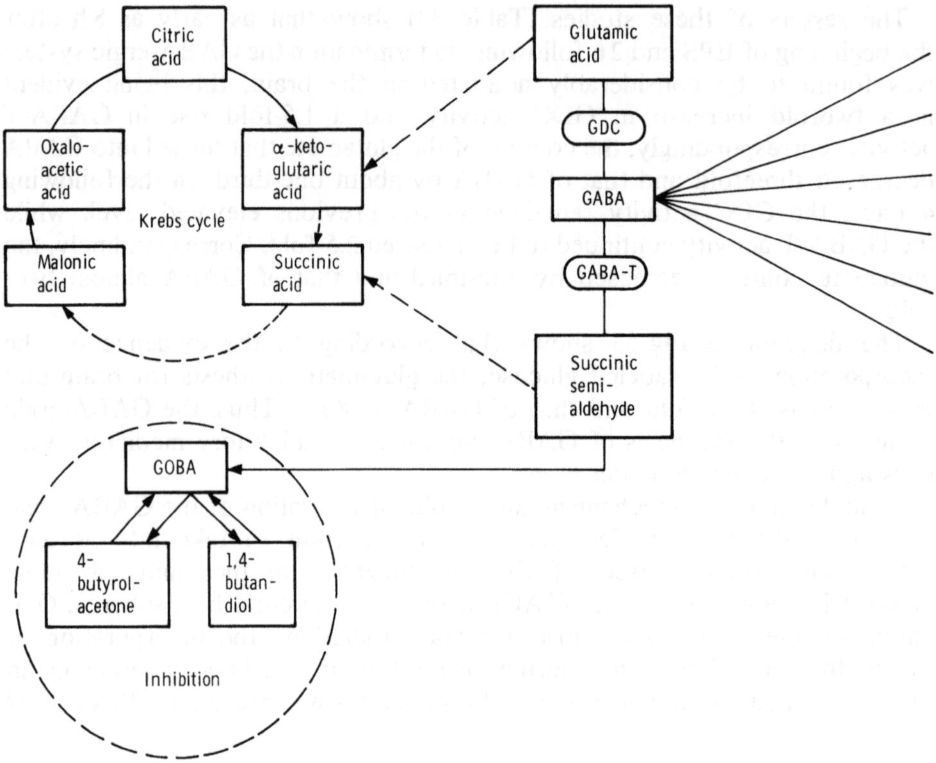

Fig. 54. Relationship between GABA-ergic system and Krebs' cycle. Explanation in text

GOBA, which possesses strong inhibitory effects. Emotional painful stress impairs utilization of succinic acid in the TAC and predetermines the transformation of increased amounts of succinic semialdehyde into GOBA.

Thus, inhibited utilization of α-ketoglutorate in the TAC, an event occurring in the brain under EPS, provides an increased glutamate influx into the GABA system, while inhibited utilization of succinic acid apparently predetermines the increased formation of GOBA, an inhibitory metabolite, from succinic semi-aldehyde. The fact that certain TAC links are limited together with activation of the GABA system results in increased amounts of GABA and GOBA in the brain during EPS. Both are inhibitory metabolites, which when acting at the level of higher control centers may restrict excitation of the stress-realizing systems.

If this assumption is correct, then administration of GOBA, which permeates well through the hematoencephalic barrier, prior to stress should reduce activation of the adrenergic and hypophysial-adrenal systems and, to some measure, prevent stress-induced injuries. Indeed, experiments have established that GOBA administered prior to and during EPS suppresses the activation of stress-realizing systems [372]. The curves in Fig. 55 show the results of an experiment in which GOBA was administered perorally (100 mg/kg) 30 min before EPS and twice during EPS. It is apparent from Fig. 55 that administration of

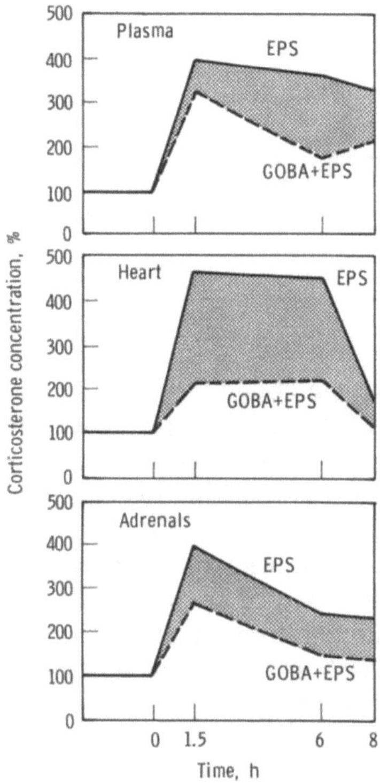

Fig. 55

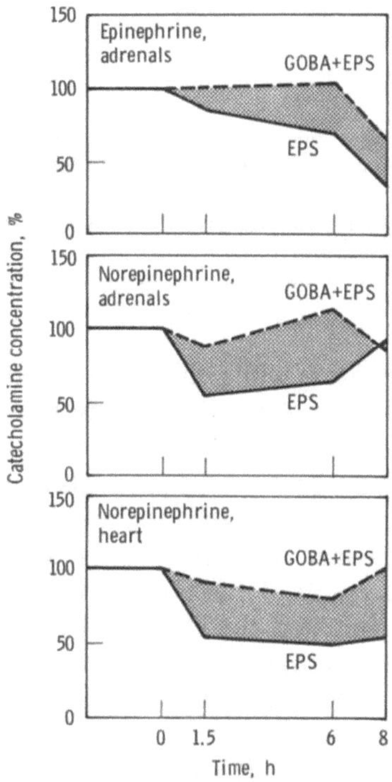

Fig. 56

Fig. 55. Effect of preliminary injection of GOBA on corticosterone concentration in the plasma, heart, and adrenals during stress (EPS); ***shaded area*** indicates GOBA effect in control
Fig. 56. Effect of preliminary GOBA injection on epinephrine and norepinephrine concentration in adrenals and norepinephrine concentration in the heart give as percentage of normal stress. *Shaded area* indicates GOBA effect

GOBA largely prevented the EPS-induced increase in corticosterone concentration in the adrenals, blood plasma, and myocardium. Figure 56 shows that GOBA also prevents the catecholamine content in the adrenals and myocardium from decreasing.

Further studies showed that by thus restricting excitation of the stress-realizing systems, GOBA largely prevents ulcerous lesions of the mucous coat of the stomach, which usually develop under EPS. Table 35 shows that EPS resulted in ulcerous lesions of the mucous coat of the stomach in all rats without exception (average ulcer length per stomach, 10.5 mm). After preliminary administration of GOBA, 28% of the animals developed no ulcers, while the remaining ones had an average ulcerous length three times less than under EPS without GOBA.

Thus, there are real grounds to conclude that activation of the inhibitory GABA-ergic system in the brain, which occurs simultaneously with stress, may restrict excitation of stress-realizing systems due to inhibitory meta-

Table 35. Effect of GOBA on development of ulcerous lesions in the stomach

Indicator	Animal group		
	EPS (n = 30)		EPS + GOBA (n = 14)
Average length of ulcerous lesions per stomach, mm	10.5 ± 1.15	$p < 0.01$	3.3 ± 1.43
Number of animals, without ulcerous lesions, %	0		28.6

bolites, while the administration of one of these metabolites (GOBA) prior to the stress action may prevent stomach ulcers from developing. Since the antistress effect of GOBA is achieved at the level of the higher vegetative brain centers and prevents excessive excitation of stress-realizing systems, one could, with the aid of GOBA, probably reduce or prevent stress damage to very different organs, specifically stress-induced injuries of the heart. As a result of the above evidence on the pathogenic chain of such injuries, we evaluated the effects of preliminary GOBA administration on the activation of peroxidation, lysosomal state in the myocardium, level of poststress fermentemia, and poststress disturbances in the contractile function of the heart.

The evidence in Table 36 shows the previously described activation of LP under the influence of EPS and indicates that the accumulation in the myocardium and skeletal muscles of lipid hydroperoxides and Schiff's bases, the major evidence of such activation, is largely prevented by the prior administration of GOBA in the doses mentioned above.

Since the following links in stress damage are caused by the action of

Table 36. GOBA effect on activation of lipid peroxidation in the myocardium and skeletal muscle during EPS

	Number of animals	Subject	Accumulation of hydroperoxidases according to UV spectra, nmol/mg lipids	Intensity of Schiff's base fluorescence, relative units
1. Control	5	Myocardium	16.7 ± 1.4	10.5 ± 3.5
		Skeletal muscle	19.1 ± 2.4	10.3 ± 0.9
2. EPS	5	Myocardium	47.6 ± 2.4	20.6 ± 2.8
		Skeletal muscle	38.1 ± 2.4	27.6 ± 4.2
3. GOBA	4	Myocardium	14.3 ± 1.4	8.3 ± 1.1
		Skeletal muscle	14.3 ± 2.4	9.1 ± 1.6
4. GOBA + EPS	5	Myocardium	26.2 ± 2.4	12.9 ± 1.2
		Skeletal muscle	16.7 ± 2.4	6.8 ± 0.5
p_{1-2}			< 0.01	< 0.01
p_{1-4}			> 0.05	> 0.1

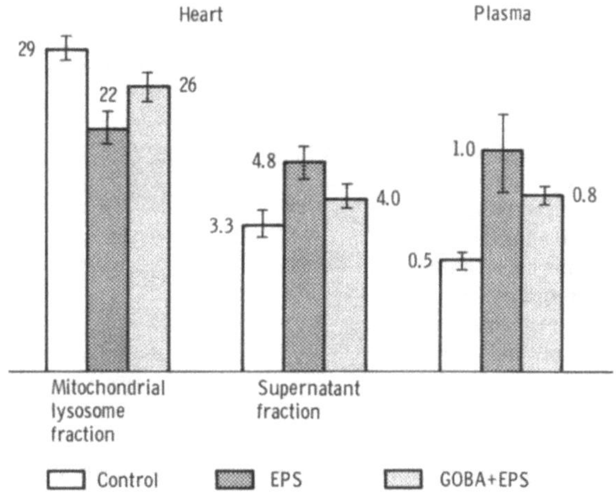

Fig. 57. Effect of preliminary GOBA injection on activity of acid cathepsins (μg tyrosine/mg protein/h) in the heart and plasma during stress (EPS)

hydroperoxides, the preventive effect of GOBA should likewise be expected to occur. Indeed, Fig. 57 shows EPS to result in lysosomal labilization, which is evident in the reduced activity of acid cathepsins in the myocardial lysosomal and mitochondrial fractions, higher activity of these enzymes in the supernatant, and a twofold increase in cathepsin activity in blood plasma. Prior administration of GOBA largely prevents acid cathepsins from passing from the lysosomal-mitochondrial fraction into the supernatant and it decreases the EPS-induced rise in the activity of these enzymes in the blood plasma.

From Fig. 58 it can be seen that pronounced fermentemia developed 2 h after EPS: Aspartate transaminase, alanine transaminase, lactate dehydrogenase, and malate dehydrogenase activity in the blood plasma increased approximately twofold. Prior GOBA administration virtually eliminated this phenomenon.

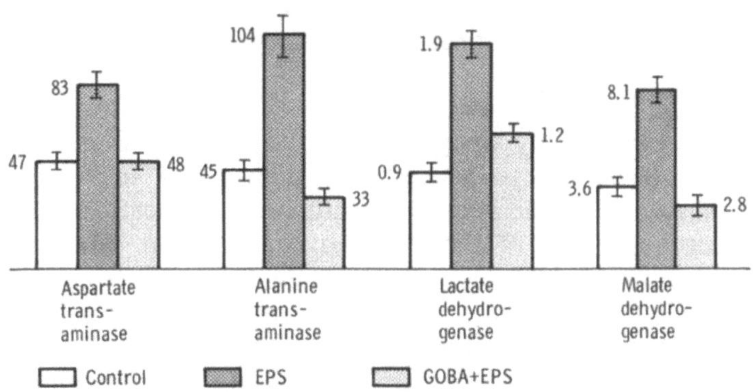

Fig. 58. Effect of preliminary GOBA injection on activity of aspartate and alanine transaminase (μmol/ml/h) and lactate and malate dehydrogenase (μmol/min/mg protein) in plasma during stress

Table 37. GOBA effect on contraction and relaxation of papillary muscle in EPS

	Control ($n = 10$)	GOBA + control ($n = 10$)	EPS ($n = 11$)	GOBA + EPS ($n = 10$)
Contraction amplitude, % of initial length	6.99 ± 0.81	9.33 ± 2.14	4.72 ± 0.44	8.03 ± 0.74
Contraction velocity, unit musc. length/s	1.086 ± 0.049	1.73 ± 0.06	0.518 ± 0.06	1.56 ± 0.092
Relaxation velocity, unit musc. length/s	0.891 ± 0.062	1.655 ± 0.084	0.474 ± 0.068	1.422 ± 0.088
Significance of differences for first three indices	$p_{1-2} < 0.01$	$p_{1-3} < 0.01$	$p_{3-4} < 0.01$	

Significantly, disturbances in the contractile functions of the heart resulting from EPS, like the metabolic disorders described above, may be prevented by prior administration of GOBA. We performed experiments on rats using the Sonnenblick method of studying the contractile function of the papillary muscles of the left ventricle [335]. The muscles were taken from controls, from animals receiving GOBA, from EPS-exposed animals, and from animals that received GOBA in the dose cited above (100 mg/kg) prior to stress. The evidence in Table 37 compares the results of these four series of experiments and leads to at least three substantial conclusions.

First, the administration of GOBA to intact animals (GOBA + control in Table 37) caused the contraction amplitude and velocity and the relaxation velocity of papillary muscles to increase.

Second, in animals that had not received GOBA, EPS reduced the contraction amplitude and also the velocities of relaxation and contraction by 30%–50%. In other words, this confirms previously cited evidence that stress impairs the contractile function of the heart muscle. Finally, the third and principal conclusion in terms of our presentation is that GOBA administration prior to EPS prevented the parameters of the contractile function of the papillary muscle from decreasing to the quite low values normally observed in stress-exposed animals, and in this sense the preventive effect of GOBA evokes no doubts.

Thus, prestress administration of GOBA restricts the excitation of the adrenergic and hypophysial-adrenal systems and as a result either prevents or decreases stress-induced activation of LP in various organs, lysosomal labilization, fermentemia, disturbances in heart contractile function, and ulcerous lesions of the stomach mucous coat. These facts agree with the concept that GABA-ergic system activation, which signifies increased formation of GOBA, is a natural physiologic mechanism whereby the stress reaction is restricted and stress-induced injuries are prevented.

In assessing this concept, one should keep in mind that the metabolic chain that forms the GABA system is represented not only in the brain, where it performs an inhibitory function, but also in other organs that in stress play the role of targets of the catecholamine and glucocorticoid effects.

Table 38. EPS effect on activity of GABA-ergic system in rat myocardium

Series	Content, % mg		Activity	
	Glutamate	GABA	GDC, μmol; GABA, h/g	GABA-T, μmol GDC, h/g
Control	92.85 ± 4.0	1.77 ± 0.15	3.0 ± 0.3	15.0 ± 1.0
EPS, 8 h	70 ± 3.4[b]	3.26 ± 0.2[b]	3.8 ± 0.2[d]	35.6 ± 2.1[b]
EPS, 24 h	64.8 ± 1.6[b]	3.53 ± 0.5[c]	5.75 ± 0.3[b]	31.2 ± 1.0[b]
EPS, 72 h	58.0 ± 2.4[b]	3.1 ± 0.2[b]	6.25 ± 0.4[b]	36.7 ± 2.1[b]
EPS, 96 h	58.0 ± 2.6[b]	4.17 ± 0.5[b]	5.0 ± 0.5[c]	26.0 ± 0.7[b]

[a] Ten animals were used in each series. [b] $p<0.001$. [c] $p<0.01$. [d] $p<0.05$.

From this viewpoint, we assessed the EPS effect on the GABA system in the myocardium and certain parameters of liver metabolism (Table 38).

From Table 38 it follows that in normal conditions, the glutamate content in the myocardium is close to that in the brain, and the GABA content is lower by one order of magnitude. Correspondingly, the GDC activity in the myocardium is also several times lower than in the brain and the GABA-T activity is about the same. The 4 days immediately following EPS showed a gradual activation of the GABA system; this was apparent in a more than twofold increase in GDC and GABA-T activities, lower glutamate content, and a somewhat increased GABA content, the cause of which is obscure.

To assess the significance of GABA activation in the myocardium during stress, it is expedient to study the evidence in Table 39 on the effect of EPS on the transition of labeled ^{14}C to CO_2 from the principal TAC intermediates and from ^{14}C-glutamate, the principal transamination product.

Since the ^{14}C transition to CO_2 characterizes the oxidation rate of the substrates used, from Table 39 it follows that EPS reduces the oxidation of α-ketoglutarate but significantly increases the oxidation of succinic acid in myocardial section. The impaired α-ketoglutarate oxidation observed under EPS apparently does not depend on OAA shortage, since reduced formation of α-ketoglutarate from isocitric acid should in itself lead to a lower dilution of the label added to the sections and to accelerated labeling of CO_2. The inverse result shown in Table 39 may most probably have been caused by the reduced activity of NAD-dependent ketoglutarate dehydrogenase, which achieves decarboxylation and dehydrogenation of α-ketoglutarate.

From Table 39 it follows that glutamate oxidation in myocardial sections increases twofold under EPS. Evidently, this shift cannot be explained by accelerated deamination of the glutamate to α-ketoglutarate, since oxidation of the ketoglutarate itself is delayed and, consequently, ^{14}C-glutamate cannot rapidly move to CO_2 via α-ketoglutarate. An alternative way of accelerated glutamate utilization is to activate its decarboxylation to GABA. Above, we showed this possibility to have indeed become reality: EPS is instrumental in the pronounced activation of the GABA system in the myocardium, which should logically lead to the increased formation of succinic acid semialdehyde.

Table 39. EPS effect on oxidation of TAC intermediates and glutamate by myocardial sections (by incorporation of ^{14}C in CO_2, impulses/min/mg)

Substrate	Control ($n=9$)	EPS ($n=8$)	p
1,4 ^{14}C-succinate	2,100 ± 1,870	35,000 ± 2,886	<0.05
1,2 ^{14}C-α-ketoglutorate	15,000 ± 1,233	9,700 ± 1,428	<0.05
5 ^{14}C-glutamate	4,160 ± 495	8,250 ± 1,750	<0.05

As was already indicated when explaining GABA activation in the brain, this compound may be transformed, first, into succinic acid and, second, into GOBA. The considerably increased utilization of succinic acid (Table 39) allows us to assume that after EPS, the myocardium utilizes the first possibility: Succinic semialdehyde is transformed chiefly into succinic acid and the activation of flavin adenine dinucleotide (FAD)-dependent succinate oxidation plays the role of a factor that compensates for the impairment of NAD-dependent oxidation of α-ketoglutarate and other substrates. Under stress-induced mobilization of the contractile function of the heart and the regeneration of poststress injuries, this mechanism may prove very important.

In terms of our presentation, it is noteworthy that *activation of the GABA system in the brain and heart under EPS apparently has a different physiologic significance in these organs.* In the brain, the GABA system is activated against a background of inhibition of FAD-dependent oxidation of succinic acid in the TAC and, hence, leads to increased accumulation of GOBA, the inhibitory metabolite that plays an important role in restricting the excitation of stress-realizing systems. In the heart, the GABA system is activated with a higher FAD-dependent oxidation of succinic acid and it plays the role of a shunt that provides additional amounts of succinic acid notwithstanding the inhibition of NAD-dependent oxidation of α-ketoglutarate in the TAC. This shuntlike functioning of the GABA system and the resultant high intensity of FAD-dependent oxidation may play a substantial role both in providing sufficient ATP resynthesis in the myocardium and in forming additional amounts of the NADPH utilized in biosynthesis. Thus, *both in the brain and heart, the GABA system comes into action under stress as a modulatory mechanism that prevents potential injuries. However, in the brain, its role is incommensurably more significant for the organism as a whole, for it is precisely at level of the brain that the GABA-ergic inhibitory system limits potentially dangerous excitation of the stress-realizing systems.*

The fact that the GABA-ergic system is activated and that increased amounts of glutamate are utilized under stress in such rapidly acting organs as the brain and heart should be kept in mind when analyzing the facts cited below on the metabolic changes induces by EPS in the liver. The data characterizing the transition rate to CO_2 of the labeled carbon from various substrates utilized by liver sections are presented in Table 40. It can be seen in Table 40 that the oxidation of lactate, α-ketoglutarate, and succinic acid, and also of aspartate and glutamate is essentially inhibited in EPS-exposed

Table 40. Effects of EPS and preliminary GOBA administration on oxidation of principal TAC intermediates with liver sections (by incorporation ^{14}C from TCA intermediates in $^{14}CO_2$, counts/min/mg)

	^{14}C-lactate ($n=6$)	^{14}C-stearate ($n=8$)	^{14}C-aspartate ($n=6$)	^{14}C-d-ketoglutarate ($n=7$)	^{14}C-succinate ($n=6$)	^{14}C-glutamate ($n=7$)	^{14}C-glucose ($n=9$)
1. Control	5,400 ± 500	1,550 ± 210	31,166 ± 1,720	18,400 ± 2,249	39,000 ± 509	12,800 ± 1,275	1,020 ± 165
2. EPS	1,225 ± 193	6,635 ± 986	1,205 ± 133	5,240 ± 370	15,800 ± 1,933	7,500 ± 935	4,200 ± 460
3. GOBA + EPS	3,450 ± 270	1,540 ± 259	15,000 ± 1,820	14,166 ± 400	16,500 ± 612	10,800 ± 1,200	1,400 ± 87
Significance of differences							
p_{1-2}	< 0.001	< 0.05	< 0.001	< 0.001	< 0.001	< 0.05	< 0.05
p_{1-3}	> 0.5	> 0.05	< 0.001	< 0.05	< 0.01	< 0.05	> 0.5

animals. Conversely, the oxidation of glucose labeled with the first carbon atom and that of stearic acid is increased.

Assessing the possible mechanism of these events, one should bear in mind that the liver is one of the major targets for glucocorticoids, whose concentration in the blood during stress may increase by four to ten times. Acting on a genetic level, glucocorticoids are known to activate the biosynthesis of the enzymes of gluconeogenesis, transamination, and pentosophosphate shunt in the liver cells. Activated gluconeogenesis leads to the utilization of large amounts of OAA, the principal intermediate and regulator of the TAC for the formation of phosphoenol pyruvate. The considerable efflux of OAA and its decreased availability in the TAC should logically result in reduced intensity of the citrate synthetase reaction and consequently in a lower acetyl-CoA involvement rate in the TAC. This causes diminished utilization and oxidation of pyruvate and, consequently, of lactate by the liver tissue. Aspartate oxidation is preceded by its conversion to OAA; increased removal from the TAC of the thus formed OAA causes the oxidation of the aspartate carbon skeleton in the TAC to slow down; correspondingly, this also slows down the transition of ^{14}C from aspartate into CO_2. Similarly, OAA shortage results in the retarded transition of ^{14}C from succinate to CO_2, since decarboxylation of the succinate carbon skeleton takes place in the TAC after it has been transformed through a number of stages into isocitric acid. This transformation is impaired by OAA deficiency. In full conformity with this argumentation, the incorporation of ^{14}C in the CO_2 excreted by liver sections proved to be reduced under EPS – fourfold for lactate, over twofold for succinate, and 2.5-fold for aspartate.

The disturbed α-ketoglutarate oxidation observed in EPS apparently does not depend on OAA shortage, since the reduced formation of α-ketoglutarate from isocitric acid should have in itself led to lower dilution of the label added to the sections and to accelerated labeling of CO_2. The inverse result observed in these experiments may most probably be caused by reduced activity of the NAD-dependent α-ketoglutarate dehydrogenase, which brings about the decarboxylation and dehydrogenation of the α-ketoglutarate. Significantly, irrespective of the cause, inhibition of α-ketoglutarate decarboxylation increases the possibility of its transamination to glutamate, which naturally accumulates in the liver when transamination is activated under the influence of glucocorticoids. This should have led to dilution of the label of the exogenous ^{14}C-glutamate and fully explains the inhibited transition to CO_2 of the glutamate labeled carbon observed in our experiments. In a whole organism, such glutamate accumulation may apparently have other, considerably more important consequences. Among other things, glutamate may abundantly enter the blood from the liver. From the blood it may reach various organs, specifically the brain and heart, where, as was just indicated above, glutamate utilization is sharply increased.

Another well-known metabolic shift occurring during stress is the activation of the pentose phosphate shunt and the apotomic oxidation of glucose under the influence of glucocorticoids; this shift is the most probable cause of a multiple increase in the supply to CO_2 of ^{14}C glucose labeled by the first carbon atom. An important result of these experiments was that the entire set

Table 41. Effects of EPS and GOBA on glycogen content and biosynthesis in rat liver

	Content, % mg		Specific activity, counts/min/10 mg glycogen				Total activity, counts/min/amount of glycogen in 1 g of tissue			
			^{14}C-glucose		^{14}C-glycin		^{14}C-glucose		^{14}C-glycin	
1. Control ($n=22$)	2,400	±185	3,033	±181	683	±65	7,272	±615	1,623	±136
2. EPS ($n=24$)	1,355	±101	2,250	±227	766	±61	2,825	±196	1,000	±101
3. GOBA + EPS ($n=20$)	2,837	±273	1,416	±160	1,500	±77	4,126	±305	4,200	±318
Significance of differences										
p_{1-2}	<0.05		>0.05		>0.05		<0.05		>0.05	
p_{1-3}	>0.05		<0.05		<0.05		<0.05		<0.05	

of EPS-induced metabolic shifts characteristic of the glucocorticoid effect was largely removed by preliminary administration of GOBA. Table 41 shows the results of the next experiment, in which we studied the effects of EPS and GOBA on the content and synthesis of glycogen. From Table 41 it can be seen that EPS leads to a twofold decrease in the glycogen concentration in the simultaneous inhibition of glycogen resynthesis by incorporation of uniformly labeled ^{14}C-glucose and glycin. Mobilization of the liver glycogen reserve under the action of stress is quite a well-known event; it results from the action of catecholamines which, via the phosphorylase system, activate glycogenolysis to cause glucose efflux into the blood.

The fact that under EPS mobilization of the glycogen reserve accompanies the inhibition of glycogen resynthesis is apparently because glucose 6-phosphate, which forms due to activation of the initial glycolysis stages and the arrival of glucose from the blood, is not utilized as it normally is for glycogen resynthesis and as a result of dephosphorylation this glucose 6-phosphate is converted to glucose and released into the blood. Glucose 6-phosphate can also be oxidized via the pentose-phosphate pathway, which, as was indicated above, is sharply activated under EPS. As a result, the incorporation of labeled glucose in glycogen proves to be 2.5 times less. The incorporation of glycin labeled by carbon in glycogen is considerably reduced under EPS, despite the activated reactions of gluconeogenesis, which involves the transformation of glycin into pyruvate. This probably takes place because utilization of the glucose 6-phosphate that formed from pyruvate is restricted by the factors just examined.

Consequently, changes in EPS-induced glycogenolysis (mobilization of glycogen reserve and inhibition of glycogen resynthesis) may very probably be explained by the combined action of glucocorticoids and catecholamines. A significant fact seen in Table 42 is that the preliminary administration of GOBA fully eliminates all the EPS-induced changes in the glycogen content and resynthesis.

Table 42. Activity of GABA-ergic system in cerebral hemispheres 2 days after four different stress actions

	Content, % mg		Activity	
	Glutamate	GABA	GDC, μmol; GABA, g/h	GABA-T, μmol; GDC, g/h
Rats				
Control	104.8 ± 2.1	14.3 ± 0.2	13.8 ± 0.8	13.0 ± 0.7
EPS	68.1 ± 2.0	10.6 ± 0.6[c]	31.6 ± 1.2[c]	35.4 ± 1.0[c]
Injection of formalin	70.0 ± 0.6[c]	11.14 ± 0.4	20.0 ± 0.8	47.7 ± 0.84
Rabbits				
Control	62.5 ± 2.0	15.25 ± 1.2	9.5 ± 0.6	9.25 ± 0.3
Ischemic myocardial necrosis	25.5 ± 1.7[b]	14.75 ± 1.1[b]	18.0 ± 2.0[c]	45.0 ± 2.4[d]
Aorta stenosis	26.5 ± 1.9[b]	14.0 ± 2.0	20.0 ± 2.2[c]	33.75 ± 2.3[d]

Ten rats and six rabbits were used in each series.
[b] $p < 0.001$. [c] $p < 0.01$. [d] < 0.05.

In assessing the results, it should be emphasized that the above EPS-induced metabolic changes in the liver are regulation-conditioned and play a substantial role in providing an adequate response to extreme environmental stimulators. This role is determined by the fact that mobilization of the glycogen reserve and activation of gluconeogenesis and transamination cause glucose, glutamate, and possibly other substances to enter the blood from the liver. These substances can be utilized in the brain, heart, and skeletal muscles, i.e., in systems that play a direct part in the organim's adaptive responses. In this respect, it is significant that the increased formation of glutamate in the liver and its excretion into the blood under EPS which was previously established and indirectly demonstrated in our experiments, occurs together with increased utilization of glutamate and with its decrease in the brain and heart, where it is intensely utilized to form either the inhibitory metabolites GABA and GOBA or succinic acid, a major oxidation substrate.

In assessing the formation of inhibitory mediators in the brain as the organism's defense reaction, one may assume that the increased glutamate transport from the liver to the brain constitutes a substantial prerequisite for realizing this protective mechanism. Similarly, increased glutamate transport from the liver to the heart may play its positive role in compensating for stress-induced disturbances in myocardial metabolism.

The definitely proven role of the GABA-ergic inhibitory system in the brain as one of natural prevention of stress damage logically leads to a highly important question: Is activation of the GABA-ergic inhibitory system a specific feature of the EPS model that we used or does it naturally occur in any stress action on the organism? To decide this issue, we compared the basic characteristics of the metabolism of the GABA-system in the cerebral hemispheres of four groups of animals exposed to various stress actions [383]. Rats from the first series were exposed to EPS and those from the second series received two formalin injections into the dorsal skin (24 h interval,

1.0 ml 2% solution); the control group included intact rats. In the third series, we used rabbits and ligated the descending branch of the left coronary artery to induce a morphologically documented ischemic necrosis of the myocardium. In the fourth series, also carried out on rabbits, surgery was used to create stenosis of the aortic ostiam; this provided a threefold decrease in the ostial gap resulting in acute overload and intense compensatory hyperfunction of the heart. The control group included animals subjected to thoracotomy with subsequent stitching of the chest. All experimental and control animals were examined 2 days after the stress action.

Table 42 shows that with all the strong stress actions used, the GDC activity in the cerebral hemispheres increased approximately twofold; correspondingly, the glutamate concentration in the brain tissues decreased by 30%–60%. The activity of the enzyme that degrades GABA during its inhibitory postsynaptic effect (GABA-T) increased with all the stress actions employed to a somewhat greater extent (three to four times); correspondingly, the GABA concentration showed a certain tendency to decline.

On the whole, these facts are unambiguously indicative of activation of the GABA-ergic brain system under various stress actions. They also show that GOBA, one of the final metabolites in the GABA system, prevents excessive activation of the hypophysial-adrenal system and injuries to internal organs in stress. This affords grounds to assume that the GABA-ergic system plays the role of a nonspecific inhibitory mechanism, which restricts the stress reaction and prevents injuries from various stress situations and harmful environmental factors. This system represents an inevitable satellite of the stress reaction. This nonspecific (like stress itself) inhibitory mechanism restricts the stress reaction and plays a role in the natural prevention of stress damage.

Hence, by the logic of the above concept, administration of GOBA, the final product of the GABA-ergic system, may prove useful in all pathologic states in whose development an excessively intensified and prolonged stress reaction, specifically during myocardial infarction (see Chap. 6), plays a substantial role. Another major cause for utilizing GOBA during infarction is that besides its antistress effect, it enhances the resistance of the heart and brain to hypoxia, this being highly significant with cardiogenic shock accompanying serious infarction. Accordingly, Popov et al. [443] used GOBA to treat 20 patients with serious infarction and pronounced painful syndrome.

When injected intravenously (20% solution, 80–150 mg/kg), GOBA removed the painful syndrome in all patients and 10–15 min later induced sleep, which lasted about 2 h. Pains occurred again after 4–8 h, but were much weaker and readily eliminated by agents usually employed in myocardial infarction clinics. At the same time, in some patients GOBA removed the arrhythmias inherent in infarction, stabilized arterial pressure, and increased the arterial pressure during large-focal infarction. As a result of this, Popov et al. [443] successfully administered GOBA per os (about 6 g daily) to other patients suffering from stenocardiac pains. Essentially similar possitive results were obtained by Yuldashev and Babadzhanov [594] in treating myocardial infarction with GOBA. Thus, the GABA-ergic system and its final product constitute real antistress and antiischemic factors.

The role of the GABA system in restricting the stress reaction and preventing stress damage is an example of how in extreme situations the organism either inhibits or cuts off certain physiologic systems in order to prevent their degradation and retain a reserve for future adaptational responses. Recent advances in biochemistry and cell physiology unambiguously show that such mechanisms for restricting the organism's response are widely represented not only at higher levels of the regulatory hierarchy but also at cell and tissue level as well. Indeed, to prevent fully injuries during protracted stress reactions, central inhibitory mechanisms are apparently insufficient. This follows from the fact that major excitation of the adnergic and hypophysial-adrenal systems, which is evident in a 7–13-fold rise in active catecholamine and glucocorticoid concentrations in the blood [294, 465], takes place in response to a stress situation almost instantly, before the central inhibitory systems that subsequently restrict the intensity and duration of stress become increasingly active.

In an optimal situation in which adaptation develops rapidly, the tremendous initial mobilization of stress systems proves short-lived and does not produce injuries in target organs, which the basic catecholamine and glucocorticoid effects essential for adaptation take place. The very fact that target organs possess such resistance leads to the inference that in addition to the receptor apparatus through which the catecholamine action is achieved, the cells must contain sufficiently specialized metabolic mechanisms which play the role of modulators that restrict hormonal effects and prevent cell injury.

Below, we shall examine three examples of such modulatory mechanisms involving (a) limitation of myocardial function by blocking the calcium influx to cardiomyocytes, (b) prevention of energy shortage and blockade of adrenergic effects by an adenine nucleotide system, and (c) blockade of adrenergic effects by the prostaglandin system. In the subject matter that follows, it will be shown that all these mechanisms of cellular self-control can play a major role in preventing stress-induced and ischemic damages.

Blockade of Physiologic Effects of Calcium: Its Role in Preventing Ischemic and Stress Damages[2]

As was shown above, excess calcium in cells constitutes the common final link in the pathogenic chain of stress-induced and ischemic injuries. It is evident, therefore, that the cell control mechanisms capable of preventing this link are of paramount significance. Despite the fact that the structure of these mechanisms is largely obscure, contemporary pharmacology has spontaneously, but quite effectively, reproduced their adaptive effect, and as a result the so-called calcium antagonists are widely used in medicine [113, 499]. The organism most vividly shows its ability to switch off a given function by blocking calcium transport or expelling calcium from points where its physiologic effect is usually achieved during acute ischemia. The essence of this pheno-

---

2 Written jointly with A. A. Ugolev.

menon is following clamping of the coronary artery, the contractile function of the ischemic myocardial site ceases virtually instantaneously. The site becomes readily expandable and moves paradoxically, i.e., bulges out during systolic contraction of the heart [546].

With biopsies of ischemic myocardia in dogs immediately after contractility had dropped because of ischemia, the total ATP and CP content in the cardiomyocytes was shown either to remain normal or to change very insignificantly [68]. Neither is ischemic depression of the myocardial contractile function the result of changes in contractile proteins, for even protracted ischemia does not cause changes in the structures of actin [208], myosin [17], actomyosin [204], and myofibrillar ATPase [4].

In full conformity with the above evidence, it was established that the contracting, passively expanding myocardial site retains the ability to contract for at least the first 10 min of ischemia in the presence of sufficiently strong stimulation. For instance, Boden et al. [39] used postextrasystolic potentiation to show that the contraction of the myocardium in the ischemic zone noticeably increased 10 min after the coronary artery was ligated.

The above combined data signifies that during the appearance of depressed contraction, the ischemic myocardium still has sufficient energy reserve and an intact contractile apparatus and also that the absence of contractions can most probably be explained by certain rapidly developing regulatory shifts in the system of excitation-contraction coupling. Calcium transport is known to play the central role in this coupling since during excitation, when the Ca^{2+} concentration in the sarcoplasma increases to 10^{-5}, it combines with myofibrillar troponin and thereby actuates actomyosin interaction, i.e., contraction. Increasing calcium concentration during excitation is based on quite a complex system of processes. First, during the action potential (AP) plateau, a slow calcium flux directed inside the cardiomyocytes and dependent on the AP voltage takes place via kinetically slow sarcolemmic channels. The flux source is the intercellular fluid (to which Ca^{2+} is supplied from the blood), unique calcium storages situated in the glycocalix and apparently bound therein with sialic acids [227, 255], and also the plasmolemma, where Ca^{2+} is bound with certain phospholipids [285]; the calcium entering the cell during AP was termed 'trigger,' since it causes Ca^{2+} to emerge from the intracellular storages, namely from the SR and mitochondria [411]. In addition to the electrogenic influx of Ca^{2+} into the cardiocytes, there is also an electroneutral Ca^{2+} flux through the sarcolemma – one Ca^{2+} enters the cell in exchange for two Na^+. This flux is achieved at the expense of the so-called Na^+-Ca^{2+} exchange mechanism involving specific transmitter proteins. In addition, Ca^{2+} can passively penetrate into the cells by the concentration gradient via channels that are present in the sarcolemma in a state of rest [256, 412]. However, these last two Ca^{2+} transport mechanisms to myocardiocytes do not substantially contribute to the forming of a 'trigger' calcium pool and, hence, their role in the excitation–contraction coupling is apparently less significant than that of the AP-dependent slow calcium flow. This in itself makes it probable that inhibition of the electrogenic, AP-dependent calcium flux causes instantaneous ischemic depression of the contractile function of the myocardium.

Indeed, it has been established that from the very outset, this depression is accompanied by a shortening of the AP plateau, an important electrophysiologic phenomenon [576]. The fact that in this case, a normal rest potential is retained indicates that the K^+-dependent membrane properties are essentially unchanged; so are the gradients of ascent and the amplitudes of the AP peak, which depend on Na^+ influx into the cardiomyocytes. In this situation, shortening of the AP plateau signifies that in ischemia, the processes that chiefly suffer are those responsible for the slow calcium flux. Taking into consideration the important role of slow calcium flux through the sarcolemma in excitation–contraction coupling in the myocardium, one may assume that the diminished Ca^{2+} influx along slow channels, observed immediately after the occurrence of ischemia, does not simply accompany ischemic depression of myocardial contractility, but is the basic cause of this depression.

This means that myocytes posses a certain mechanism that can forecast impending energy shortage, preblock Ca^{2+} influx to the cardiomyocytes, and thus in advance of rapid ATP-deficient degradation of myocardial structures. In case of short-term or spatially limited ischemia, the adaptive significance of such a mechanism is obvious; in case of long-term and general ischemia, it may constitute a pathogenic mechanism leading to termination of the contractile function of the heart and accelerate death, which in these conditions is unavoidable. Understandably, therefore, the blockade mechanism of electrogenic calcium transport in ischemic depression of the heart contractile function has become the subject of many forms of research. An analysis of the obtained results permits the distinction of at least three hypotheses formulating the essence of this mechanism in different ways. Since an instantaneous and for some time reversible depression of the contractile function of the myocardium is typical not only of ischemia but also of considerable hypoxia, below we shall regard these concepts as hypotheses of early reversible depression of the heart contractile function in both ischemia and hypoxia.

The first energy-deficient hypothesis of early reversible depression of heart contractile function in ischemia and hypoxia still remains an item for discussion despite the fact noted above that there is no ATP shortage in the myocardium when its contractile function starts to diminish. To activate the slow calcium channels, the channel-forming proteins must be phosphorylated; it is, therefore, possible that impaired conduction over slow calcium channels may be caused by a shortage of ATP as the source of phosphate groups. However, the energy shortage in the myocardium in ischemia and hypoxia occurs later than the contractile function decline and ATP shortening [68]. In addition, the protein kinase responsible for the phosphorylation of membrane proteins has a high affinity for ATP and, hence, the intracellular ATP concentration should drop to very low values – to about 1% of the normal amount – until the energy shortage produces a considerable effect on the phosphorylation of slow calcium channels [530]. And yet, it is possible that energy shortage may play a role in suppressing slow calcium flux [530], since the ATP reserves in the cell are not distributed uniformly but are divided among various structures [411]. Hence there is the possibility that during ischemia and hypoxia, the first to become exhausted is the ATP pool connected with the sarcolemma and

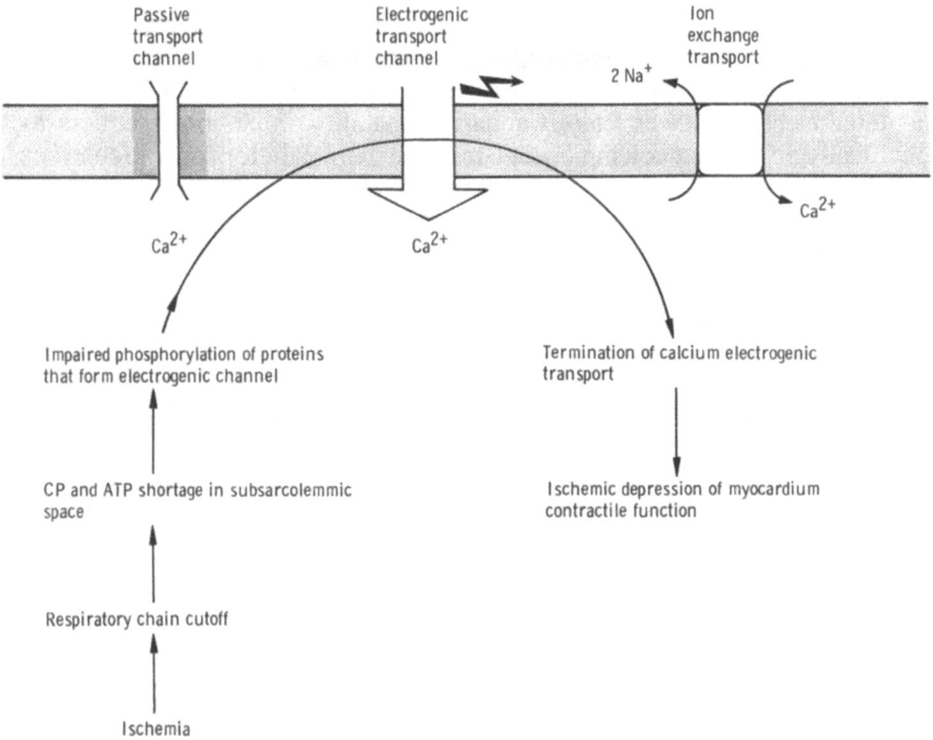

Fig. 59. Energy-shortage hypothesis of early reversible depression of contractile function of the heart during ischemia and hypoxia

this restricts the possibilities of the membrane to accumulate and transport Ca^{2+}. Localization of the ATP pool near or within the sarcolemma has still to be experimentally proven; however, some facts indicate that this may be highly probable. For example, the rapidity with which ATP leaves the myocardium and the myocardial cell culture for the extracellular fluid during hypoxia may indicate that it is released from a surface pool, or possibly from the cell membrane itself, not from the cytosol. There is also evidence that under hypoxia and ischemia, the CP reserves deplete to a greater extent and more quickly than the ATP reserves [411]. If one takes into account that CP is a transport form of energy in the myocardial cell, the above fact presumes a possible disturbance of the energy supply to membrane systems, including the sarcolemma. This may also cause a local shortage of energy needed for slow calcium channels to function.

These largely indirect arguments are diagrammatically summarized in Fig. 59, which shows how the energy-deficient mechanism of calcium influx blockade could theoretically function.

The second so-called lipid hypothesis of early reversible depression in the contractile function of the myocardium in ischemia and hypoxia explains the blockade of Ca^{2+} influx and the depression of the contractile function by the action of fatty acids, whose content in the myocardium naturally increases during ischemia [490]. Shortening of the AP plateau is one of the best known fatty acid effects on the myocardium [284]. It may be explained by the fact that a slight increase in fatty acid concentration, which apparently takes place in the initial ischemic stages, causes enhanced stability of membrane structures, specifically of the sarcolemma, and reduced permeability for various ions, including Ca^{2+} [2]. This may possibly be caused by membrane swelling due to the incorporation therein of amphiphylic lipid molecules with subsequent conformational changes in the membrane structural proteins [57].

Recently, evidence has appeared to show that the role of excess fatty acids in depressing the heart contractile function may go beyond fatty acid effect on the sarcolemma. It was shown, for instance, that the relatively low concentrations of free fatty acids, specifically oleic acid, suppress Ca^{2+} efflux from the SR [210]. Palmitinic acid [209] increases Ca^{2+} deposits in isolated SR vesicles. This is indicative of a possible restriction of calcium discharge from the SR during ischemia with subsequent depression of the contractile function. The lipid hypothesis of ischemic depression of the heart

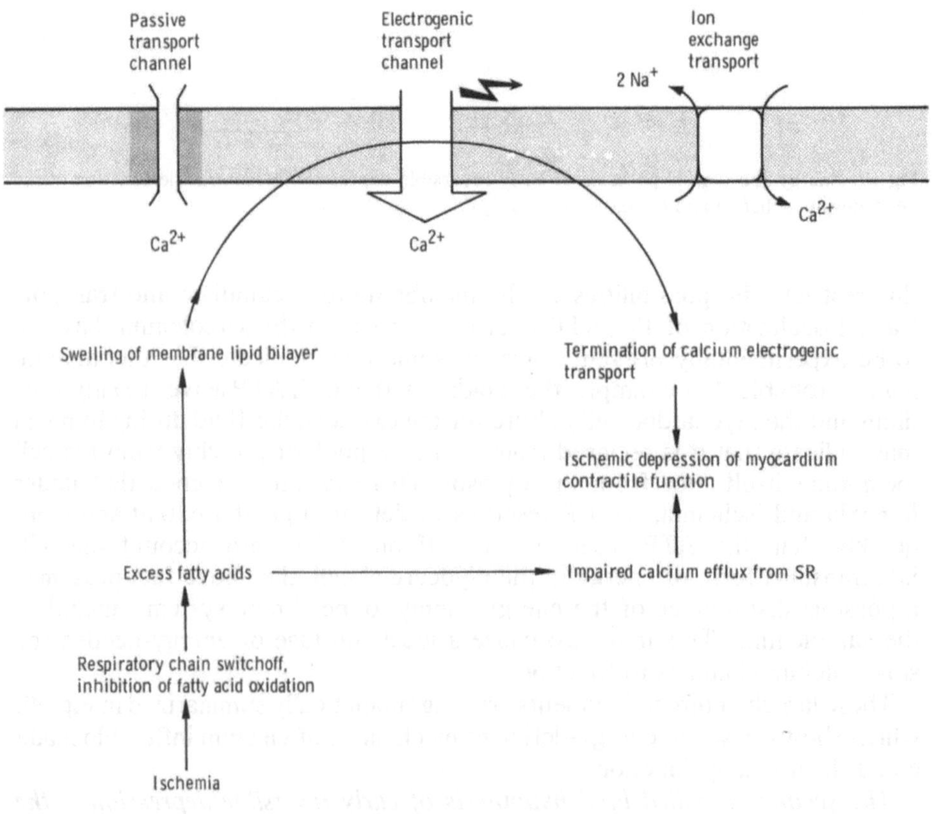

Fig. 60. Lipid hypothesis of early reversible depression of contractile function of the myocardium during ischemia and hypoxia

contractile function is schematically shown in Fig. 60. Here, in conformity with the above facts, fatty acids are shown to be capable of simultaneously blockading the Ca^{2+} influx via the slow sarcolemma channel and releasing Ca^{2+} from the SR; this may possibly result in depressed contractile function.

The third possible and, in our view, most workable mechanism of rapid reversible depression of the heart contractile function in ischemia may at present be represented within the framework of the so-called ionic hypothesis. According to this concept, schematically summarized in Fig. 61, the two following shifts, occurring on account of a disconnection in the respiratory chain, play a major role in ischemic and hypoxic depression of the heart contractile function: (a) acidosis, i.e., increased proton concentration, and (b)

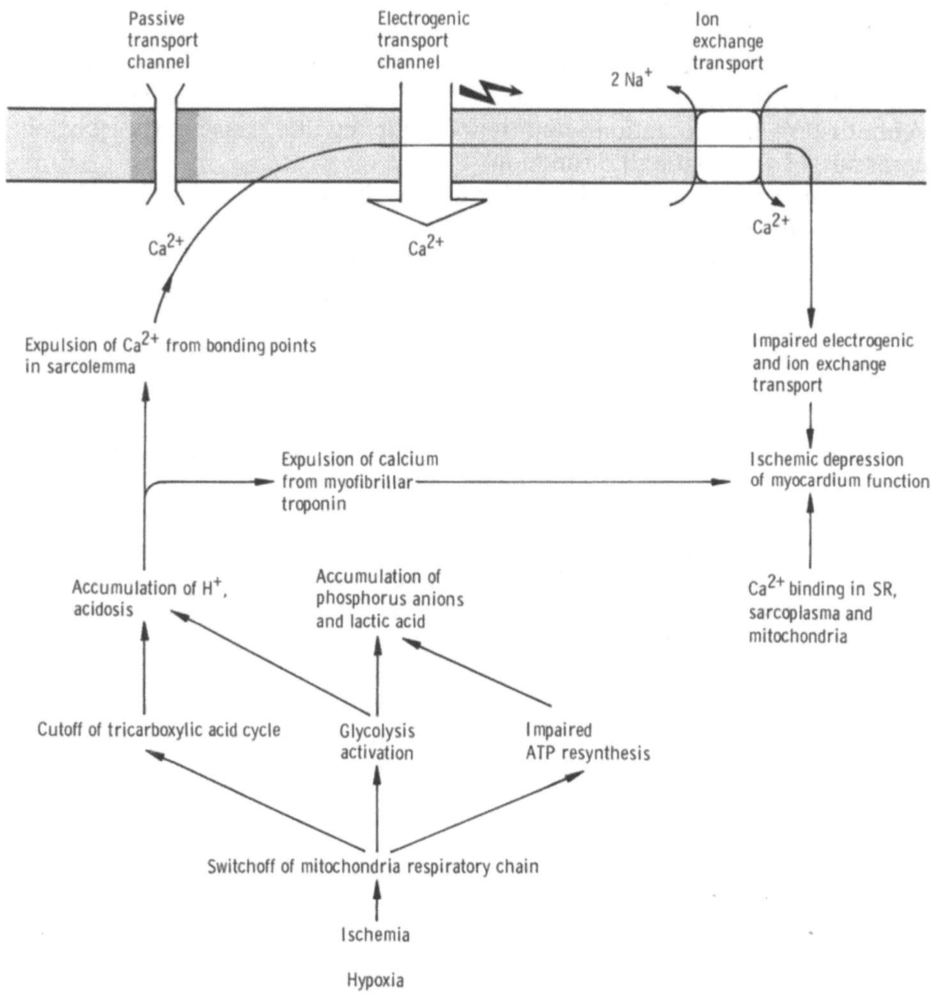

Fig. 61 Ionic hypothesis of early depression of contractile function of the myocardium during ischemia and hypoxia

phosphate anion excess, caused by disturbed aerobic resynthesis of ATP. The significance of acidosis is that H^+ expel Ca^{2+} from their binding sites in the sarcolemma and myofibrils and reduces the SR sensitivity to the trigger effect of the Ca^{2+} influx from outside [221]. The increase in the sarcoplasma of phosphate concentration in addition to that of other anions, e.g., lactate, may reduce the amount of free Ca^{2+} in the sarcoplasma that can interact with contractile and regulatory proteins. This phenomenon is designated the 'Kübler-Katz phosphate trap' [247]. These investigators calculated that ischemia may involve an eightfold increase in the free phosphate concentration in the cytosol of cardiomyocytes. In this case, the inorganic phosphate sources disintegrate ATP and CP molecules. The amount of inorganic phosphate incorporated in adenosine monophosphate (AMP) molecules, whose content in this case rises, is slight and may be ignored. With such a multiple increase in the inorganic phosphate concentration, the free calcium is bound in the form of poorly soluble salts in all the cytosol sections and forms precipitates. Since inorganic phosphate diffuses freely through the membranes of the SR and mitochondria, precipitates form in these organelles as well. These events disturb the excitation–contraction coupling and cause early ischemic depression of the contractile function.

Irrespective of what form the molecular mechanism of reversible ischemic depression of the heart contractile function adopts, it is now already clear that it is based on rapid regulatory blocking of the slow calcium channel. This important concept signifies that the wide use in modern cardiology of blockers of the slow calcium channel, viz verapamil, D 600, nefedipin, and diethiazeme, directly imitates the control mechanism that actually functions in the natural conditions of a living organism. When used in therapeutic doses, these blockers possess moderate, negatively inotropic, and pronouncedly vasodilatory effects; they reduce the amount of oxygen utilized by the heart to prevent the ATP concentration from declining and protect the heart from adrenergic and ischemic injuries. Today, this effect has not only been proven experimentally and clinically but it is also theoretically sound, since it is excess calcium in the cardiomyocytes that constitutes the final link in stress-induced and ischemic damage (see scheme in Chap. 6). At the same time, it signifies that the natural self-control mechanism that provides complete or partial blockade of the slow calcium channel either to restrict or shut off the natural function-determining signal may in extreme situations have great protective significance. At the expense of functional decrease, this mechanism prevents the principal functioning structures from disintegrating.

Role of the Adenine Nucleotide System and Its Product Adenosine in Preventing Energy Shortage and Limiting Adrenergic Effects[3]

The adenine nucleotide system is known to play a decisive role not only in supplying energy to the cell, but in controlling the basic intracellular pro-

3 Studied jointly with M. G. Pshennikova.

cesses. Indeed, with any significant increase in function, even a slight lagging of ATP aerobic resynthesis behind its increased utilization entails higher [ADP] + [AMP] + [CP]/[ATP] + [CP] ratios and, consequently, activated ATP resynthesis both in the mitochondria and the glycolysis system. At the same time, greater amounts of adenosine, a powerful vasodilator, form from AMP and the resultant hyperemia causes the supply of oxygen and oxidation substrates to the mitochondria to increase. As a result, the adenine nucleotide system proves to be a mechanism that adapts O_2 supply and ATP synthesis in the cell to its functional level, i.e., in the end to the requirements of an intact organism. Adenosine plays the role of an important link in that mechanism.

Adenosine is produced by myocardial cells to be released when the oxygen level in the blood is reduced and its utilization by the myocardium increased [30]. In normal conditions, the adenosine concentration in interstitial fluid is low; however, even when hypoxia is very brief, say 5 s, the adenosine level rises about threefold [31]. It is significant that this reaction is fully achieved in a whole organism. For example, Watkinson et al. [575] showed that in dogs with catheters permanently sewn into the pericardial cavity, the adenosine level during rest and after a physical load was the same and increased 2–3-fold under exertion in a treadmill.

Adenosine is synthesized from 5'-AMP by means of dephosphorylation catalyzed by 5'-nucleotidase. This enzyme is connected with many membranes in the myocardial cell, including the sarcolemma, insert disks, SR membranes, etc. [483]; in addition, in the sarcolemma it is present with adenyl cyclase [501], a significant factor in the presentation that follows. Adenosine synthesis depends on the level in the cell of 5'-nucleotidase-inhibitors and activators. The inhibitors of the enzyme are ATP, ADP, and CP, the latter being the most powerful inhibitor [538]. Hence, normally, the enzyme in the cell is 100 times less active than in vitro [419]. The enzyme is activated by free Mg^{2+}, which in the cell is bound with ATP and released when the latter disintegrates. With an increased load on the heart or under the action of stress-determined excess catecholamines, the rate of ATP and CP resynthesis is less than the rate of their utilization; as a result, the level of CP and ATP, both 5'-nucleotidase inhibitors, is reduced and the level of AMP, the substrate of 5'-nucleotidase, and Mg, its activator, increases. This causes activated adenosine synthesis. Increased amounts of adenosine are released into the interstitial space; it reaches the arterioles, causes dilation, and increases blood flow, oxygen supply, and the generation of energy in the myocardium increasing the CP and ATP content; in other words, adenosine reduces its synthesis in correspondence to the cell's energy status.

At present, adenosine has been shown to act in two ways on the heart: It causes the vasodilation mentioned above and restricts the effect of excess catecholamines. A number of facts show that in achieving both of these actions, adenosine emerges as a factor that blocks excessively intense control signals. It has recently appeared probable that the vasodilatory effect of adenosine is based on its ability to block calcium influx to coronary vessel myocytes and thus to cause their relaxation. This is evidenced by the fact that the addition of adenosine (10^{-6} M) removes the AP in a guinea pig auricle, in which rapid sodium channels are blocked and a slow Ca^{2+} flux is clearly pronounced [500]. This means that adenosine inhibited the Ca^{2+} influx, which in these conditions was responsible for the generation of AP. Addition of adenosine deaminase to the perfusion solution reduced the AP. Further, the studies of Belardinelly et al. [26] revealed that like lanthanum, adenosine displaces the AP curve of the guinea pig auricle downward without changing the ascent gradient, i.e., inhibits the Ca^{2+} influx without influencing the sodium flux. The action mechanism is still obscure, but is presumably due to a lesser number of Ca channels or their lower permeability to Ca^{2+} [30].

The second aspect of adenosine action is its ability to restrict the mobilizing, possitively inotropic effect of catecholamines on the heart. It was demonstrated by Lammerant and Becsei [252], who in experiments on dogs established that adenosine does not affect the heart contractile function when the effect of the adrenergic system on the heart is either lower than normal or normal, and it considerably affects the cardiac function when this system is activated. In this case, adenosine reduces the functional increment. Later, similar evidence was obtained by Schrader et al. [501], who showed that adenosine prevents the more rapid isoproterenol-induced contraction velocity of an isolated guinea pig heart. These results agree with those of Dobson [87], who used rat hearts to establish that adenosine suppresses the catecholamine-induced increase in contractility as well as the activation of cAMP-dependent protein kinase and glycopene phosphorylase. These facts confirm that adenosine can reduce the positively inotropic effect of catecholamines.

In examining the mechanism of this phenomenon, Schrader et al. [501] showed that it apparently takes place independently in cardiomyocytes and not through the vasodilatory effect of adenosine. They also showed that when used without catecholamines, adenosine itself does not affect the cAMP level and the phosphodiesterase (PDE) and adenyl cyclase activities in membrane preparations. True, it may suppress adenyl cyclase activity, but then its concentration must exceed the physiologic activity (10^{-3} M) by two orders of magnitude. This evidence correlates with the facts obtained in other tissues subjected to catecholamine action [311]. Thus, we obtain the impression that adenosine produces a modulating action on catecholamine effects not by directly affecting the adenyl cyclase – phosphodiesterase system.

Increased calcium influx into the cardiomyocytes under the action of catecholamines has been proven and the role of this phenomenon in the positive inotropic and cardiotoxic effects of catecholamines cannot be doubted; at the same time, it has been proven that adenosine inhibits the calcium influx into cardiomyocytes (see above). Hence, it would be logical to assume that the adrenal-inhibiting and essentially antistress effect of adenosine is achieved by blockading the calcium influx into the cells, i.e., through the same mechanism as that instrumental in its vasodilative effect. Taking this into account, a hypothesis for the mechanism by which adenosine protects the heart from excessive adrenergic effects and oxygen shortage is given in Fig. 62.

Figure 62 shows that any shortage of energy-rich phosphorous compounds, irrespective of whether it is due to a stress-induced adrenergic effect or to the direct action of ischemia, activates adenosine synthesis. Adenosine blocks the calcium influx to coronary artery myocytes (vasodilation) and produces the same effect in the cardiomyocytes (restriction of positive inotropic effect of catecholamines). This blockade of the basic control signal proves sufficient either to limit or prevent a stress- or ischemia-induced harmful energy shortage. This hypothesis corresponds to the general idea that the control mechanisms of a living organism are organized quite economically and that one and the same control contour is normally utilized to solve several tasks. At the same time, it is quite probable that in real conditions the situation is more complex than shown in Fig. 62. For example, the adrenoblocking effect of adenosine is possibly achieved by inhibiting the catecholamine-β-adrenoceptor interaction or by preventing the receptor and catalytic parts of the adenyl cyclase complex from coupling. The evidence of Rodbell and Londos [473], who showed on liver cells that adenyl cyclase has adenosine-binding sites, is in favor of this hypothesis. Inasmuch as 5'-nucleotidase, the adenosine synthesis enzyme, is situated in the sarcolemma, i.e., near the adenyl cyclase, this interaction is spatially quite possible. These arguments correlate with the evidence that phosphodiesterase inhibitors (theophylline and methyl xanthine) are simultaneously antagonists to the adenosine action on the myocardium [501].

In this context, it is unimportant whether adenosine protection is achieved by blockading the adrenoceptors or by calcium influx into the cells. The significance of the presented facts is that a timely blockade or simple restriction of the central control signal can, through self-control mechanisms of the

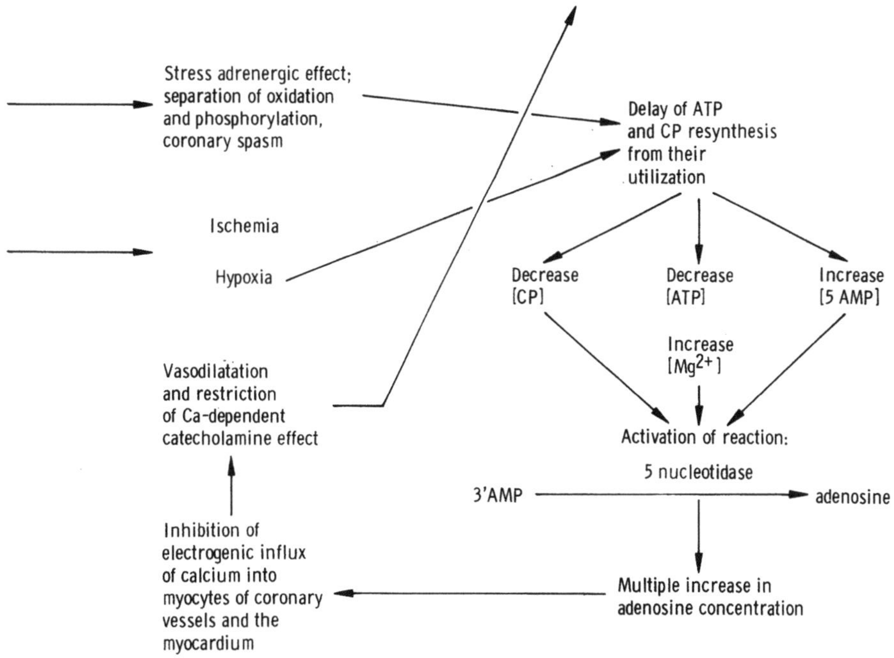

Fig. 62. Protective vasodilative and adrenolytic effect of adenosine during hypoxia and ischemia

executive organ, prevent its vital structures from disintegrating in extreme environmental situations that quite often cause diseases. It is, therefore, understandable that the organism possesses highly specialized mechanisms for achieving such a blockade and these mechanisms, which become established in the course of evolution, play an important role in adaptation to stress situations.

Activation of the Prostaglandin System: A Factor for Preventing Stress and Ischemic Damages[4]

Being powerful vasodilators and blockers of catecholamine action, prostaglandins are low-molecular substances that form from arachidonic acid, the nonsaturated fatty acid contained in cell membrane lipids. A basic action mechanism of this modulatory system involves an increase in the catecholamine level activating the formation and release of prostaglandins E (PGE), which in turn limits catecholamine action. Let us examine the principal links of this feedback system.

Prostaglandins (PG) form chiefly in the external plasma membrane from unsaturated fatty acids, arachidonic acid being the direct PG precursor. Prostaglandin synthesis is achieved by means of several enzymes. In addition, several forms of PG may develop in different tissues and even in the same cell since PG can transform from one form into another. A simplified scheme of PG synthesis is shown in Fig. 63. It shows that PG formation is based on the formation of cyclic endoperoxides from arachidonic acid by means of cyclodioxygenase or PG synthetase; this is followed by the formation

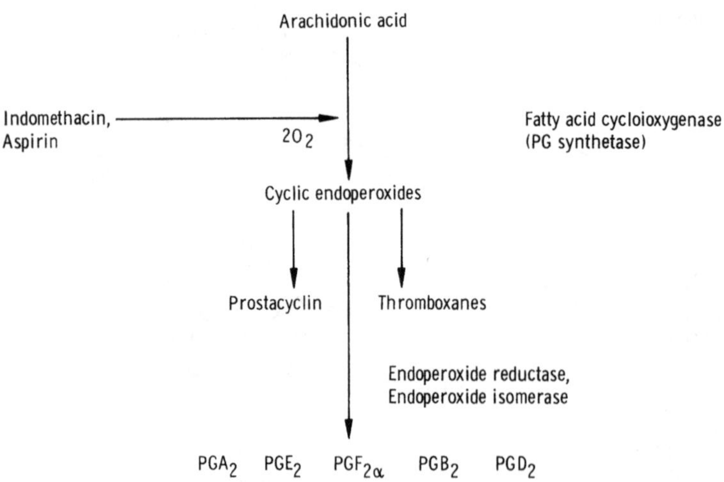

Fig. 63. Scheme of prostaglandin (PG) synthesis from arachidonic acid

4 Written jointly with M. G. Pshennikova.

of a number of intermediate products, biologically active endoperoxides, which under the influence of corresponding enzymes are in turn transformed into the main PG forms [128, 558].

To understand the PG-forming mechanism, it is essential to know that synthesis is possible only with sufficient amounts of free precursors. Inasmuch as arachidonic acid is almost absent in the cell in a free state and on release is instantly incorporated either in PG or back in the lipids, an essential condition for PG synthesis is the preliminary splitting of lipids containing the compound. Hence, it is clear that the disturbed activity of cell phospholipases, which release arachidonic acid from phospholipids, and of triglyceride lipases, which release fatty acids from neutral fats, leads to the suppression of PG synthesis. The activity of these enzymes is largely dependent on the Ca^{2+} level in the cell; hence, the intensity of PG synthesis also depends on Ca^{2+} transport [120].

In addition to the presence of free precursors, successful PG synthesis also requires the activity of PG synthetase and other enzymes responsible for PG synthesis. There are a number of factors that cause the activity of PG synthetase to be either suppressed or stimulated. Several pharmacologic compounds are also specific inhibitors of PG synthesis; the most effective ones are cyclodioxygenase inhibitors, viz indometacin and aspirin [526].

Experiments in vitro showed that the antioxidants examined below, e.g., ascorbic acid, α-tocoferol, which in small doses activate PG synthetase but in large doses suppress it, have an essentially stimulative effect on PG synthesis [541, 557]. The glutathione sulfhydryl group, a cofactor in the reaction in which intermediate endoperoxides are transformed into PG, also substantially affects PG synthesis [592].

Figure 64 shows a scheme that approximately reflects the formation of PG in the cell and characterizes the presumed PG activation mechanisms under the action of catecholamines. As is apparent from Fig. 64, catecholamine effects on the cell may involve four presumed activation mechanisms of PG synthesis. The first two mechanisms have been studied best; they appear to be the most probable and are connected with the activated release of arachidonic acid from cell depots – from cell triacyglycerides and membrane phospholipids. The two other mechanisms have been studied very little and are largely hypothetical.

The activation mechanism I of PG synthesis is due to the effect of catecholamines on β-adrenoceptors, activated cAMP formation, and stimulated nonspecific lipases, which cause the release of fatty acids from triacyglycerides. In this case, different fatty acids are released, not only arachidonic acid. These processes are accompanied by the incorporation of arachidonic acid in PG and by fatty acid release from the cell. This activation pathway is not specific for PG formation since many other reactions in the cell are also activated under the influence of growing cAMP concentration [183].

Pathway II appears to be the most specific as it is achieved through the activation of specific phospholipases (for example, phospholipase A_2), which provide a rapid release of arachidonic acid from membrane phospholipids. Phospholipases may supposedly also be activated in pathway I of PG synthesis

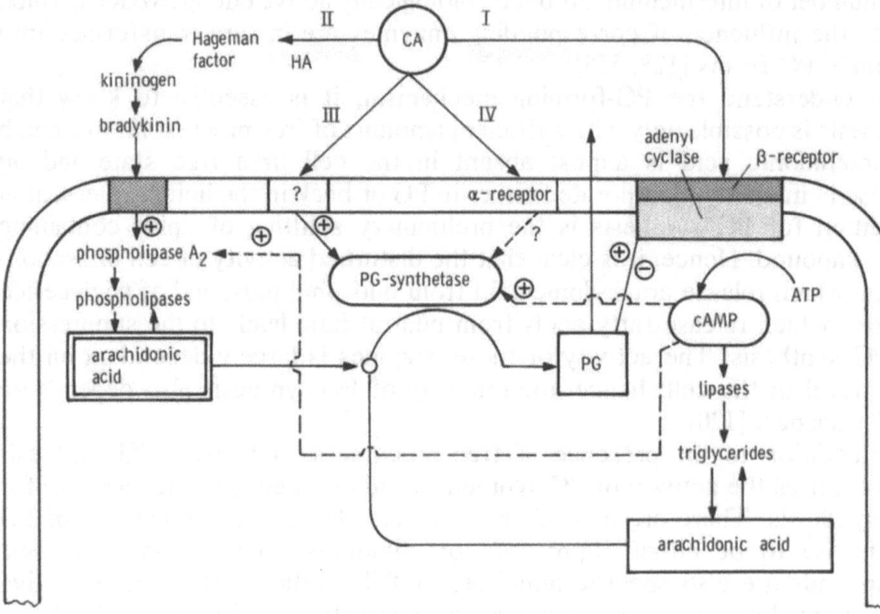

Fig. 64. Catecholamine *(CA)*-activated prostaglandin *(PG)* synthesis. Activating effect is designated by +, and suppressing effect by −. Further explanation in text

activation, due to the effect of cAMP on phospholipases; however, this process appears to be chiefly activated by means of a kinin system. The process essentially involves an increased catecholamine content leading to the activation of the kallikrein-kinin system and to the formation of bradikinin. Bradikinin, in turn, has a specific effect on the membrane, where activation of phospholipase A_2, release of arachidonic acid, and its incorporation in PG occur. This reaction takes place in the membrane and is highly effective in utilizing arachidonic acid [183].

Unlike the principal pathways of activation of PG synthesis by catecholamines cited above, the other two are largely hypothetical and, if they really do exist, their role is more probably ancillary. For example, pathway III presumes direct catecholamine molecular involvement in PG synthetase activation as the enzyme cofactor [274]. This catecholamine effect would probably activate PG formation only when the catecholamines achieve their principal action and activate the release of arachidonic acid. Pathway IV, which supposes an activating catecholamine effect on PG synthesis via α-adrenoceptors, appears equally hypothetical. This suggestion stems from studies in which the catecholamine effect stimulating PG release in certain organs was suppressed by α-adrenoblockers and remained unchanged during β-adrenoblockade [584].

Despite the hypothetical nature of some of the links in the above scheme, it is unquestionable that PG synthesis and release are really activated under the influence of catecholamines. This is evidenced by at least three groups of facts: (1) PG release into the blood from various tissues increases under the influence of catecholamines; (2) the catecholamine-induced PG release is blocked by inhibitors of catecholamine action; and (3) the catecholamine-induced increase in PG release is blocked by inhibitors of PG synthesis. These findings were obtained in various organs. For example, it was shown in an isolated perfused rabbit heart with normal innervation that the stimulation of sympathetic nerves causes the intensive release of PG into the perfusate. The blockade of heart adrenoceptors by propranolol and phentolamine prevented PG release under the influence of sympathetic stimulation [582]. Further, by using ^{14}C-arachidonic acid, Wennmalm [583] established that PG release from the heart due to the effect of sympathetic stimulation is caused by activated PG synthesis from arachidonic acid; stimulation resulted in an increase in ^{14}C-labeled PG. Boroyan [45] showed on a dog heart in situ that administration of adrenaline and noradrenaline sharply increased the PG content in the venous blood running back from the heart. In this case, indometacin, the PG synthesis inhibitor, fully suppressed both spontaneous PG release in the myocardium and PG release in response to catecholamine administration. Similar evidence showing that PG synthesis is activated and that PGs are liberated during the stimulation of sympathetic control and in response to the administration of exogenous noradrenaline was also obtained in other organs [159, 534].

These facts are unambiguously indicative of activation of the PG system under the action of catecholamines and allow us now to examine the question

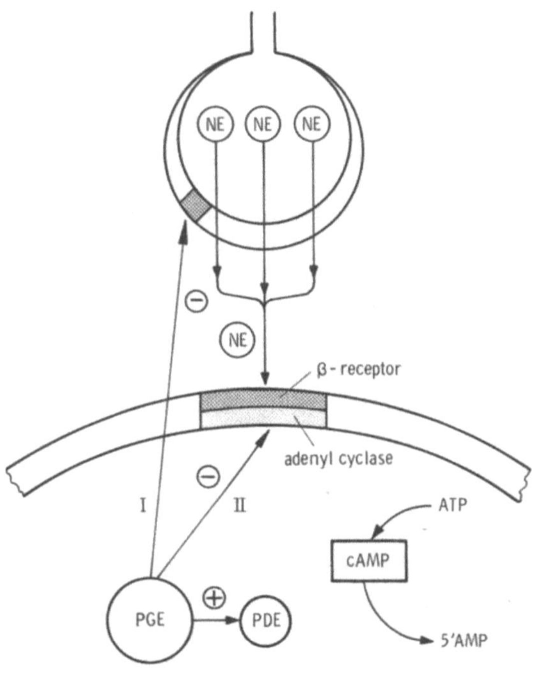

Fig. 65. Effect of prostaglandins on noradrenalin *(NE)* release from nerve ending *(I)* and on catecholamine effect in effector cell *(II)*. Explanation in text

concerning the PG influence on catecholamine effects. The PG system modulates the catecholamine effect. This is achieved in a feedback manner in two principal ways: (1) the PGs that had formed under the action of catecholamines influence the noradrenaline release from sympathetic terminals and thus decrease or increase the mediator level in the zone of influence of effector cells; and (2) the catecholamine-induced PGs either restrict or enhance the direct effect of catecholamines on the effector cells. The result of the modulating action of PGs on catecholamine effects is determined by the dynamic ratio of the PG forms and their positive or negative action and also by the nature of the effector cells. The scheme in Fig. 65 approximately shows how the above pathways I and II of modulating PG action upon the catecholamine effect are achieved. Let us now examine the two pathways.

Most of the now available evidence shows that PGs of group E are released in a whole series of organs during activation of sympathetic control. They restrict the latter's effect due to suppressed mediator release from sympathetic nerve endings. This suppression is achieved by the PGE effect on a calcium-dependent noradrenaline secretion mechanism in the ending membrane; in this case, the noradrenaline release is diminished, while the neuronal reuptake does not change. As a result, the amount of noradrenaline in the action zone decreases and correspondingly restricts the mediator action on effector cells; and, on the whole, the sympathetic control excitation proves reduced [171].

As is shown in Fig. 65, apart from influencing the release mechanism of the sympathetic mediator, PGs can also influence the catecholamine effect in organs and tissues. Most relevant studies indicate that PGs influence the catecholamine effect at the levels of the adrenoceptors and the adenyl cyclase system of the effector cells [53, 584]. In Fig. 65, this influence is shown by the arrow from PGE to adenyl cyclase; it is achieved in the cell as feedback in the form of suppressed cAMP formation. The sign of this influence is determined both by the PG form and the nature of the tissue. Relevant investigation results are highly contradictory.

At present, the question regarding the role of PGE in modulating the effect of catecholamines and sympathetic regulation on fatty tissue and the stomach appears to be the best studied. It seems that in both cases, PGE limit the adrenergic effect by suppressing cAMP accumulation. Indeed, it was established that in the fatty tissue, the PGs that form under the influence of catecholamines in the course of cATP-dependent lipolysis restrict the lipolytic effect of catecholamines and reduce the release of free fatty acids into the blood. This restriction is achieved by the feedback principle by suppressing the interaction between catecholamines and the adenyl cyclase complex in the cell membrane [53, 536] and, possibly, by activating cAMP phosphodiesterase [423]. Figure 66 shows a scheme of this mechanism, which prevents excessive increase in fatty acid concentration in the blood under strong adrenergic effects and, therefore, plays an important role in the antistress protection.

Presently available data indicate that in the same manner, PGE restrict the effect of adrenergic control excitation on the stomach [53, 212] and, as will be shown below, protect the organ from ischemic injuries caused by

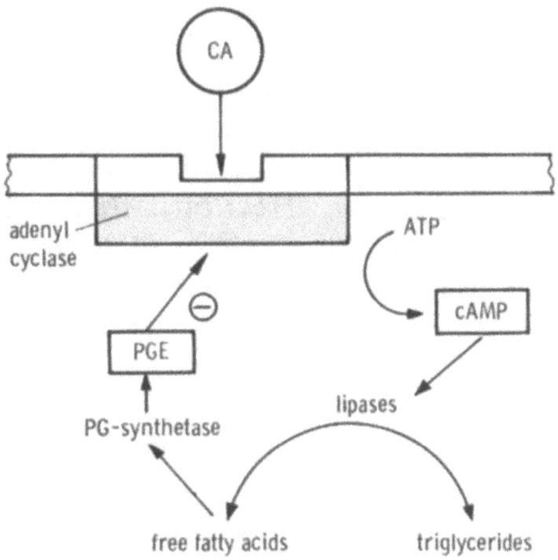

Fig. 66. Prostaglandin-limited lypolytic effect of catecholamine (after Jacquemin, 1976)

adrenergic spasms during stress situations. At the same time, evidence provided by many authors shows that in most tissues, PGs activate the adenyl cyclase complex and boost cAMP synthesis; the highest effect is produced by PGE and lesser effects by other PG forms. More often, PGE have no effect on adenyl cyclase whatsoever and, on the contrary, stimulate cAMP synthesis [14, 584].

On the whole, the above evidence shows that stronger adrenergic effects on various organs and the increased release therein of noradrenaline activates the synthesis and release of PGE, which by feedback may restrict the catecholamine effect. They may do so at the expense of suppressed noradrenaline release from sympathetic endings, as was shown for the heart, spermatic cord, pancreas, spleen, and vessels of numerous organs, and also by suppression of the interaction of catecholamines with effector cell adrenoceptors, as was established for fatty tissue and the stomach. This mechanism plays an important role in preventing stress damage to internal organs.

Another no less important mechanism, through which PGs play a role in preventing stress damages or stress-potentiated diseases is caused by the fact that these compounds possess direct natriuretic and vasodilative effects and the resultant pronounced antispastic and hypotensive effects. Lee [260] established that PGE_1, PGE_2, and PGA have direct natriuretic and vasodilative effects. The most active are PGAs: The PGAs synthesized in the kidney medullar layer are apparently the only PGs circulating in the blood since PGEs disintegrate in the lungs [160].

In the light of these concepts, we can now examine the facts that directly characterize the action of PG as one of a factor that prevents stress damages or stress-provoked diseases. The prophylactic effect of PGEs is most distinctly evident in their ability to prevent ulcerous lesions in the stomach mucous coat upon various stress actions. For instance, Kawarada et al. [212] subjected rats to a 2-h immobilization stress. This resulted in stomach ulcers in 90% of the cases. Prior injection of PGE_1 (1–5 mg/kg) prevented the ulcers, but this depended on the PG dosage: With 1 mg/kg, ulcers developed in 65% of the cases, with 2 mg/kg it was 43%, and with 4 mg/kg they were totally absent. As was shown in Chap. 6, ulcerous lesions of the stomach during stress are caused by the following chain of events: Stress → vasoconstriction in the stomach → focal ischemia of the mucous coat → impaired resistance to the digestive effect of gastric juice. Hence, Kawarada et al. [212] suggest that the protective role of PGE_1 is in removing vasoconstriction, the principal link in the chain. Whereas, according to Segawa et al. [504], PGs do not affect the norepinephrine turnover in the stomach, it may be assumed that their effect on adrenergic control of the stomach is achieved not by restricting norepinephrine release, but rather by influencing the adenyl cyclase complex of the stomach cell membranes. Indeed, Butcher and Baird [53] showed that PGs reduce the cAMP content in the stomach, which according to Ruoff [486] considerably increases with the administration of catecholamines.

Thus, the protective action of PGs against stress-induced ulcers is apparently determined both by the direct vasodilative effect of these compounds and by their effects as factors that restrict the catecholamine influence. A number of studies [65, 504] present considerable relevant interest in that they show that suppression of PG synthesis in the stomach due to the administration of indometacin or aspirin naturally causes ulceration. Segawa et al. [504], for instance, established that when indometacin administered to rats caused PG synthesis to decrease by 86%, stomach ulcers developed in 90% of the cases. Inasmuch as inhibitors of PG synthesis may also cause stomach ulceration without stress, it may be assumed that under stress the PG protective function is determined not only by the action that limits the vasoconstrictive effect of catecholamines, but also by some specific influence, possibly connected with the stabilization of cell membranes.

The hypotensive and preventive effects of PGE are very strikingly evident in hypertension, for example, in hereditary animal hypertension, the development of which we managed to delay by adaptation to hypoxia (see Chap. 3). As another example, Cohen et al. [66] showed that the protracted administration of arachidonic acid, a PG precursor, to rats with pronounced hereditary hypertension leads to a quite stable hypotensive effect. This may be prevented by administering indometacin, which inhibits PG synthesis from arachidonic acid and, hence, depends on activated synthesis of PGE and PGA, which possess vasodilative and antiadrenergic effects. Similar results were obtained by Somova [528], who showed that a diet enriched with linolenic acid, another PG precursor, prevents salt hypertension from developing in rats and causes PG synthesis to be activated; at the same time, the renin activity and catecholamine concentration in the blood plasma decline. In full conformity with these experimental data, Lee [260] noted the direct hypotensive and natriuretic effects of PGE and PGA in patients with hypertension. It cannot, therefore, be doubted that an *alimentary-induced increase in PG biosynthesis may turn into a hypertension-preventing factor.* The possible role of PG in preventing heart injuries in stress has been studied less and is less apparent since PGs, as was shown above, can under certain conditions themselves activate adenyl cyclase and intensify the adrenergic effect.

In our view, the investigation results of Gudbjarnason et al. [152] are of great interest. In their work, they compared the effect of the long-term administration of vitamin E and cod liver oil (in the diet) on rat mortality from isoproterenol administerd twice (20–40 mg/kg) with a 24 h interval between doses. Vitamin E caused a considerable increase in arachidonic acid, a PG precursor, in heart phospholipids, while cod liver oil conversely caused more than a twofold decrease in the arachidonic acid content in heart phospholipids. It appeared that in animals receiving isoproterenol without vitamin E, mortality was 50%; on the other hand, mortality among those that did receive vitamin E was only 25% and among those whose diet included cod liver oil it was invariably 100%.

These data indicate that the protective effect of vitamin E was due to the fact that by creating an increased reserve of the direct precursor of PGs in the myocardium, it helped to activate synthesis of PGE, which as was shown above can restrict the effect of catecholamines. The aggravating influence of cod liver oil on the lethal effect of isoproterenol was apparently due to the fact that cod liver oil caused the arachidonic acid content in the myocardial phospholipids to decrease, thus inhibiting PGE synthesis and, correspondingly, its anticatecholamine protective action. Thus, PGs can play a substantial role in preventing adrenergic injuries, specifically stress-induced injuries of internal organs.

This fact makes probable the important role of the PG system in protecting the heart from ischemia for, as was shown above, the stress reaction plays an essential role in the pathogenesis of myocardial infarction (see Chap. 6). Indeed, recent studies by Lepran et al. [266] showed the role of the PG system in limiting ischemic injuries to be highly significant. In their experiments, rats were given a diet containing 12% sunseed oil for 3 months so as to change the fatty acid composition in the myocardial phospholipids. This resulted in a substantial increase in the level of arachidonic and similar fatty acids in the phospholipids. A special method was used to ligate the descending branch of the left coronary artery in closed-chest conscious rats. This method of inducing experimental myocardial infarction was previously used on dogs to show that it leads to a considerably higher mortality than arterial occlusion under narcosis [296].

In terms of our presentation, it is significant that the high mortality was caused by the concomitant stress reaction, which aggravates heart injury on account of the mechanisms described above (Chap. 6). In the experiments mentioned above of Lepran et al. [266], the experimental infarction reproduced in conscious animals resulted in the death of 81% of control rats. Preliminiary enrichment of the myocardium with phospholipids containing arachidonic acid resulted in a fivefold reduced rat mortality and sharply decreased the duration and frequency of arrhythmias usually observed with experimental infarction. For instance, in the initial 20 min after the artery was ligated, the average heart fibrillation lasted one-sixth of the previous time. This enormous protective effect was completely absent in animals that prior to the infarction had received indometacin, a PG synthesis inhibitor, and was, therefore, determined by mobilization of the PGS, which provided coronary dilation and blocked adrenergic impulses. Generally speaking, the above evidence unambiguously shows that in limiting the mobilizing adrenergic signal, the PG system plays a major role in protecting the organism from stress and ischemic injuries.

In examining the antistress and antiischemic mechanisms that function at the cell and tissue level, we constantly encountered a situation in which these protective mechanisms either blocked or restricted catecholamine action, i.e., the mobilizing central signal. This may create the false impression that protective control mechanisms always restrict the function of a given organ and prevent it from satisfying the requirements that the whole organism makes of it. Actually, it is obvious that in most situations, adaptation is based on the perfect transmission and realization of central control signals and the role of the examined mechanisms is to provide timely specification, i.e., modulation of responses to these signals by the organs in question. It is only in extreme or so-called 'hopeless' situations that the modulating effect may reach the level of blockade, an alternative to the disintegration of the structure of vitally important organs; this chapter has been specifically concerned with adaptation to such situations.

However, one should keep in mind that in extreme situations, control mechanisms may block not only control signals that mobilize the function of a given organ but also, conversely, block the inhibiting signals that could prevent a vitally important organ from hyperfunctioning and, thus, they keep the whole organism alive. We reproduced this situation by means of our previous method, which permitted us to perform dosed narrowing of the ascending aorta and thereby reproduce heart hyperfunction in conscious rabbits with closed chests [327]. The experiments performed using this method [206, 326] were as follows: Prior to aortal narrowing and at various stages of increasing narrowing, i.e., during the growing intensity of heart hyperfunction, we stimulated the peripheral end of the vagus nerve dissected at the neck. The high, negatively chronotropic effect that normally occurs when the vagus nerve is stimulated was found to decrease naturally and disappear altogether with increasing load on the heart. An analysis of this phenomenon showed it to be based on the occurrence of a ventricular pacemaker, which assumes the usual function of the sinoatrial node, i.e., the ventricular pacemaker determines the contraction frequency of the heart ventricles. Prior to stimulation of the vagus nerve, this heterotropic focus of excitation is evident, at considerable narrowing of the aorta, only in individual ventrical extrasystoles. During stimulation of the vagus nerve, automatism of the sinoatrial node is suppressed; as a result, the ventricular pacemaker fully assumes its function. Yet, this never happens in the absence of big loads on the heart. In effect, it emancipates the heart from the effect of the vagus nerve and prevents mortally dangerous disturbance of the cardiac pump function under increased resistance to expulsion. Enhanced automatism of the ventricular pacemaker forms the basis of this vitally important emancipation and it was found to result from the combined effect of cardiac self-control mechanisms and the central adrenergic effect that takes place during narrowing of the aorta.

In terms of our presentation, the phenomenon of emancipation is important as it demonstrates the efficiency of control mechanisms, which may block the inhibitory control signal in order to provide a stably increasing function of a vitally significant organ and thus guarantee the interests of the whole organism, its perfect adaptation to injuries that develop. The presence in the organism of evolutionarily established control mechanisms capable, in extreme conditions, of blocking both mobilizing functions and, conversely, inhibitory control signals is possibly a reflection of the simple fact that some extreme environmental situations demand maximal mobilization of functions, while others require their restriction. As a result, we can conclude that *control modulation and, in extreme situations, complete blockade of central controlling signals is one of the important adaptation mechanisms; in other words, in certain conditions, blockade of information may be no less important for survival than its timely transmission.*

Chapter 8
Antioxidant Factors in the Organism: A System of Natural Prevention of Stress-induced and Ischemic Injuries

> **"Before I go over to these experiments, permit me to express the hope that none of you would do mischief in trying to repeat them for fun."[1]**
>
> Faraday

Activation of the GABA system in the brain, as well as activation of modulatory systems in target organs and other inhibitory and modulatory systems of central and peripheral action undoubtedly limit the stress reaction and may play a role in preventing stress damage. However, the experimental evidence cited above unambiguously shows that despite the functioning of natural prevention systems, sufficiently serious stress actions involve considerable and prolonged LP activation, which is evident in the accumulation of fatty acid hydroperoxides that damage cell membranes. Facts indicate that the role of the stress reaction in the etiology of circulatory diseases and tumor growth is achieved precisely through LP activation. This leads to the concept that the chain of events involving stress → LP activation → injury constitutes a major common link in the pathogenesis of basic diseases that restrict human life expectancy. In this conection, we shall briefly examine the nature of lipid peroxidation and its activation mechanisms and then dwell on the use of antioxidants exogenously administered to prevent stress-induced and ischemic injuries.

Nature and Harmful Effect of Lipid Peroxidation

Lipid peroxidation (LP) represents a physiologic process continously taking place in cell membranes. In addition to renovating the membranes and controlling the activity of membrane-bound enzymes, it plays a role in the biosynthesis of prostaglandins (PGs) and leucotrienes, as well as in phago- and pinocytosis and in breaking down intracellular membranes [201]. The role of LP activation, i.e., of the transition of this physiologic process into a pathologic one has, until very recently been studied in cases in which LP was activated by hyperbaric oxygenation, ionizing radiation, and avitaminosis E. Despite the significance of these factors, taken together they are encountered immeasurably less often than stress-induced LP activation, which seems to occur repeatedly in each individual life. It is, then, obvious that not all stress situations causing LP activation result in desease. This in itself is indicative of the existence in the organism of mechanisms that restrict the duration and intensity of LP activation and prevent stress-induced injuries at the membrane level. Such LP-regulating antioxidant systems really do

1 from *The Story of the Candle.*

exist and to assess their role in preventing stress damage and understand the role of antioxidants in the prevention of these injuries, it is essential to dwell briefly on the very essence of the LP process itself.

The theory of free-radical chain reactions of lipid peroxidation developed on the basis of investigations by Bach [15], Semenov [508], and Emanuel et al. [102]. The role of these processes in normal and pathologic states was the subject of compretensive studies by Tarusov [543], Vladimirov and Archakov [566], Spirichev [532], Burlakova et al. [51], Kozlov [237], Kagan et al. [199], and others. Free-radical oxidation is known to constitute an essential link in such vitally important processes as electron transport by numerous flavinic enzymes, oxidative phosphorylation in mitochondria, nervous impulse conduction, and cell division. Lipid peroxidation always takes place in cell membranes and changes their lipid composition and, thereby, the activity of the lipid-dependent membrane-bound enzymes, to which all the principal enzymic systems in the organism essentially belong.

In this context, we shall briefly examine two questions, namely, the nature of LP and its activation mechanism during ischemic and stress damage, and we shall dwell in greater detail on our data on the use of antioxidants, i.e. LP inhibitors, to provide protection from stress-induced and ischemic injuries.[2] To understand the nature of LP, one should have a clear idea that there are two ways in which oxygen is utilized in the organism. The first is the widely known and studied *oxidase pathway* connected with the oxidation of energy substrates and achieved by cytochromoxidase, the terminal link in the respiratory chain. This reaction is characterized by the fact that oxygen attaches four electrons to form water as a result of this tetraelectron reduction of oxygen. This oxidase oxidation does not stipulate that oxygen be incorporated in the molecule of the oxidized substrate; in normal conditions, it is coupled with oxidative ATP resynthesis and is the principal energy source in living systems. It has been the subject of major attention for many years.

The other much less studied pathway of oxygen utilization, the *oxygenase pathway*, is characterized by the fact that complete tetraelectron reduction of oxygen does not take place; the so-called active forms of oxygen occur due to the attachment of one, two, or three electrons to oxygen. These active forms of oxygen, which are essentially free radicals, possess the ability to react with endogenous substrates that comprise the organism's structures; in such cases, one of the atoms or the entire oxygen molecule is incorporated in the oxidized substrate. Significantly, polyunsaturated fatty acid residues of phospholipids that form the lipid bilayer of all cell membranes occupy an important place among the endogenous substrates of free-radical reactions. The so-called peroxide compounds form during the free-radical oxidation of these lipids; hence, the entire process is called lipid peroxidation (LP).

The diagram in Fig. 67 compares the oxidase and oxygenase pathways of oxygen utilization in the organism; at the same time, it emphasizes the substantial fact that LP in a whole organism is, on the one hand, a source

---

2 This section to p. 385 inclusive was written jointly with V. E. Kagan.

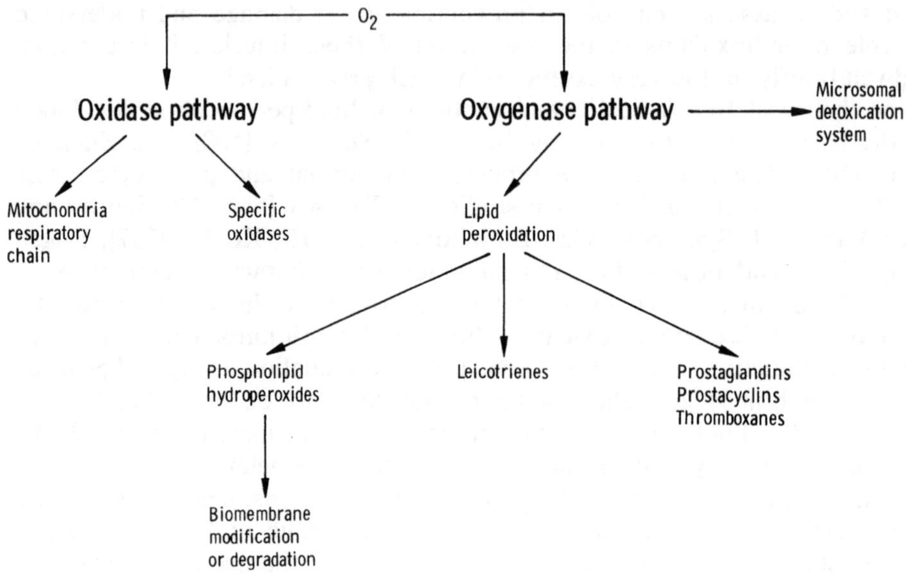

Fig. 67. Correlation between oxidase and oxygenase pathways of oxygen utilization by the organism

of major biologic compounds, viz PGs and leucotrienes, and on the other, is instrumental in renovating and modifying biomembranes and, with excess activation, in disintegrating them as well. To realise how important the mechanism of such excessively destructive LP activation is in pathology, one should keep in mind that in a healthy organism, specialized enzymic systems providing oxygenase utilization of oxygen are chiefly present in the liver. Here there is a specialized oxygenase system containing cytochrome P-450 and providing the detoxifying disintegration of steroid hormones, barbiturates, etc. In other organs, specifically in the myocardium, this type of specialized system of oxygenase oxidation is rare. However, studies performed in recent years have shown a considerable switchover from the oxidase respiratory pathway of utilizing molecular oxygen to the oxygenase pathway constituting the foundation of CP to be feasible. The tetraelectron reduction of O_2 to H_2O that takes place via the cytochromoxidase is possible only with the faultless coordination of the ensemble of transmitters in the respiratory chain. With blockading of the respiratory chain in the final link, i.e., under the action of ischemia, anoxia, and cyanides, the inevitable reduction of NAD to NADH and of intermediary transmitters in the chain may cause the molecular oxygen dissolved in the lipid matrix of the membrane to be reduced. However, this will not be tetraelectron reduction to the final product, H_2O, but incomplete one-, two- and three-electron reduction accompanied by the development of active forms of oxygen, viz free radicals.

Equations(1)–(3):

$$O_2 + O_2^{\bullet -} \rightarrow O_2^{\bullet -} \quad (1)$$

$$O_2^{\bullet -} + O_2^{\bullet -} \rightarrow H_2O_2 \quad (2)$$

$$O_2^{\bullet -} + H_2O_2 \rightarrow OH$$
$$Me^{n+} \cap Me^{n+1} \quad (3)$$

show that one-electron reduction results in the anion-radical of the superoxide [Eq. (1)]; two-electron reduction results in hydrogen peroxide [Eq. (2)]; and, finally, three-electron reduction produces a hydroxyl radical [Eq. (3)]. These processes may take place both on flavin-containing components and on heme-containing components of the respiratory chain; their existence has been proven in experiments, which demonstrated the formation of a superoxide radical and hydrogen peroxide in mitochondria [56].

It is highly significant that in a whole organism, all these reactions and the increased formation of free radicals may take place not only in ischemia and anoxia but also under the directly opposite situation, namely, under hyperbaric oxygenation. Here the amount of oxygen dissolved in the mitochondrial lipid matrix is sharply increased and correspondingly enhances the probability of oxygen reacting with reduced transmitters in the respiratory chain, notwithstanding their insignificant number.

In terms of our presentation, it is important that irrespective of the cause of the increased concentration of free oxygen forms, these free radicals, primarily the hydroxyl radical, react with unsaturated fatty acid residues of membrane phospholipids and this leads to activation of LP. This situation may result from two opposite changes in the oxygen conditions in the cell. In ischemia and anoxia, LP is activated due to the excess electron donors, i.e., reduced transmitters, while under hyperbaric oxygenation, it is activated because of excess electron acceptors, i.e., molecular oxygen. In addition, this concept implies that the highest LP activation may be expected when both these conditions are combined, i.e., when excess oxygen appears in the wake of protracted anoxia and the considerable accumulation of reduced transmitters. This situation, designated reoxygenation, has now been studied well and has been found to be characterized by a considerable accumulation of secondary LP products [147, 148] representing a factor that most effectively disintegrates cell membranes and causes irreversible changes in the heart and other organs [168, 169].

$$RH + OH^{\bullet} \rightarrow ROOH + H_2O \quad (4)$$

$$ROOH + M^{n+} \rightarrow RO^{\bullet} + Me^{n+1} \quad (5)$$

$$RH + RO_2^{\bullet} \rightarrow ROOH + R^{\bullet} \quad (6)$$

Equations (4)–(6) show the manner in which free oxygen forms activate lipid peroxidation. It is clear that in interacting with unsaturated RH lipids in Eq. (4), the hydroxyl radical $OH^{\bullet}$affords the primary molecular product of LP, lipid hydroperoxides ROOH. For further development of LP, complexes of reduced metals with variable valencies are needed. $RO_2^{\bullet}$ radicals result in the participation of these complexes at the expense of the radical disintegration of lipid hydroperoxides in Eq. (5); as with the hydroxyl radical, the above radicals can attack unsaturated lipids and result in Eq. (6).

Naturally, the above changes in the oxygen conditions in the cell are not the only cause of excess LP activation. One may identify at least six LP-activating factors that can act in the whole organism.

1) Increased generation of the radical superoxide $O_2^{\bullet -}$ due to excess catecholamines and the accumulation of their incomplete oxidation products. In this case, the semiquinone forms of catecholamines, namely semiquinone adrenaline and semiquinone noradrenaline, are electron donors for O_2. This situation is more probable during the stress reaction with which we are concerned, i.e., when the oxidation of excess adrenaline to adrenochrome results in semiquinone adrenaline, which as is shown in Fig. 68 may discharge the electron onto oxygen and generate the superoxide radical O_2^- [46], an important LP inducer.

2) Activation of the enzymic systems of generation of active oxygen forms, e.g., activation of the generation of O_2 during the accumulation of reduced pyridine nucleotides under hypoxia, ischemia, and subsequent reoxygenation corresponds to Eq. (1) (see above).

3) Increased oxygen partial pressure (pO_2) in the cell under hyperbaric oxygenation (see above) or by cyanide poisoning whereby O_2 accumulates as a result of its disturbed utilization.

4) Enhanced concentration of reduced metals with variable valencies during the accumulation and disintegration of iron- and copper-containing proteins [Eqs. (3) and (5)], e.g., in hemolytic anemia and other states.

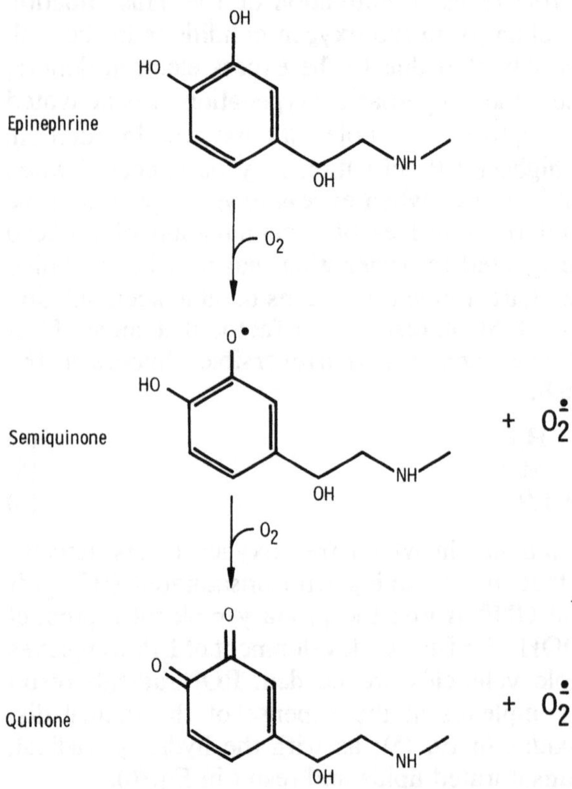

Fig. 68. Formation of superoxide radical during epinephrine oxidation

5) Increased content in biomembranes of unsaturated polyene lipids attacked by peroxide radicals and the sharply accelerated LP in Eq. (6). This situation is observed up on enriching the diet with unsaturated fatty acids and with obesity.

6) Finally, increased LP intensity may be observed during the inactivation of specialized antioxidant systems, which inhibit and thereby limit LP in a whole organism; all this is discussed below.

It is noteworthy that irrespective of the factor that causes excessive LP activation, its significance in the organism is that it entails modified and, subsequently, also harmful effects in cell membranes. It has already been shown that in conformity with the accepted fluid-mosaic model, the biomembrane structure is based on a bimolecular phospholipid bilayer with incorporated membrane proteins. In addition, the phospholipids form an intermittent bilayer in which the polar heads are turned outwardly toward the so-called aqueous hydrophilic phase and the hydrocarbon tails of the fatty acids toward the internal, hydrophobic phase of the membrane.

The reaction of the unsaturated fatty acid residues of phosphatidyl ethanol amine and of other phospholipids from the lipid bilayer with active forms of oxygen (Eqs. (4)–(6)] is the foundation of all the modifying or injurious LP effects in the membrane. As a result of this reaction, a polar hydroperoxide group occurs in phospholipid molecules containing a fatty acid in the second site. The scheme in Fig. 69 reflects this initial phenomenon of LP. The phospholipid hydroperoxides resulting from incorporation of the polar group are unstable; correspondingly, their accumulation leads to a chain of further transformations, i.e., to several quite important modifying effects [11, 203].

The first effect of this kind is an accumulation of lipid hydroperoxides and a decrease in the unsaturated lipids in the lipid microsurroundings of integral proteins, which as already stated shown are the main functional proteins in the membrane. Figure 70 shows that with moderate LP activation involving not more than 2%–5% membrane phospholipids, the occur-

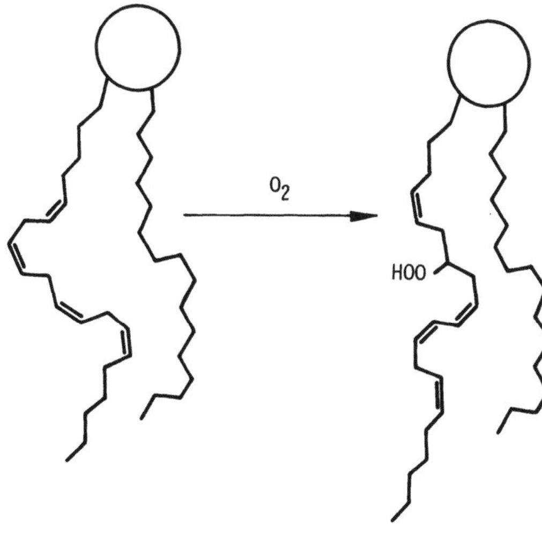

Fig. 69. Formation of phospholipid hydroperoxide, initial stage of lipoprotein (LP)

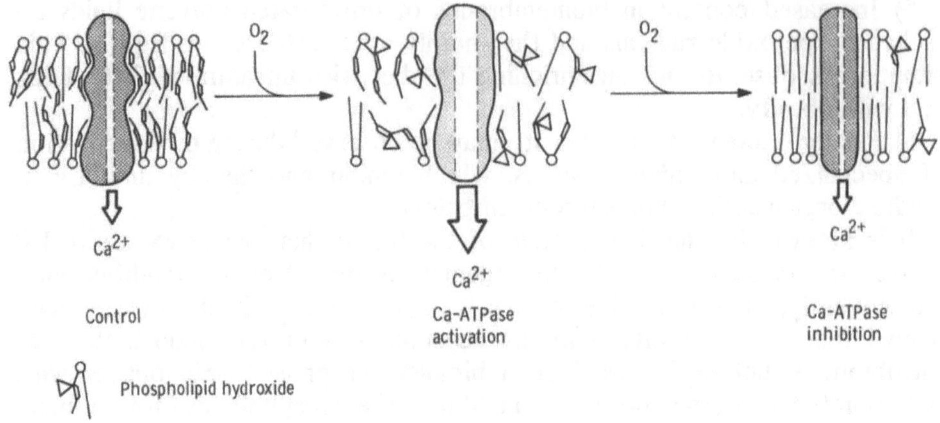

Fig. 70. Variations in Ca-ATPase activity in SR membranes due to modification of lipid surrounding by LP. Explanation in text

rence in the microsurroundings of integral proteins of polar LP products possessing a detergent effect causes the polypeptide chain to become more mobile and as a rule this is accompanied by increased enzyme catalytic activity. With pronounced LP activation, when considerable amounts of membrane phospholipids are subjected to oxidative degradation, a decrease in the amount of unsaturated phospholipids, which causes integral proteins to be 'frozen' in a more rigid matrix, becomes most significant. In this case, the lipid microsurroundings restrict the conformational mobility of the polypeptide chain needed for the normal functioning of enzymes, receptors, and channel-forming proteins. As a result, their functional activity is inhibited. This development, most exhaustively proven for Ca^{2+} ATPase of the SR of cardiomyocytes in vitro [203], should in a whole organism naturally lead to impaired calcium removal from the sarcoplasma and to the harmful effect of the cation.

The second modifying effect is that the LP-caused accumulation of oxidized phospholipids form regulated groups, the so-called peroxide clusters. The scheme in Fig. 71 shows the forming of clusters. As is apparent from Fig. 71, at first the membrane contains no oxidized phospholipids; then, as a result of an attack by active oxygen forms these oxidized phospholipids appear. This is followed by the so-called lateral diffusion of oxidized phospholipids along the membranes to form associates of these phospholipids, the associates being fixed by the interaction of phospholipid molecules with one another and with water molecules. Unlike the invariable bilayer, these membrane sites are not hydrophobic, but hydrophilic, i.e., accessible to water. Further, as a result of the lateral diffussion of whole associates, in each of the membrane monolayers the associates prove to be opposite one another and form channels of permeability that are passable for, e.g., calcium ions [200, 258]. It may be assumed that the formation of such clusters during LP activation caused by ischemia, stress, or other factors may play an essential

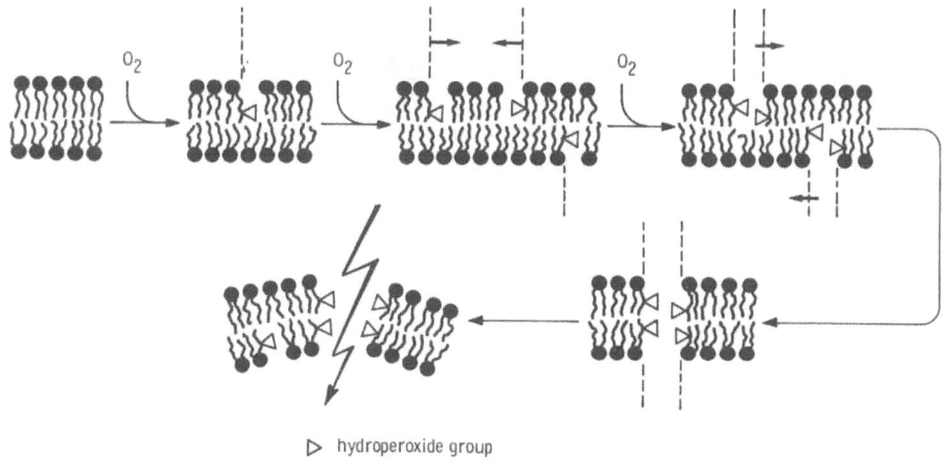

Fig. 71. Formation of "peroxide clusters" and fragmentation of membrane upon LP induction. Explanation in text

role in generating excess Ca^{2+} in the myocytes and in producing the harmful effect of that cation. A further increase in LP products and clusters may become the foundation of fragmentation and degradation of sarcolemmal and sarcoplasmic reticular membranes. Thus, this effect first modifies the myocytic membrane function and then becomes the cause of its degradation.

The third injurious effect of LP is that the degradation products of phospholipid hydroperoxides, namely dialdehydes, interact with free amino groups in membrane proteins, form intermolecular cross-links, and inactivate these proteins. This event is naturally reproduced in vitro when incubating membranes with malonic, glutaric, and other dialdehydes [232] and in vivo forms the basis for the formation of so-called Schiff's bases of the lipofuccin wear pigment. Figure 72 diagrammatically shows how the development of this process leads to cross-links. It is apparent that the hydroperoxide that formed as a result of a reaction between phospholipids and molecular oxygen is degraded to a phospholipid with a shortened hydrocarbon chain in the second site and to dialdehyde, a short hydrocarbon fragment. The simultaneous interaction of the essentially bifunctional dialdehyde molecule with the amino groups of two protein molecules leads to the formation of a cross-link on account of the inactivation mechanism of cation pumps. Ion conduction channels, the systems responsible for transmembrane signal transmission, may play a significant role in injuring the myocardium during all the processes for which LP activation has been proven.

The fourth and, in this context, the last harmful effect of LP is the oxidation of sulfhydryl groups, quite often localized in enzymic centers active to sulfones. This event is known to be accompanied by inactivation of bound enzyme membranes [203] and by increased membrane permeability [203]. It is significant that the first two LP modifying effects represent changes in membrane phospholipids and are to some extent reversible due to the replacement of

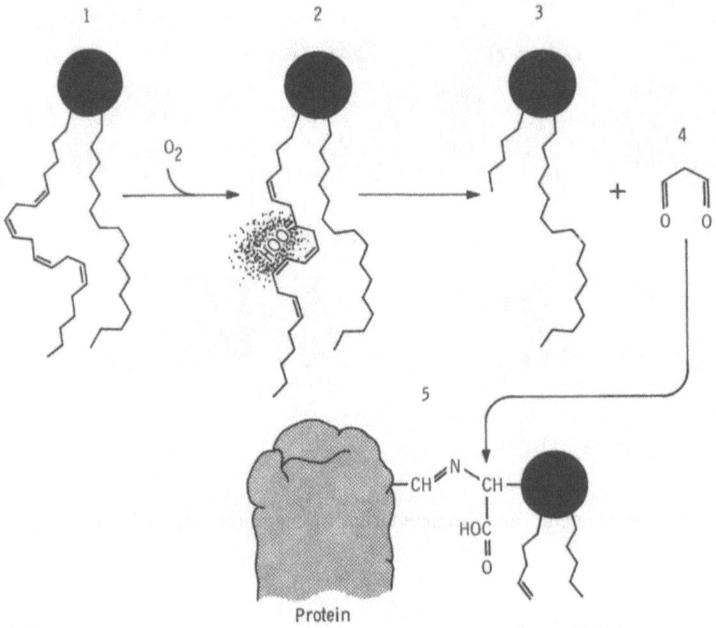

Fig. 72. Inactivation of membrane enzyme proteins through forming of structures during LP activation. Explanation in text

oxidized hydrocarbon chains in oxidized phospholipids by means of a special system of enzymes, i.e., by phospholipases and acyl transferases. At the same time, the last two effects are connected with injuries to the protein components of membranes; they are less reversible, since the injuries may be eliminated only by replacing the damaged molecules with proteins synthesized de novo.

To understand the role of LP activation in injuring cell membranes and in the death of cells, specifically cardiomyocytes, during stress, ischemia, and reoxygenation, one should keep in mind that this component of the lipid triad is organically associated with its other components and pathogenic links of both stress-induced and ischemic heart injuries. As was already shown, LP activation causes the system of lysosomes, in which the larger moiety of the cell phospholipases is concentrated, to be labilized. The phospholipases released from the lysosomes may play an important role in degrading the membrane lipid bilayer and in forming lysophosphatides [210]. As with the high fatty acid concentrations observed in the blood in stress and myocardial infarctions, lysophosphatides, by means of their detergent action, impair the orderly arrangement of the phospholipids in the membranes, and this so-called chaotropic effect has now been shown to activate LP [166]. The increased membrane permeability to calcium resulting from this chain reaction, and the excess calcium in the cell formed as a result in turn activate phospholipases and, possibly, LP [566]. As a result, the vicious circle that plays an important part in injuring cell membranes, in making these injuries irreversible, and in causing the death of cells, is broken closes.

Any cell that achieves aerobic metabolism has the necessary conditions for an LP reaction to take place. Indeed, polyene lipids, the LP substrates, are one of the major components of plasma and intracellular membranes. Metalloproteins, the initiators and catalysts of LP and the generators of active oxygen forms, are also permanently present in the cell. Yet, in spite of this, in a healthy organism harmful LP activation and cell membrane degradation do not take place. During normal stress situations and relative hypoxia caused by big loads, this activation proves to be highly limited and does not result in irreversible injuries. This situation, which makes it possible to preserve life and active behavioral responses in complex environmental conditions is provided by the existence of a continually functioning and sufficiently reliable antioxidant system, which restricts LP virtually in all its links.

Antioxidant Systems and Action Mechanisms of Exogenous Antioxidants

The reaction of free-radical lipid oxidation in the cell is controlled by various control systems. The multistage character of LP permits it to be controlled in the cell and, under specific conditions, to be restricted at different stages; different authors [51, 123, 415] have shown this to be achieved by various systems, both of an enzymic and nonenzymic nature.

The scheme in Fig. 73 provides a generalized view of the concept on the generation mechanism of active oxygen forms and lipid peroxides mentioned above and, based on this, shows the principal points of application of the cell antioxidant systems. From Fig. 73 it follows that various steps of the cascade in the antioxidant cell mechanism have a specific specialization. For instance, the formation of $O_2^{\bullet}$ may be prevented by electron acceptors competing with $O_2^{\bullet}$, e.g., by vitamin E in quinone form, by vitamin K_3. Transition from $O_2^{\bullet}$ to H_2O_2 is blocked by acceptors of the superoxide anion radicals, for example, by sulfur-containing amino acids, methionine and cysteine.

The consistent effect of two intercoordinated enzymic systems, i.e., superoxide dysmutase and catalase, inhibits the formation of hydroxyl radicals. Inasmuch as the effectivity of the interaction between hydroxyl radicals and polyene lipids is exceptionally high and is only diffusively limited, these two enzymic systems play the central role in controlling LP at the initiation stage. Despite the fact that these systems are powerful in physiologic conditions, reliable antioxidant protection generally requires the presence of inhibitors specialized in limiting the subsequent LP stages. They are 'traps of hydroxyl radicals,' namely aliphatic alcohols, and also natural antioxidants, the traps of peroxide radicals α-tocopherol, steroid hormones, vitamin C, and ubiquinones.

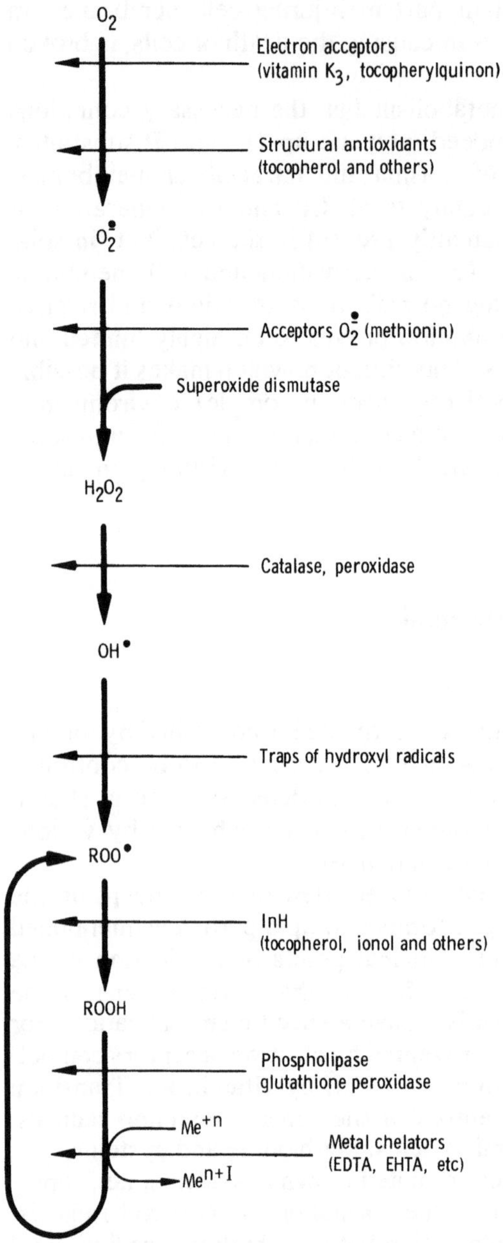

Fig. 73. Basic antioxidant systems of the organism. Explanation in text

The final LP stage involves reactions catalyzed by glutathione peroxidase, which converts phospholipid hydroperoxides to oxyderivatives. The latter, in turn, are attacked by phospholipase A_2 and this results in the release of oxy acids, intermediates of β-oxidation of fatty acids. However, in the case of insufficiently active glutathione peroxidase and in the presence of metal ions of variable valency, new radicals may form from the phospholipid hydroxide to cause the chain reaction and the ramified nature of lipid peroxidation. To prevent the reaction from developing in this manner, there is yet another mechanism of antioxidant protection, which, as shown in Fig. 73, involves the removal of metal ions of variable valency from the membrane hydrophobic phase by means of chelating compounds, viz organic acids, histidine.

Finally, a substantial nonspecific factor of LP control shown in the scheme is the so-called structural antioxidant [283]. The latter should be understood as a set of membrane properties that restrict the accessibility of polyene fatty acyls of phospholipids, the LP substrates, for molecular oxygen and also for LP catalysts.

The fact that the LP is sharply accelerated through the action of 'chaotropic agents', which disturb the orderly organization of the lipid and protein components of the membrane and enhance their molecular mobility, is major proof of the significance of the structural antioxidant. Apart from high fatty acid and lisophospholipid concentrations, chaotropic factors also include anions of the rhodamin, bromide, and perchlorate types possessing detergent-like effects [166].

Inasmuch as the conditions needed to activate LP are constantly present in a living cell, a defect in any of the links of the above system of antioxidant protection should in itself understandably lead to LP activation. This situation is actually observed in avitaminosis E, acatalasemia, and genetically determined decreased activity of superoxide dysmutase [122] and glutathione peroxidase; as will be shown below, it is also observed in hypoxic and reoxygenative stress-induced injuries of the heart. Conversely, increased power of the antioxidant system, caused, for example, by administration of natural of synthetic antioxidants, provides a certain guarantee against LP activation.

In terms of our presentation, the action mechanism of tocopherol, the principal nonenzymic antioxidant, is of the highest interest, since the exogenously administered synthetic antioxidants presently used for protection against stress and ischemia act on a similar mechanism.

It has now been established that α-tocopherol, a phenol-type antioxidant, virtually does not interact with lipid peroxides, but does effectively react with free $RO^{\bullet}$ radicals that form in the disintegration of hydroperoxides in Eq. (5); these hydroperoxides are themselves capable of attacking unsaturated lipids to lead the peroxidation chain.

The reaction of LP inhibition may be written down in a general form, keeping in mind that InH is the LP inhibitor molecule, $RO_2^{\bullet}$ the peroxide radical, $R^{\bullet}$ the radical of the unsaturated fatty acid, ROOH the hydroperoxide, and $In^{\bullet}$ the inhibitor radical.

$$InH + RO_2^{\bullet} \rightarrow In' + ROOH \tag{7}$$

$$InH + R^{\bullet} \rightarrow In' + RH \tag{8}$$

It is clear that binding of peroxide radicals $RO_2^{\bullet}$ in these reactions sharply limits the LP intensity. This inhibitory effect is deepened by the reactions in Eqs. (9) and (10), in which the free radicals of the inhibitor react with the fatty acid radicals $RO_2^{\bullet}$ and with one another. This results in the binding of the additional amount of peroxide radicals and in neutralization of the radicals of the inhibitor itself.

$$In' + RO_2^{\bullet} \rightarrow RO_2In \qquad (9)$$
$$In' + In' \rightarrow In_2 \qquad (10)$$

As a result, the LP intensity proves substantially restricted so as to prevent considerably the harmful effects of LP and the lipid triad as a whole. This is exactly how α-tocopherol, ionol (a synthetic antioxidant of the phenol type), and antioxidants from the oxypyridine group act; their use in preventing stress-induced, ischemic and reoxygenative injuries is discussed below.

Use of Antioxidants in Preventing Stress-induced, Reoxygenative, and Ischemic Injuries

The use of exogenous antioxidants in preventing stress damage is, in effect, simulating the way the organism naturally reacts – just as in all the previously examined examples of antistress protection – since in the stress reaction the organism produces antioxidants, i.e., glucocorti-coids, simultaneously with proantioxidants, i.e., catecholamines. In addition to its pragmatic aspect, antioxidant protection by the organism based on the factors cited above possesses substantial theoretical significance. Indeed, the presence of intense and protracted activation with stress does not in itself strictly prove the LP role in the pathogenesis of stress-induced injury, for abstract reasoning would assume LP to be the effect, not the cause, of such injuries. It would appear that in order to ground thoroughly the concept that LP activation is an essential or decisive link in stress-induced injuries of the organism, one should, prior to the stress situation, block LP with a sufficiently powerful antioxidant to prevent the considerable accumulation of lipid hydroperoxides and clarify whether or not stress-induced injuries could occur in such conditions.

Correspondingly, in the first stage of a set of our experiments, we studied the effect of the preliminary administration of ionol (2,6-ditert-butyl-4-methyl phenol) on LP activation in stress; in the second stage, we studied the effect of ionol and other antioxidants on stress-induced injuries assessed by the degree of fermentemia, extent of ulcerous lesions of the stomach, and injury to the brain and retina; finally, in the third stage, we studied the effect of the preliminary administration of ionol on stress, and on hypoxic, ischemic, and reoxygenative injuries of the heart. All these experiments were performed on Wistar male rats. Ionol was injected intraperitoneally (120 mg/kg body weight) daily for 3 days prior to the EPS and decapitation [367].

Table 43 shows the results of our experiments. From Table 43 it follows that preliminary ionol injection substantially reduced the accumulation of hydroperoxides and fully prevented the accumulation of Schiff's bases in the brain, skeletal muscles, and myocardium of stress-exposed animals. This means that post-EPS LP activation may be prevented or substantially reduced by means of ionol. In assessing, on the basis of this fact, the ability of ionol to protect the organism from stress-induced injuries, one may ostensibly decide what role LP activation plays in the pathogenesis of those injuries.

Table 43. Effect of ionol on LP activation in the myocardium in EPS

	Number of animals	Accumulation of lipid hydroperoxides (according to UV spectroscopy), nmol/mg lipids	Fluorescence of Schiff's bases, relative units
1. Control	5	16.7 ± 1.4	10.5 ± 3.5
2. EPS	7	47.6 ± 2.4	27.6 ± 4.2
3. Ionol	7	16.7 ± 0.5	10.6 ± 2.2
4. Ionol + stress	7	28.6 ± 2.4	13.2 ± 1.8
Significance of differences			
p_{1-2}		<0.001	<0.01
p_{2-4}		<0.001	<0.01

A reliable criterion of cell membrane injury is fermentemia, which, as shown above, is fully achieved under the stress action we applied. Hence, in the second stage of the experiment, we studied the effect of the preliminary administration of the antioxidants ionol, α-tocopherol (50 mg/kg intraperitoneally for 5 days), and sodium selenite (30 μg/kg subcutaneously for 3 days, including the day of stress), and the β-blocker Inderal (0.2 mg/kg intravenously, once before the beginning of stress action) on fermentemia, which was observed 8 h after the start of EPS [378].

The results of these experiments (Table 44) show that just as in the previously described experiments, EPS caused pronounced fermentemia in rats: The activity of lysosomal enzymes and also of alanine and aspartate trans-

Table 44. Prevention of EPS-induced fermentemia by preliminary administration of Inderal and Antioxidants

	Activity		Content, μmol/ml/h	
	Acid cathepsin, μg tyrosine/mg protein/h	Leucine aminopeptidase, μmol NH_3/h	Alanine transaminases	Aspartate transaminases
1. Control	0.74 ± 0.05	0.055 ± 0.002	1.58 ± 0.3	2.39 ± 0.29
2. EPS	1.82 ± 0.06	0.099 ± 0.006	3.17 ± 0.4	6.10 ± 0.10
3. Inderal + EPS	0.83 ± 0.06	0.073 ± 0.003	1.82 ± 0.3	3.17 ± 0.12
4. Ionol + EPS	0.80 ± 0.01	0.063 ± 0.001	1.84 ± 0.4	2.90 ± 0.16
5. α-tocopherol + EPS	1.22 ± 0.02	0.074 ± 0.001	2.57 ± 0.4	4.13 ± 0.12
6. Na-selenite + EPS	1.38 ± 0.02	0.080 ± 0.001	2.52 ± 0.6	3.73 ± 0.11
Significance of differences				
p_{1-2}	<0.001	<0.1	<0.001	<0.001
p_{1-3}	>0.1	>0.1	>0.1	>0.1
p_{1-4}	<0.05	>0.1	>0.1	>0.1
p_{1-5}	<0.05	<0.05	<0.05	<0.05
p_{1-6}	<0.05	<0.05	<0.05	<0.05

Ten animals were used in each series.

aminases in the blood plasma increased 2–2.5-fold. Injection of the β-blocker, Inderal, prevented fermentemia almost completely. This highly significant fact agrees with the above concept that noradrenaline, which in high concentrations acts via adenyl cyclase, is the principal cause of cell damage in stress. From Table 44, it can further be seen that antioxidants that block the next link in the pathogenic chain of stress injury, namely LP, also prevent fermentemia. However, the extent of this preventive effect for various antioxidants varies. Ionol prevents fermentemia as effectively as Inderal, or even better. The activity of all four enzymes determined in the blood plasma of rats that received ionol prior to stress was significantly the same as in control animals not exposed to stress. The two other antioxidants, namely α-tocopherol and sodium selenite, also possessed certain preventive effects, which however were less pronounced than those of ionol and Inderal.

Thus, prior administration of ionol almost completely prevented the stress injury as assessed by fermentemia; other antioxidants possessed less pronounced effects.

The evidence presented in Table 45 shows that in our experiment with EPS, 75% of the rats developed ulcerous lesions of the mucous coat of the stomach. The average length and number of lesions per stomach were 5.4 and 6.6 mm, respectively. Preliminary administration of ionol resulted in a twofold reduction in the number of animals with ulcers, a 2.5-fold decrease in the average number of ulcers, and a fourfold decrease in the average ulcer length. Thus, ionol largely prevented ulcerous lesions of the stomach in stress.

In the next series of experiments, we studied the effect of EPS and ionol on LP activation in the brain, and also the appearance in the blood of autoantibodies against brain tissue, a significant indication of cerebral injury [41]. A study of the accumulation dynamics for products of LP in brain tissues after EPS showed that as early as 2 h after EPS, the number of fluorescent LP products had increased by 2.7-fold and continued to rise for 2 days; control values were attained on day 10. Prior administration of ionol significantly reduced LP activation in the brain tissues of EPS-exposed rats. For instance, the stress-induced increased intensity of fluorescence of Schiff's bases in the lipids extracted from the brains of animals that did and did not receive ionol was equal to 39% ± 25% ($p < 0.05$)

Table 45. Effect of preliminary administration of ionol on development of ulcerous lesions of the stomach during EPS

	Stress ($n = 16$)	Ionol + stress ($n = 16$)	p
Average number of ulcers per stomach	5.4 ± 1.1	2.1 ± 0.8	< 0.05
Average ulcer length per stomach, mm	6.6 ± 1.0	1.6 ± 0.7	< 0.05
Number of animals with ulcers (against total number of animals), %	75	36	

and 165% ± 51% ($p < 0.001$), respectively, i.e., it decreased fourfold under the action of ionol.

In determining the antibodies against two forms of brain antigens in the serum of EPS-exposed animals, we revealed the antibody titer to be reliably increased both against water-soluble and membrane-bound antigens. The content of antibodies against membrane-bound antigens was reliably higher than against water-soluble antigens (18, 19, and 5.71 CH_{50} hemolysis units, respectively; $p < 0.001$). In this connection, it should be noted that the LP products are surface-active substances and can solubilize not only peripheral but also integral membrane proteins. The administration to rats of ionol, an inhibitor of free radical reactions, eliminates the EPS-induced accumulation of LP products and totally prevents antibodies against brain antigens appearing in the blood. In control experiments, ionol injections do not affect the formation of antibodies in response to a single administration of the antigen.

It was, thus, shown that the use of antioxidants prevents the stress-induced LP activation in brain tissues and prevents the titer of antibodies against the brain antigens from rising in the peripheral blood. Since the antioxidant concentrations used in the study had no effect on the organism's ability to generate antibodies in response to antigen administration, the data obtained show that LP was involved in the mechanisms of antigen discharge by the stress-injured brain via the hematoencephalic barrier. One may expect that with other LP-inducing effects too organospecific antigens would appear in the blood; as a consequence, antibodies would form against these antigens. For instance, hyperoxia is known to cause the appearance of brain-specific proteins, including S-100 proteins, in animal blood [523].

Hence, the experimental results indicate that LP activation is the actual cause of the release from the brain and appearance in the blood of antigens via the hematoencephalic barrier. In view of this, it would appear advisable to study the role of LP in injuring the brain during stress and the initiation of its autoimmune diseases.

Similar results were obtained by Shvedova et al. [518] in studying the effect of Inderal in EPS-induced injury of rat retina. The experiments were performed on male Wistar rats (170–200 g). Emotional painful stress was produced by the procedure described above. Antioxidants, i.e., ionol and oxypyridine (OP-6), were injected intraperitoneally for 3 days once a day prior to EPS (doses: 100 mg/kg and 120 mg/kg, respectively). Inderal was administered once (0.5 mg/kg) prior to EPS. The retina electrogram was recorded by means of a cored electrode (active electrode) on an encephalograph. Light test-stimulation was achieved by a photostimulator (C-1, flash energy 0.3 J). The pupil was widened with 1% homotropin, and the eye was anesthetized with 1% dicain. The rats were decapitated 2 h after the termination of stress. Lipid peroxidation activity in the retina was estimated by the above method (p. 268). The α-tocopherol content in the retina was determined by a modified method [544], recording characteristic fluorescence (λ maximum excitation = 240 nm, maximum emission = 340 nm). The superoxide dysmutase activity in the retina was measured by the method described by Kagan et al [202].

Table 46. Content of LP endogenous products in rat retinas 2 h after EPS

	Lipid hydroperoxides (Optical density unit $\lambda_{max} = 232$ nm)	LP final products (fluorescence intensity with $\lambda_{maximum}$ excitation = 365 nm, $\lambda_{maximum}$ emission = 420 nm, relative units
Control	0.51 ± 0.06 $n = 11$	4.7 ± 0.3 $n = 11$
EPS	0.83 ± 0.07 $n = 11$	11.4 ± 0.7 $n = 11$
Antioxidants + EPS:		
a) 4-methyl-2,6-ditertbutyl phenol	0.52 ± 0.03 $n = 8$	5.2 ± 0.6 $n = 8$
b) OP-6 from oxypyridine class	0.55 ± 0.06 $n = 8$	4.8 ± 0.7 $n = 8$
Inderal + EPS	0.67 ± 0.10 $n = 8$	5.6 ± 0.3 $n = 8$

From Table 46 it is apparent that 2 h after EPS, the amounts of primary hydroperoxides of polyene lipids and of fluorescing Schiff's bases, the LP final products in the retina, had increased by 1.5- and 2.5-fold, respectively. This means that as in the other previously studied tissues, EPS in the retina leads to activation of endogenous LP.

The occurrence of disturbances in the LP control system, whose main components in the retina are tocopherol and superoxide dysmutase, may be one of the causes of the accumulation of LP endogenous products.

Table 47 shows that EPS does not cause changes in the activity of superoxide dysmutase in rat retinas. At the same time, the post-EPS content in the retina of α-tocopherol (vitamin E), a natural inhibitor of free-radical oxidation, was reduced by more than 35% against the initial level. Thus, EPS causes a shortage of this natural antioxidant in the retina. Replenishment of this shortage by synthetic antioxidants should have, with great probability, prevented the EPS-induced accumulation of LP products in the retina. Indeed, the preliminary administration of ionol or OP-6 from the class of oxypyridines completely eliminated the EPS effect: LP products did not accumulate (Table 46). Hence the question: Is the natural antioxidant shortage, i.e., its decreased concentration, the primary mechanism of LP activation in the retina

Table 47. Effect of EPS on superoxide dysmutase activity and α-tocopherol content in rat retina

Antioxidant system	Control	After EPS
Activity of superoxide dysmutase (unit/mg protein)	0.58 ± 0.03 $n = 10$	0.63 ± 0.04 $n = 10$
α-tocopherol content, μg/kg protein	0.47 ± 0.05 $n = 8$	0.29 ± 0.07 $n = 8$

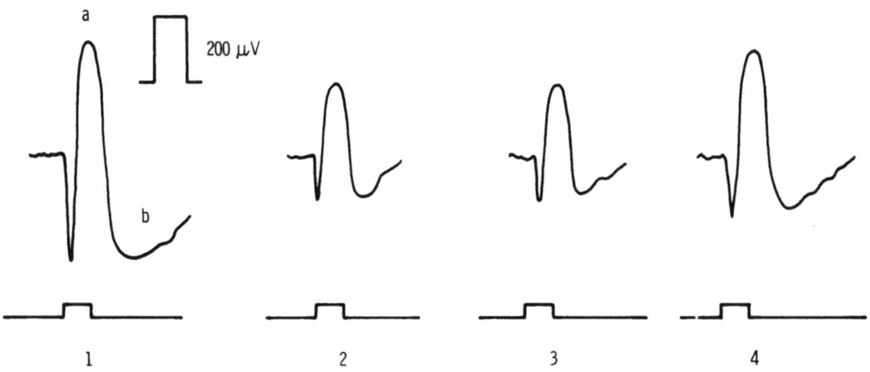

Fig. 74. Effect of stress on typical readings of electrogram of rat retina. *1,* prior to stress; *2–4,* 2, 24, and 48 h after stress action

during EPS, or is it a secondary outcome resulting from the interaction of α-tocopherol with radical LP intermediates? The answer may be found in the results of experiments with Inderal, a β-adrenoblocker (Table 46). Inderal, which in itself is not an LP inhibitor either with nonenzymic (Fe^{2+} + ascorbate) or enzymic (Fe^{2+} + NADPH) pathways of inducing LP in vitro, causes virtually complete inhibition of LP activation in the retina during EPS. Thus, Inderal prevents LP activation during EPS, acting by means of a mechanism that basically differs from the antioxidant action mechanism, i.e., it most probably inhibits the adenyl cyclase system or reduces the Ca^{2+} influx into the cells. In studying the EPS effect on retinal electric activity (Fig. 74), we found the amplitude of waves a and b in the electroretinogram (ERG) to be reduced; this decrease was observed 2 h after EPS and persisted even after 20 h. Lipid-peroxidation induction in retinas isolated in vitro is known to result in the suppression of ERG. A comparison of this fact with the above result indicates activation of endogenous LP in vivo to be the mechanism that inhibits retinal electric activity. The data in Table 48 show the protective effect of both antioxidants, preliminarily injected, in preventing EPS-induced retinal injury and experimentally corroborate the above hypothesis. Thus, EPS results

Table 48. Changed amplitudes of waves a and b ERG for rat retina during EPS. Protective effects of antioxidants

	ERG wave	Control	After EPS			
			2 h	24 h	48 h	72 h
Control	a	267 ± 17	109 ± 8	120 ± 10	130 ± 20	260 ± 30
	b	625 ± 47	256 ± 18	300 ± 23	340 ± 37	630 ± 50
Ionol	a	267 ± 17	175 ± 17	260 ± 19	260 ± 18	–
	b	625 ± 47	510 ± 28	617 ± 34	620 ± 35	–
OP-6	a	267 ± 17	180 ± 19	253 ± 22	257 ± 19	–
	b	625 ± 47	500 ± 26	610 ± 38	613 ± 28	–

Table 49. Parameters of enzyme system of Ca^{2+} transport in SR membranes (M ± m)

	Control	Ionol	EPS	Ionol + EPS
Activity, nmol, P_{inorg} per 1 mg protein in 1 min:				
ATPase	1,267 ± 163 (12)[a]	1,220 ± 181 (6)	1,026 ± 249 (11)	1,199 ± 105 (8)
ATPase	694 ± 113 (12)	688 ± 93 (6)	642 ± 153 (10)	640 ± 187 (8)
Ca ATPase	502 ± 68 (12)	488 ± 71 (6)	359 ± 127[b] (10)	469 ± 68 (8)
Velocity, nmol Ca^{2+} per 1 mg protein in 1 min				
Ca^{2+} accumulation	37.2 ± 4.0 (10)	36.2 ± 3.2 (6)	23.1 ± 4.2[b] (8)	29.0 ± 3.8 (8)
Ca^{2+} binding	3.80 ± 0.42 (10)	4.11 ± 0.85 (6)	2.23 ± 1.14 (8)	3.85 ± 0.31 (8)
Amount of Ca^{2+} bound, nmol Ca^{2+} per 1 mg protein	7.20 ± 2.78 (10)	7.35 ± 3.01 (6)	3.90 ± 1.22 (7)	8.37 ± 3.31 (8)

[a] Number of experiments in brackets.
[b] $p < 0.05$.

in the accumulation of LP products in the rat retina causing the suppression of retinal electric activity. Hence, the suppression of LP activation by antioxidants prevents stress-induced retinal injury no less effectively than it prevents stress-induced injuries of the brain or internal organs.

As was indicated above, the third stage of the investigation was devoted to antioxidant protection of the heart. At the outset, the protective effect of antioxidants was estimated by the prevention of stress-induced disturbances in metabolism and structure and then by the prevention of stress-induced impairment of the function of an isolated heart and by the loss of enzymes from that heart.

Table 49 characterizes the effects of EPS and ionol administered prior to stress on the ability of myocardial SR elements to bind and accumulate calcium. Relevant data in the previously mentioned investigation [381] unambiguously show that for the microsomal fraction of the myocardium of stress-exposed rats, the basic parameters of active calcium transport are essentially reduced and the preliminary administration of ionol eliminates this phenomenon. Indeed, the calcium accumulation rate in the presence of oxalate,

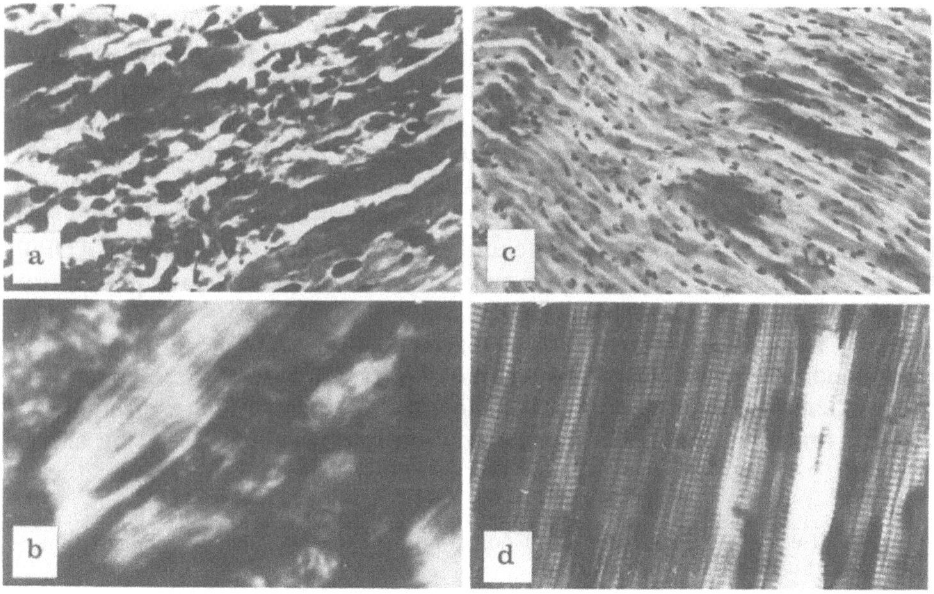

Fig. 75 a–d. Effect of ionol on morphological changes in rat myocardium during stress. **a** Rat myocardium 45 h after sustained stress. Focal myocardial necroses with early inflammatory infiltration; intensive reaction to iron. Pyknotic changes in nuclei and fragmentation of muscle tissue perceptible in necrotized myocardial tissues. Perls' reaction, × 126. **b** Rat myocardium 45 h after stress. Sharply increased anisotropy with merger of myofibrillar A disks forming continous anistropic conglomerate. Third and fourth degrees of contracture. Polarized light, × 295. **c** Rat myocardium 45 h later against ionol background. Positive Perls' reaction to iron weakly pronounced in individual muscle tissues. Perls' reaction, × 79. **d** Rat myocardium 45 h after stress sustained against ionol background. Individual myocardial tissue with intensified anisotropy. Insignificant (first degree) myofibrillar contracture. Polarized light, × 126 its binding rate, and the amount that can be bound by the SR in the presence of oxalate were reduced after EPS by 40%–45%; in animals that were given ionol prior to stress, the impaired function of the calcium pump was not significantly evident.

In effect, this signifies that the suppression of excess LP activation prevents the membrane system of calcium transport from being damaged during serious stress action. The excess Ca^{2+} resulting from impaired calcium transport in the cardiomyocytic SR is in itself a harmful factor and constitutes an important link in the pathogenic chain of stress-induced injuries. Hence, one may a priori expect that the antioxidant that prevents the given link from becoming a reality prevents injury itself [387]. Indeed, morphological studies have shown EPS to cause pronounced focal myocardial lesions characterized by the presence of micronecroses in myocardial cells with a sharply increased Perls' test for iron, disappearance of transverse striation, continuous or partial homogenization, and fragmentation of muscle fibers. The cell nuclei are subjected to lysis or pyknotic changes. Cells with deep contractural changes (degrees 3 and 4), evident in enhanced anisotropy with the merging of myofibrillar A disks prevail among the injured cells (Fig. 75 a, b). Animals that were administered ionol prior to EPS were found to have only a few muscle fiber sites with weakened transverse striation, swelling, and positive Perls' test for iron. The myocytic nuclei were preserved, albeit somewhat swollen. Individual myocardial cells showed enhanced anisotropy and myofibrillar contractures (degrees 1 and 2) (Fig. 75 c, d). As in the previously described experiments, the following indices were determined in studying mitochondrial functions: V_3, the oxygen utilization rate when adding ADP (natoms 02/mg protein/min); V_4, the oxygen utilization rate after exhausting the added ADP; RC_C, the coefficient of respiration control after Chance (the energy-linked reation of calcium with mitochondria) (V_3/V_4); ADP/0, the index of phosphorylation efficiency; and ADP/t, the oxidative phosphorylation rate (μM ADP/mg protein/min).

From Table 50, it can be seen that oxidation and phosphorylation in mitochondria isolated from the hearts of rats after EPS were essentially independent of one another. For instance, the ADP-activated respiration rate was 1.5 times less, and the phosphorylation rate was suppressed by almost twofold.

Table 50. Effect of EPS and antioxidant (ionol) on oxidative and phosphorylating functions of mitochondria isolated from rat hearts

	Indices of respiration and oxidative phosphorylation of mitochondria				
	V_3	V_4	RC_C	ADP/O	ADP/t
Control	253.3 ± 22.9	91.3 ± 8.3	2.86 ± 0.2	1.95 ± 0.1	522.2 ± 40.9
Ionol	273.5 ± 18.1	104.0 ± 8.5	2.75 ± 0.09	1.8 ± 0.13	486.2 ± 43.0
Stress	161.7 ± 13.8^b	112.5 ± 6.5	1.7 ± 0.1^c	1.3 ± 0.2^a	275.8 ± 39.2^c
Ionol + stress	213.8 ± 14.7	82.7 ± 7.4	2.6 ± 0.2	1.9 ± 0.2	419.8 ± 38.2

Significant differences: [a] $p < 0.05$; [b] $p < 0.01$; [c] $p < 0.001$. [d] Twelve animals in each group.

The coefficient of respiratory control decreased by 40% and phosphorylation efficiency decreased by more than 30%. Ionol prevented these disturbances in respiration and phosphorylation. For instance, the ADP-activated respiration rate proved to be reduced only by 16%, and the phosphorylation rate by 20%. The coefficient of respiratory control and phosphorylation efficiency remained essentially the same.

A similar protective effect was obtained in our laboratory in studying the effect of the preliminary administration of Inderal and ionol on stress-induced injuries and on subsequent reparation of the DNA in the myocardial cell nuclei. Inderal, the β-blocker, and ionol, the inhibitor of lipid peroxidation, were found to prevent the decrease in DNA polymerism and the activation of its reparative synthesis that usually arise in the heart muscle and orther organs during EPS [341]. These facts agree with the above concepts that during stress, excess catecholamines induce LP activation in the heart and that free-radical LP products injure the DNA in the myocardial cell nuclei.

Multiple-injury–DNA-reparation cycles accompanying repeated stress actions may, in the course of ontogenesis, lead to the accumulation of genetic faults and become a factor in aging. The fact that antioxidants prevent the occurrence of these cycles agrees well with the gerontoprotective effects of antioxidants proven by Emanuel et al. [103]. Further studies of the peroxidation inhibitor's ability to prevent stress-induced and hypoxic injuries of the myocardium involved a comparison of isolated hearts from four groups of animals – controls, stress-exposed animals, animals administered with the protective factor, and stress-exposed animals after preliminary administration of the peroxidation inhibitor. The contractile function of the isolated hearts was studied by two different methods, namely, that of Neely et al. [413], which allows the stroke volume of a working heart and the oxygen it utilizes to be measured, and that of Fallen et al. [107]; the latter was used on an isovolumic heart contracting isometrically around a latex bulb filling the cavity of the left ventricle.

In experiments performed on an isovolumic heart, samples of the perfusate that had passed via the coronary bed were taken parallel with recording the contractile function, and the Bergmeyer [27] method was used to determine the CPK content. Release of CPK from the myocardium is known to be one of the most reliable criteria of myocardial cell injury, and under hypoxia it increases proportionally with the degree to which ATP concentration decreases in the myocardium [531]. A program of experiments on an isovolumic heart required hypoxic tests, during which the oxygen pressure in the perfusate dropped from 600 to 150 mm in Hg in 20 min, and glucose was absent in the perfusate. During hypoxic tests, the contractile function was naturally increasingly depressed; at the same time, increasingly larger amounts of CPK were released from the myocardium.

By comparing the evidence obtained in the four series mentioned above, it was possible to resolve two separate issues, namely, the effect of ionol on heart resistance to stress-induced injury and on animal heart resistance to hypoxia and reoxygenation. Taking into account the information content of this approach, we consistently used it to evaluate the protective effect of four factors, namely the β-blocker, Inderal, the principal natural antioxidant, α-tocopherol, the ionol already mentioned, and the water-soluble synthetic antioxidant from the class of oxypyridines, the so-called OP-6 [338, 369].

A study of the protective effect of Inderal using the criterion of CPK loss by the heart gave the results shown in Table 51.

From Table 51 it can be seen that 30 min after the beginning of isolated heart perfusion, the latter lost a considerable amount of CPK to the perfusate. At the outset of perfusion, this loss was observed in the experiments of other investigators [396] and was the result of myocardial injury by hypoxia and subsequent reoxygenation. Emotional painful shock led to a twofold increase in CPK loss. This result was apparently caused by the fact that EPS substantially lowered the heart resistance to hypoxia. Inderal-induced β-blockade fully prevented this potentiating effect of stress on hypoxic injury. Table 51 further shows that in the 90th min of perfusion under good oxygenation in the 2nd min causes further increase in the CPK loss to the perfusate. as was approximately the same amount from the hearts of animals in the other series. In control animals, 20-min hypoxia leads to an almost fourfold increase in the CPK release from the myocardium, and subsequent reoxygenation in the 2nd min causes further increase in the CPK loss to the perfusate. In the 20th min of hypoxia, the hearts of stress-exposed animals lost 1.5 times more CPK than controls, and in the 2nd min of reoxygenation the CPK release from the hearts of these animals was also higher than in controls. Inderal β-blockade led to lower CPK losses during hypoxia, and also decreased them twofold during reoxygenation. Thus, Inderal prevents the potentiating effect of stress on hypoxic and reoxygenation injuries to the heart by such a substantial factor as CPK loss by the myocardium.

Table 51. Effects of EPS and Inderal on CPK release from myocardial cells

	CPK activity in perfusate, units/l			
	30 min perfusion	90 min perfusion	20 min perfusion	2 min reoxygenation
I $n = 14$	27.5 ± 2.4	9.0 ± 2.0	34.0 ± 2.4	49.5 ± 2.8
II $n = 14$	50.0 ± 6.7	11.4 ± 1.3	51.9 ± 5.2	60.3 ± 4.1
III $n = 11$	22.0 ± 3.1	7.3 ± 0.7	28.8 ± 3.4	24.8 ± 5.8
IV $n = 11$	24.4 ± 4.0	7.1 ± 0.7	39.1 ± 3.2	32.2 ± 2.3
Significance of differences				
	$p_{I-II} < 0.001$	$p_{I-II} < 0.1$	$p_{I-II} < 0.001$	$p_{I-II} < 0.05$
	$p_{I-III} > 0.1$	$p_{I-III} > 0.1$	$p_{I-III} > 0.1$	$p_{I-III} < 0.001$
	$p_{I-IV} > 0.1$	$p_{I-IV} < 0.05$	$p_{I-IV} > 0.1$	$p_{I-IV} < 0.001$
	$p_{II-IV} < 0.001$	$p_{II-IV} < 0.05$	$p_{II-IV} < 0.001$	$p_{II-IV} < 0.001$

I, control; II, EPS; III, control + Inderal; IV, EPS + Inderal; *n* number of animals in series.

Physiologic studies showed the contraction force, estimated by the developing pressure, the contraction velocity, and the relaxation velocity in the hearts of stress-exposed animals, to have been 22%–30% lower than in controls. Beta-blockade, induced in intact animals by Inderal, had absolutely no effect on these characteristics of the conractile function of the heart. The principal experimental series in which Inderal β-blockade was created prior to EPS, was found not only fully to prevent poststress disturbances of the contractile function of the heart but also gave rise to a somewhat unexpected increase in the basic parameters of contraction and relaxation compared with controls.

Since the Inderal injected into intact animals has no positive inotropic effect, one may state that stress sustained under β-blockade led to an increase in the developing pressure at low frequencies by 9%–13%, and at high physiologic frequency equal to 330 beats/min by almost 30%. The contraction velocity proved to have increased by 15%–22% and the relaxation velocity by 19%–32%. Thus, β-blockade not only prevented poststress disturbances in the contractile function of the heart and the stress-caused excess loss of the enzyme but also indicated the positive effect of stress on the contractile function of the myocardium, an effect that was previously concealed by these disturbances.

The data shown in Table 52 allows the effect of the β-blockade on the resistance of heart function to be evaluated in intact animals and in animals exposed to hypoxic and reoxygenation stress. As is apparent from Table 52, in response to hypoxia control animals, in the wake of rapidly arising depression of systolic pressure, gradually develop contracture which is evident in the rise of diastolic pressure from 5.4 ± 0.6 to 39.6 ± 3.2 mm Hg. The developing pressure decreased from 105.8 ± 3.8 to 22.9 ± 2.2 mm Hg. The contraction velocity decreased by 4.5-fold and the relaxation velocity by more than fivefold. Upon reoxygenation, systolic pressure slowly rose and the diastolic pressure declined. However, even after 20 min neither parameter attained the initial level. The developing pressure remained reduced by 25% and the contraction and relaxation velocities by 30%–40%. Similar but more pronounced phenomena were observed in the hearts of stress-exposed animals. In this case, the absolute values of developing pressure at the 20th min of hypoxia and the 20th min of reoxygenation were essentially lower than in controls. This was due not only to a lower starting systolic pressure in the hearts of stress-exposed animals but also to the more pronounced hypoxic contracture, i.e., to a higher rise in diastolic pressure, which at the 20th min of reoxygenation, for example, was significantly higher than in controls.

From Table 52 it further follows that β-blockade had no effect whatsoever on the initial hypoxic depression of systolic pressure; however, it substantially decreased the extent of hypoxic contracture. As a result, it proved possible to retain a higher developing pressure during the effects of hypoxia and to provide a more rapid restoration of the contractile function of the heart during reoxygenation. This protective effect of the β-blockade was achieved both for the hearts of intact animals and those exposed to stress.

Thus, in intact animals, the Inderal-induced β-blockade does not affect the contractile function level under normal oxygenation but enhances its

Table 52. Effect of Inderal β-blockade on resistance of isolated rat hearts to hypoxia and reoxygenation after EPS

	Series	Prior to hypoxia	Hypoxia		Reoxygenation	
			1 min	20 min	1 min	20 min
Developing pressure, mm Hg	1. Control, $n = 14$	105.8 ± 3.8	57.8 ± 3.5	22.9 ± 2.2	49.0 ± 3.5	76.1 ± 4.6
	2. Control + Inderal, $n = 11$	102.7 ± 3.6	56.9 ± 6.3	35.1 ± 4.0[c]	90.5 ± 5.7[c]	89.0 ± 6.0
	3. EPS, $n = 14$	68.4 ± 4.9[c]	48.0 ± 4.3	14.9 ± 1.2[b]	44.6 ± 3.8	65.2 ± 3.0[b]
	4. EPS + Inderal $n = 11$	120.9 ± 4.8[b]	60.6 ± 4.0	34.0 ± 3.5[d]	93.3 ± 6.8[c]	94.4 ± 8.0[b]
Diastolic pressure, mm Hg	1. Control, $n = 14$	5.4 ± 0.6	9.4 ± 1.2	39.6 ± 3.2	27.2 ± 3.5	17.3 ± 1.3
	2. Control + Inderal $n = 11$	9.1 ± 1.1[b]	7.3 ± 1.4	30.8 ± 2.9[b]	18.3 ± 2.3[b]	13.7 ± 1.3[b]
	3. EPS, $n = 14$	8.1 ± 0,8[b]	9.1 ± 0.1	44.5 ± 4.5	33.2 ± 3.5	23.4 ± 3.0[b]
	4. EPS + Inderal, $n = 11$	10.1 ± 1.7[b]	8.3 ± 1.8	29.2 ± 3.0[b]	15.0 ± 1.4[b]	9.9 ± 0.9[b]
Contraction velocity, mm Hg	1. Control, $n = 14$	1,842 ± 75	919 ± 77	424 ± 50	904 ± 70	1,220 ± 71
	2. Control + Inderal $n = 11$	1,941 ± 85	1,112 ± 100	707 ± 51[b]	1,358 ± 116[c]	1,564 ± 114[b]
	3. EPS, $n = 14$	1,390 ± 110[b]	900 ± 87	286 ± 27[b]	686 ± 80	1,125 ± 70
	4. EPS + Inderal, $n = 11$	2,311 ± 144[b]	1,327 ± 90[b]	770 ± 70[b]	1,560 ± 75[c]	1,685 ± 150[b]
Relaxation velocity, mm Hg	1. Control, $n = 14$	1,163 ± 25	404 ± 46	230 ± 21	421 ± 67	692 ± 42
	2. Control + Inderal, $n = 11$	1,128 ± 42	535 ± 50	458 ± 50[c]	858 ± 75[c]	857 ± 85
	3. EPS, $n = 14$	840 ± 45[c]	415 ± 60	162 ± 20[b]	375 ± 66	685 ± 40
	4. EPS + Inderal $n = 11$	1,511 ± 88[c]	742 ± 68[c]	401 ± 40[c]	850 ± 85[c]	1,015 ± 70[c]

n, number of animals in experiment.
[b] Significance of differences from controls – $p < 0.05$.
[c] Significance of differences from controls – $p < 0.01$.
[d] Significance of differences from controls – $p < 0.001$.

Table 53. Effect of EPS and α-tocopherol on the coronary flow and CPK release from an isolated heart during hypoxia and reoxygenation

	30 min perfusion		90 min perfusion		20 min hypoxia		2 min hypoxia	
	Coronary flow, ml/min	CPK loss, univ units/min	Coronary flow, ml/min	CPK loss, univ units/min	Coronary flow, ml/min	CPK loss, univ units/min	Coronary flow, ml/min	CPK loss, univ units/min
1. Control $n = 10$	10.2 ± 0.6	0.281 ± 0.019	8.5 ± 0.5	0.077 ± 0.006	6.8 ± 0.4	0.21 ± 0.017	8.4 ± 0.3	0.416 ± 0.032
2. α-tocopherol $n = 10$	9.8 ± 0.6	0.203 ± 0.016	9.2 ± 0.5	0.072 ± 0.005	8.1 ± 0.7	0.209 ± 0.016	8.4 ± 0.6	0.272 ± 0.017
3. EPS $n = 10$	10.5 ± 0.7	0.525 ± 0.024	8.4 ± 0.4	0.096 ± 0.007	4.9 ± 0.4	0.252 ± 0.021	6.5 ± 0.2	0.412 ± 0.040
4. EPS + α-tocopherol	10.2 ± 0.8	0.25 ± 0.012	10.2 ± 0.7	0.076 ± 0.06	9.8 ± 0.5	0.232 ± 0.019	9.8 ± 0.5	0.308 ± 0.034
Significance of differences	$p_{1-2} > 0.1$ $p_{1-3} > 0.1$ $p_{3-4} > 0.1$	$p_{1-2} < 0.05$ $p_{1-3} < 0.001$ $p_{3-4} < 0.001$	$p_{1-2} > 0.1$ $p_{1-3} > 0.1$ $p_{3-4} > 0.1$	$p_{1-2} > 0.1$ $p_{1-3} > 0.1$ $p_{3-4} > 0.1$	$p_{1-2} > 0.1$ $p_{1-3} < 0.05$ $p_{3-4} < 0.001$	$p_{1-2} > 0.1$ $p_{1-3} < 0.05$ $p_{3-4} > 0.1$	$p_{1-2} > 0.1$ $p_{1-3} < 0.05$ $p_{3-4} > 0.1$	$p_{1-2} < 0.001$ $p_{1-3} > 0.1$ $p_{3-4} < 0.05$

n, number of animals in series.

resistance to hypoxia and restores it more completely during reoxygenation. The creation of a β-blockade prior to stress prevents all contractile functional disturbances that occur under serious EPS and discloses the positive effect of sustained stress upon the heart function despite the fact that this effect is usually concealed by the contractile functional disturbances. This positive poststress effect on the heart is evident in that the principal characteristics of the contractile function of the hearts of stress-exposed animals prove increased under β-blockade, first under normal oxygenation conditions and, second, under reoxygenation after hypoxia.

In terms of our presentation, it is significant that the results obtained in protecting the heart with Inderal agree with the concept previously discussed that natural antioxidants, glucocorticoids, which are released in increased amounts under stress, in themselves possess a specific cardioprotective effect in stress and reoxygenation. This hypothesis agrees with the above observation that antioxidants enhance the organism's resistance to acute overload associated with the stress reaction (Chap. 2); this should be taken into consideration in further studies of the cardioprotective effect of exogenous natural and synthetic antioxidants.

In studying the protective effect of α-tocopherol by the criterion of CPK loss into the perfusate of an isolated heart, we obtained the evidence shown in Table 53. From this data it follows that in the initial perfusion period, when the heart that was ischemic during preparation was subjected to perfusion, i.e., was actually subject to reoxygenation conditions, the coronary flow in all the series was the same and the CPK loss for stress-induced hearts was increased twofold. The preliminary administration of α-tocopherol fully prevented this reoxygenation-induced injury. Subsequently, under stable oxygenation, the coronary flow changed little and the CPK loss reduced three-to fivefold proved to be about equal in all the series. In the 20th min of the hypoxic test the coronary flow slightly decreased, this decrease being significant only for the hearts of stress-exposed animals, while CPK loss increased again more than threefold and, for the hearts of stress-exposed animals, proved slightly more pronounced than for controls. Finally, with subsequent repeated reoxygenation, the coronary flow for stress-exposed animals unprotected by α-tocopherol slightly increased. In the other series, it did not significantly change and the CPK loss in animals unprotected by α-tocopherol proved substantially increased. In the presence of α-topherol protection, this increase was either slight or totally absent. Thus, under reoxygenation, prior injection of α-tocopherol again prevented CPK loss from animal hearts. In this case, the loss for control animals that had received α-tocopherol and for stress-exposed animals was 26%. In general, these data show that, first, stress potentiates CPK loss from an isolated heart during reoxygenation after ischemia and, second, prior administration of α-tocopherol largely prevents CPK loss from an isolated heart during reoxygenation both after ischemia and after hypoxia.

The curves in Fig. 76 characterize the effect of stress and prior administration of α-tocopherol on the pressure developed by an isolated heart during an induced increasing frequency of contractions. Figure 76 shows that stress

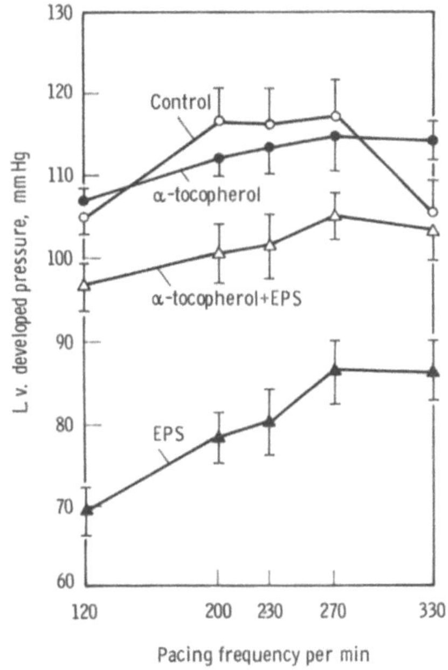

Fig. 76. Effect of EPS and α-tocopherol on left-ventricle *(L. v.)* developed pressure upon induction of various contraction rates

itself had reduced this index by 20%–30%. Alpha-tocopherol did not affect the developing pressure in intact animals and almost completely prevented the stress-induced depression. Similar effects of stress and α-tocopherol were traced in this experiment when determining the contraction and relaxation velocities. Thus, α-tocopherol treatment had no effect on the contractile function in intact animals; however, it prevented the heart contractile function from being depressed by the serious stress action.

The data in Table 54 reflects the effect of hypoxia and subsequent reoxygenation on the contractile function of isolated hearts from control animals, from stress-exposed animals, and from animals protected from stress by α-tocopherol. From Table 54 it can be seen that hypoxia leads to a rapid decrease in the developing pressure and contraction velocity in the hearts of control animals; by as early as the 1st min of hypoxic action, these indices are reduced twofold, and by the 20th min four- to fivefold. At the same time, the decline in the relaxation velocity and the increase in diastolic pressure become even more pronounced: In the 1st min, the relaxation velocity drops fourfold and the diastolic pressure increases twofold; by the 20th min, the relaxation velocity is reduced 7.5-fold, and the diastolic pressure increases to the same extent. This increase in diastolic pressure signifies the development of hypoxic contracture and is the major factor in the decrease in the developing pressure. Under subsequent reoxygenation, these phenomena are eliminated slowly and incompletely; Even in the 20th min, the developing pressure remains reduced by 16% and the diastolic pressure increased threefold over the initial level.

Pretreatment with α-tocopherol moderately limits the decrease in the de-

Table 54. Effect of EPS and α-tocopherol on heart resistance to hypoxia and reoxygenation

	Initial values	Hypoxia		Reoxygenation	
		1 min	20 min	1 min	20 min
Developing pressure, mm Hg					
1. Control, $n = 10$	105.8 ± 4.0	57.8 ± 3.5	22.9 ± 2.6	40.0 ± 4.0	76.1 ± 4.8
2. Control + α-tocopherol, $n = 10$	93.8 ± 4.1	47.8 ± 3.5	31.8 ± 3.0[b]	74.7 ± 9.4[b]	77.6 ± 6.0
3. EPS, $n = 10$	68.4 ± 4.9[b]	48.0 ± 4.3	14.1 ± 1.2[b]	44.6 ± 3.8	65.2 ± 3.0
4. EPS + α-tocopherol, $n = 10$	88.8 ± 4.4[b]	48.0 ± 4.1	28.0 ± 2.4	85.5 ± 4.4[b]	77.0 ± 5.2
Diastolic pressure, mm Hg					
1. Control, $n = 10$	5.4 ± 0.6	9.4 ± 1.3	39.4 ± 3.8	27.2 ± 2.5	17.3 ± 1.3
2. Control + α-tocopherol, $n = 10$	5.8 ± 0.8	9.1 ± 1.0	23.1 ± 3.0[b]	17.8 ± 2.0[b]	10.2 ± 1.3
3. EPS, $n = 10$	8.1 ± 0.8[b]	9.1 ± 0.8	44.5 ± 4.6	33.2 ± 3.5	23.4 ± 2.0
4. EPS + α-tocopherol, $n = 10$	7.5 ± 0.6	4.5 ± 1.2[b]	17.3 ± 1.9[b]	5.5 ± 0.6[b]	4.0 ± 0.3
Contraction velocity, mm Hg/s					
1. Control, $n = 10$	1,842 ± 75	919 ± 75	424 ± 50	904 ± 70	1,220 ± 71
2. Control + α-tocopherol, $n = 10$	1,958 ± 91	850 ± 78	469 ± 31	999 ± 88	1,200 ± 80
3. EPS, $n = 10$	1,380 ± 110[b]	900 ± 87[b]	286 ± 27[b]	586 ± 90	1,125 ± 70
4. EPS + α-tocopherol, $n = 10$	1,775 ± 81	925 ± 78	650 ± 53[b]	1,325 ± 100[b]	1,230 ± 68
Relaxation velocity, mm Hg/s					
1. Control, $n = 10$	1,163 ± 25	404 ± 46	230 ± 23	421 ± 67	692 ± 42
2. Control + α-tocopherol, $n = 10$	1,166 ± 66	412 ± 37	250 ± 25	600 ± 50[b]	700 ± 60
3. EPS, $n = 10$	840 ± 45	415 ± 60	162 ± 20[b]	375 ± 66	685 ± 40
4. EPS + α-TP, $n = 10$	1,400 ± 120[b]	568 ± 35	350 ± 19[b]	837 ± 49[b]	862 ± 35

n, number of animals in series.
[b] Significantly differs from control ($p < 0.05$).

veloping pressure and the degree of contracture during hypoxia (in the 20th minute, $p < 0.05$), but quickens the restoration of the developing and diastolic pressures during reoxygenation (in the 1st min, $p < 0.001$). Conversely, stress potentiates the depression of the developing pressure and the contraction velocity and also the development of hypoxic contracture.

Pretreatment with α-tocopherol prior to the stress action essentially prevented hypoxic contracture from developing. In the hearts of animals given α-tocopherol before stress, the relaxation velocity in the 20th min of hypoxia was two times higher and diastolic pressure was 2.5 times lower than in the hearts of animals that did not receive the antioxidant before stress. Correspondingly, the developing pressure and the contraction velocity in α-tocopherol-protected animals proved to be 2–2.5-fold higher than in unprotected ones. Under reoxygenation, the protective effect of α-tocopherol on the hearts of stress-exposed animals was no less specifically evident. This was primarily expressed in a considerably more rapid disappearance of hypoxic contractures: In the 1st min of reoxygenation, the relaxation velocity in the hearts of α-tocopherol-protected animals proved to be 2.5 times higher and the diastolic pressure six times lower than in unprotected ones. Correspondingly, the developing pressure and contraction velocity were restored much more rapidly and completely.

Hence, stress potentiates the development of hypoxic and reoxygenation injuries of the heart, the latter being evaluated by the CPK release and the degree of hypoxic contracture; on the other hand, the natural antioxidant α-tocopherol fully prevents the stress-induced potentiation of reoxygenation and hypoxic injuries.

In assessing the evidence that the effect of α-tocopherol enhances heart resistance to hypoxia and reoxygenation, one should keep in mind that it mainly coincides with the evidence obtained earlier by Guarnieri et al. [146]. The fact that pretreatment with α-tocopherol prevents disturbances in the contractile function due to serious EPS and eliminates the potentiating effect of stress on the development of hypoxic injury, evaluated by the CPK release as well as by the development of hypoxic contracture, was shown

Table 55. CPK activity in perfusate of an isolated rat heart (units/l)

	Number of animals	35 min perfusion	95 min perfusion	20 min hypoxia
1. Control	8	46.87 ± 6.7	19.65 ± 2.15	31.64 ± 3.78
2. Stress	8	86.55 ± 10.52	29.56 ± 1.89	47.26 ± 5.25
3. Ionol	8	16.11 ± 4.31	8.80 ± 2.59	10.02 ± 3.62
4. Ionol + stress	7	18.44 ± 4.85	10.57 ± 1.27	18.70 ± 1.96
Significance of differences				
p_{1-2}		< 0.01	< 0.01	< 0.05
p_{1-3}		< 0.01	< 0.01	< 0.01
p_{1-4}		< 0.01	< 0.01	< 0.02
p_{3-4}		> 0.1	> 0.1	> 0.1

here for the first time. The significance of these data for the cardiologic clinic is determined by the fact that during acute myocardial infarction, humans are found to be affected by the combined action on the heart of both ischemia and subsequent reoxygenation and adrenergic stress effects.

In studying the cardioprotective effect of ionol, we obtained the data shown in Table 55, from which it can be seen that 35 min after the beginning of perfusion, the hearts of control animals release large amounts of CPK into the perfusate). After 95 min, the amount of released CPK was reduced by more than twofold. A hypoxic test performed under such conditions again leads to considerable CPK release. For the hearts of stress-exposed animals, the dynamics of CPK release remains the same as in controls; however, in all stages, the released CPK activity showed an increase of 1.5-fold or higher. Thus, stress induces heart resistance to hypoxic injury arising during the prepration or removal of the heart or caused by a special hypoxic test. Ionol administered to intact animals reduced CPK release in early perfusion by almost threefold compared with controls; upon protracted perfusion under sufficient oxygenation, CPK release was reduced by more than two-fold and during the hypoxic test it was reduced threefold.

Hence, pretreatment with ionol undoubtedly enhances heart resistance to hypoxic injury. From Table 55 it is evident that the hearts of animals that received ionol prior to stress released CPK in all experimental stages to about the same extent as the hearts of intact animals that received ionol. In other words, the administration of ionol removed both the harmful effects of hypoxia and the potentiating effect of stress on that injury.

Figure 77 presents the data on the contractile function of a heart that contracts in isovolumic conditions, i.e., of the same heart that was studied for CPK release. As is apparent from Fig. 77, at equal electrostimulator-induced contraction frequency, the pressure developed by the hearts of stress-exposed

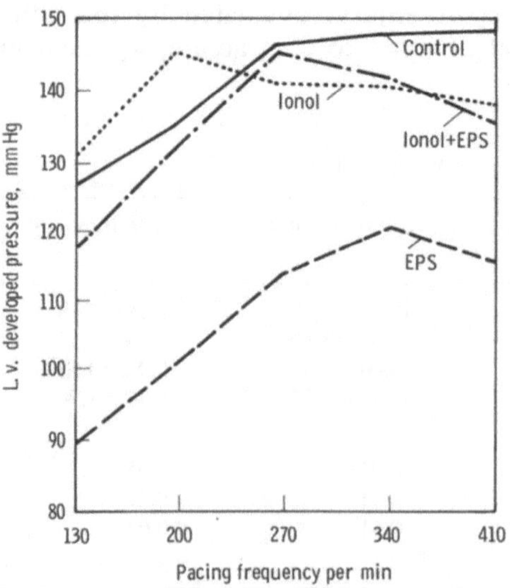

Fig. 77. Effect of stress *(EPS)* and ionol on contraction force of left ventricle *(L. v.)* of rat isolated heart upon induction of increasing contraction rate

Table 56. Effect of EPS and ionol on contractile function of an isolated working heart

	Stroke volume ml · 10^{-2}	Minute volume, ml	External work, $\frac{kg}{g \cdot min} \cdot 10^{-3}$	Efficiency of oxygen utilization by heart, %
1. Control $n=8$	17.8 ±1.6	37 ±3	35.7 ±4.2	13.5 ±1.2
2. Stress, $n=6$	12.0 ±1.0	24 ±3	21.2 ±2.5	8.0 ±0.9
3. Ionol, $n=5$	20.7 ±2.3	36 ±5	37.0 ±5.2	12.5 ±1.7
4. Ionol + stress, $n=8$	19.2 ±2.1	36 ±4	36.9 ±5.0	14.1 ±1.8
p_{1-2}	<0.05	<0.05	<0.05	<0.05
p_{1-3}	>0.05	>0.01	>0.05	>0.05
p_{3-4}	>0.02	n/s	>0.01	>0.05

n, number of animals; n/s, nonsignificant.

animals was reduced by about one-third compared with controls ($p<0.05$). In animals exposed to stress against a background of ionol, this defect in the contractile function was absent; in intact animals that received ionol, the developing pressure also did not differ from controls.

Similar results were obtained in studying the contractile function of rat hearts using the Neely method. Table 56 shows that with a filling pressure equal to 10 mm Hg, the stroke volume, minute volume, and external work and effectivity of oxygen utilization by a heart taken from a stress-exposed animal were significantly reduced compared with controls. Pretreatment with ionol fully prevents this depression in pump function and heart efficiency. The administration of ionol to intact animals had no significant effect on the pump function and heart efficiency. Thus, ionol administration itself had no effect on the contractile function of the heart; however, it prevented the usual stress-induced depression of the contraction force, stroke volume, external work, and efficiency of the heart.

The curves in Fig. 78 show the reaction of an isovolumic heart to a hypoxic test. In the left-hand side of Fig. 78, we see that after hypoxia, the developing systolic pressure sharply drops for the hearts of stress-exposed animals. This decrease, largely caused by cardiac contracture, starts from a lower level; correspondingly, the dotted curve lies below the other three curves. Starting from the 10th min of hypoxia, the developing pressure becomes very low and is virtually the same for the hearts of animals from all the series (8–9 mm Hg). Under reoxygenation, the posthypoxic contractile function is restored differently for the hearts of animals from different series. Stress-exposed animals need the longest time for the contractile function of their hearts to be restored. One minute after the elimination of hypoxia, the pressure developed by these hearts amounts only to 45 mm Hg, whereas in controls it amounts to 33–34 mm Hg and in intact animals that received ionol it is about 50 mm Hg. These differences are subsequently retained: 20 min after

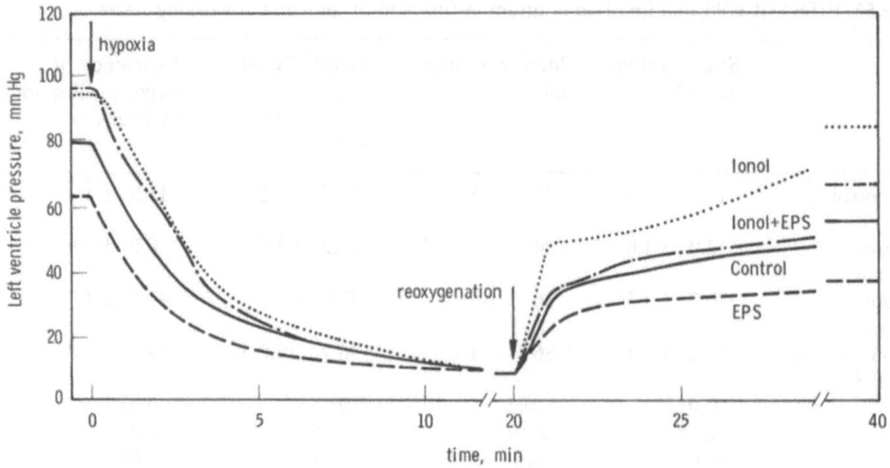

Fig. 78. Disturbances in contractile function of isolated heart due to sustained stress *(EPS)* and hypoxia; prevention of these disturbances by means of ionol (in heart)

the removal of hypoxia, the developing presure was 58% of the initial level (prior to hypoxia) for stress-exposed animals; it was 74%–75% of the initial level for 'ionol ± stress' and control animals and about 90% for the hearts of intact animals that received ionol.

Thus, judging by the enzyme release, stress potentiates hypoxic injury of the heart and simultaneously inhibits the restoration of its contractile function during reoxygenation. Pretreatment with ionol fully removes the potentiating effect of stress on hypoxic injury and also its inhibitory effect on restoration of the contractile function in the posthypoxic period. Finally, the administration of ionol to intact animals fully protects the heart from hypoxic injury, as judged by the criterion of CPK release, and considerably accelerates the restoration of the contractile function during subsequent reoxygenation.

In studying the cardioprotective effect of the water-soluble antioxidant oxypyridine OP-6, basically the same results were obtained as those established in the study of ionol. The oxypyridine OP-6 reduced CPK loss from isolated animal hearts during hypoxia and reoxygenation, prevented disturbances in the contractile functions of the hearts of stress-exposed animals, largely hindered disturbances in the contractile function during hypoxia and reoxygenation, and also prevented stress-induced uncoupling of oxidation and phosphorylation in the mitochondria. The curves in Fig. 79 show the results of experiments in which we studied the effect of preliminary injection of OP-6 (100 mg/kg, for 3 days prior to stress) on the resistance of isolated animal hearts to hypoxia and reoxygenation [380]. The curves for the systolic and diastolic pressures in Fig. 79 show that in controls, the diastolic pressure increases quite considerably to as much as 40 mm Hg. Combined with the decline in systolic pressure, this leads, in the 20th min, to a drop in the

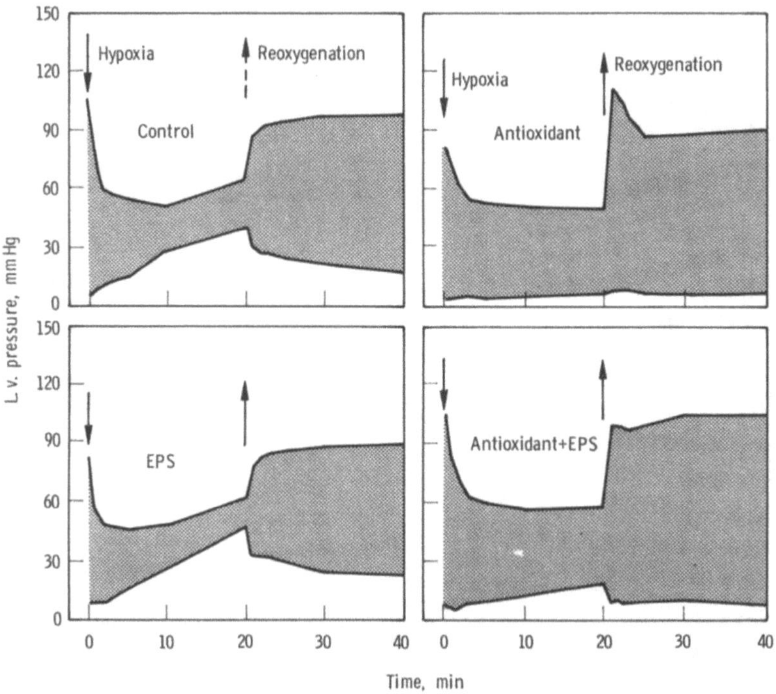

Fig. 79. Effect of preliminary administration of water-soluble antioxidant from oxypyridine class on development of hypoxic contracture of isolated rat heart under stress *(EPS). Top line,* systolic pressure; *bottom line,* diastolic pressure or pressure during rest; *shaded area,* extent of developed pressure

developing pressure from 105 to 2 mm Hg, i.e., more than fourfold compared with the initial value. Inasmuch as the developing pressure determines the stroke volume in a working heart, such a depression in the contractile function of the heart would have dangerous consequences for a whole organism.

This harmful hypoxic effect proved much stronger in the isolated hearts of stress-exposed animals (Fig. 79 lower left); contracture caused the diastolic pressure to rise to 47 mm Hg and, in conjunction with the depression in systolic pressure, this led to a decrease in the developing pressure to as low as 14.5 mm Hg, i.e. by fivefold in comparison with the initial level. Under the same experimental conditions, the isolated hearts of animals that for 3 days prior to isolation were administered a water-soluble antioxidant from the oxypyridine class did not respond to hypoxia by significant contractures (see Fig. 79 upper right, lower right); correspondingly, the developing pressure did not bring about a large decline in those hearts. This prevention of heart contracture by the antioxidant proved equally effective for the hearts of both control and stress-exposed animals.

To evaluate and rationally utilize the data on antioxidant protection, one should separately examine the two aspects of antioxidant action, viz their

Table 57. Effect of preliminary ionol administration on intensity of stress reaction

	Catecholamine content, μg/g			Noradrenaline	DOPA	Corticosterone content, % μg		
	Noradrenaline[a]	DOPA[a]	Adrenaline[b]			In plasma	In heart	In adrenals
Control	0.77 ± 0.08	0.3 ± 0.03	281.8 ± 26.4	128 ± 14.6	60.7 ± 3.1	6.45 ± 1.0	16.5 ± 1.3	1,177.4 ± 169.4
EPS	0.38 ± 0.04[c]	0.23 ± 0.05	156.9 ± 2.36[c]	72.2 ± 19.0	38.1 ± 5.3[d]	21.3 ± 2.4[d]	49.8 ± 5.8[d]	2,400.5 ± 122.5
Ionol	0.64 ± 0.06	0.3 ± 0.04	343.3 ± 46.8	117.1 ± 10.6	63.6 ± 3.9	12.7 ± 0.8	24.1 ± 1.7	2,286.8 ± 213.0
Ionol + EPS	0.59 ± 0.06	0.28 ± 0.07	234.1 ± 36.5	106.5 ± 7.1	46.7 ± 2.0	8.1 ± 0.9	17.7 ± 0.3	913.7 ± 158.0

[a] In heart.
[b] In adrenals.
[c] Ten animals were used in each series.
[d] Differences are significant in relation to controls.

antistress protection and protection against hypoxia, reoxygenation, and ischemia. Antistress protection is evident when preinjection of nontoxic doses of ionol largely prevents the activation of LP, which naturally develops under the influence of EPS. At the same time, it prevents or reduces fermentemia, ulcerous lesions of the stomach, and disturbances in heart structure, metabolism, and function. This means that LP activation is an essential link in the pathogenic chain of stress damage and that, correspondingly, the administration of sufficiently powerful LP inhibitors prevents this damage.

It would be easy to ascribe the mechanism of this protective effect to the above-mentioned prevention by the antioxidant of LP activation in the heart and other target organs, in which the effect of high catecholamine concentrations takes place. However, in reality, the protective effect of ionol and, apparently, of other antioxidants is achieved not only at the level of the target organs but also at the level of the organism's control systems. The data in Table 57 shows that the preliminary administration of ionol in the doses that we used noticeably reduces the stress reaction arising in response to EPS. Indeed, the decrease in noradrenaline, which usually develops in EPS-exposed animals, was twofold under the influence of ionol pretreatment and the post-stress rise in corticosterone concentration in the plasma and myocardium was virtually absent. This seemingly unexpected fact agrees with the concept that LP is an essential link in the function of any cell. The restriction of this process during stress reduces mobilization of the functions of nerve centers and endocrine glands in the stress-realizing system.

Thus, the ability of antioxidants to prevent stress-induced injuries is realized both through restriction of LP in the target organs and limitation of the stress reaction itself.

Antioxidant Protection of the Heart Against Hypoxia, Ischemia, and Reoxygenation

Many facts indicate that antioxidant protection against anoxia and ischemia is not achieved at all unambigously as it is against stress; it requires careful consideration. For example, in evaluating the protective action of antioxidants during hypoxic contracture, it should be underlined that the latter develops because of hypoxia, i.e., after the O_2 tension has declined from 600 to 150 mm Hg. This situation is characterized by the combined accumulation of restored transmitters in the respiratory chain and a sufficient amount of oxygen in the mitochondrial lipid matrix and, in accordance with the above concept, should lead to LP activation.

In addition to ATP shortage, LP activation plays a role in damaging or switching off the membrane calcium pump of cardiomyocytes and, as a result, in the occurrence of excess calcium and the development of the contracture. By inhibiting LP, antioxidants inhibit the development of contractures as well. In full conformity with this, an isolated heart contracture caused not by hypoxia but by anoxia conditioned by nitrogen saturation of the perfusate and evident in a diastolic pressure increase to 60-70 mm Hg by the 20th min,

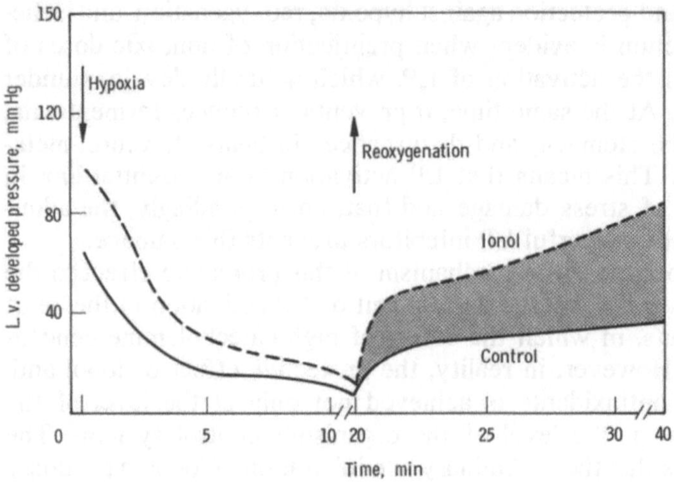

Fig. 80. Effect of ionol on contractile function of isolated heart during hypoxia and reoxygenation. *Shaded area* shows ionol effect

was unsensitive to natural and synthetic antioxidants. One may assume that this was an ATP-deficient contracture, in which the role of LP was insignificant. It is, therefore, understandable that the antihypoxic effect of antioxidants may not be realized under excessively tight hypoxia or anoxia.

Conversely, protection by antioxidants from reoxygenative damage is achieved very regularly, for this type of injury is naturally connected with LP activation. This concept has been consistently illustrated in three different experimental situations. The first situation (Fig. 80) shows how the isolated hearts of ionol-treated rats restore their contractile function following hypoxic depression more amply and quickly than the hearts of control animals.

The second situation was reproduced in a whole organism [384] and may be assessed by the curves in Figs. 81 and 82. The curves characterize the effect of preliminary administration of antioxidants on the main parameters of the contractile function of the heart during hypoxic depression of the contractile function; the depression was induced in acute experiments on rats by cutting off the respiratory system for 5 min. Figures 81 and 82 show that this kind of hypoxia naturally caused depression of the contractile function of the left ventricle, the function being retained in control animals for the first few minutes following the termination of hypoxia and beginning of reoxygenation; the contractile function was restored very slowly. Conversely, in animals that received ionol, a peroxidation inhibitor, before the experiment, the contractile function was restored instantly; moreover, super-restoration was naturally observed (Fig. 81). A similar, albeit less pronounced, effect was obtained by preadministration of another antioxidant, OP-6 (Fig. 82). Further examination showed that as was anticipated, the ability of antioxidants to prevent reoxygenation disturbances in the cardiac contractile

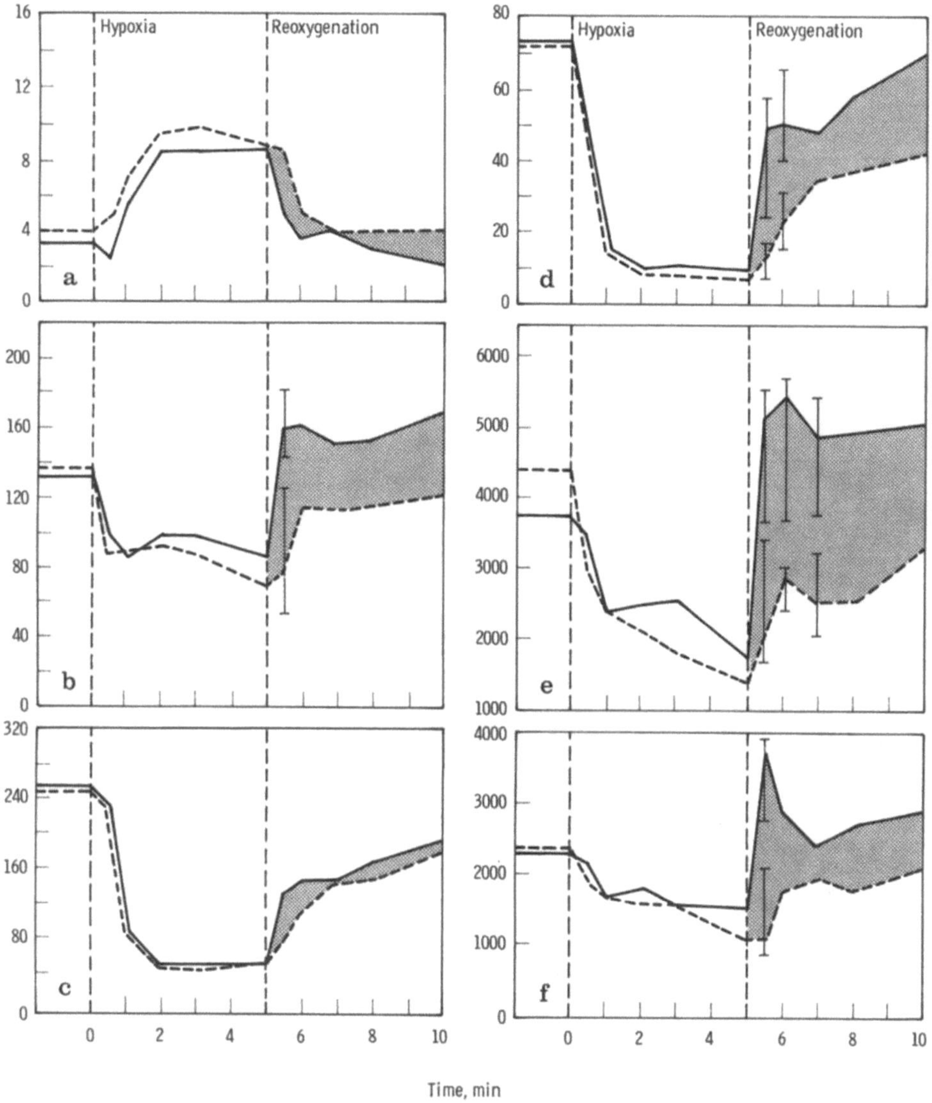

Fig. 81 a–f. Effect of ionol on restoration of heart contractile function during reoxygenation. **a** diastolic pressure, mm Hg; **b** developed pressure, mm Hg; **c** heart contraction frequency per minute; **d** structure functional intensity (SFI), mm Hg/mg/min; **e** contraction frequency, mm Hg/s; **f** relaxation frequency, mm Hg/s. *Dashed line,* control; *continuous line,* ionol, *shaded area* shows ionol effect

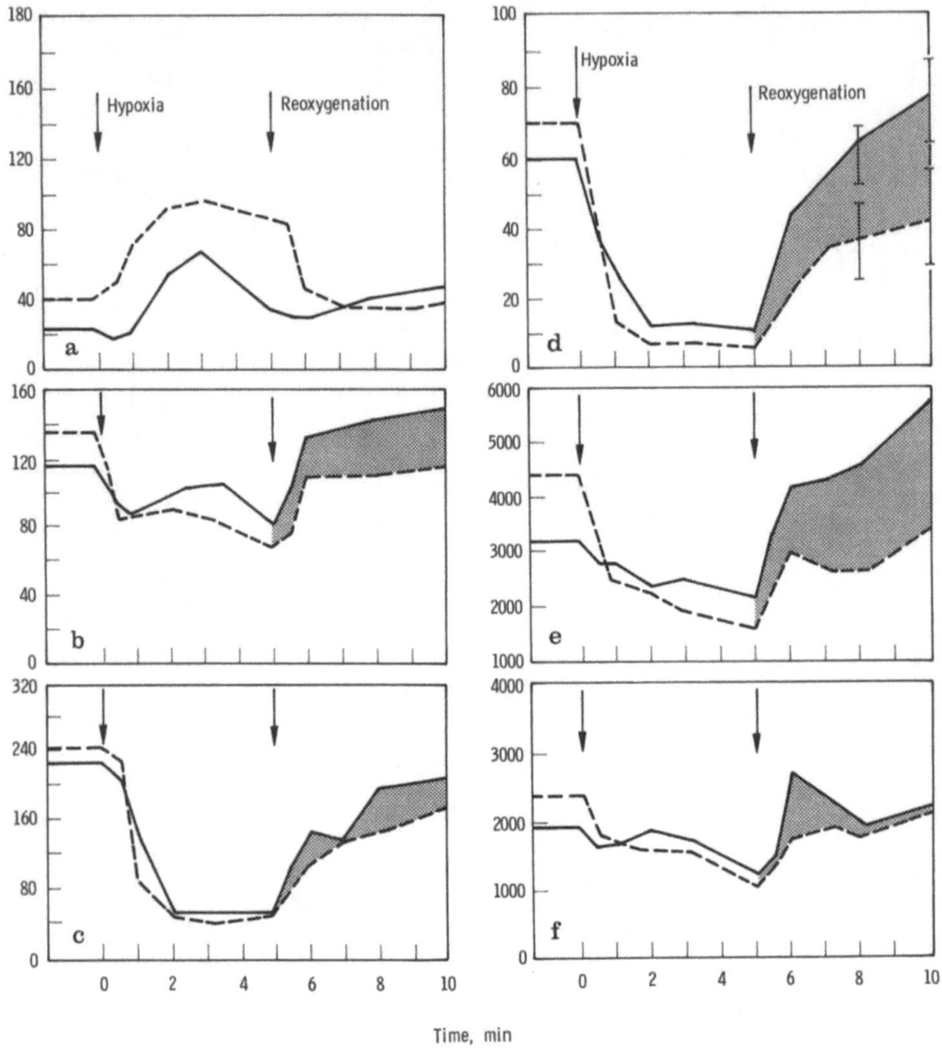

Fig. 82. Effect of antioxidant OP-6 on restoration of heart contractile function during reoxygenation. *Continuous line,* OP-6; *other symbols* as in Fig. 80

function and to provide super-restoration of that function after termination of hypoxia, is based on inhibition of lipid peroxidation.

From Table 58 it can be seen that the 5-min hypoxia itself causes a nonsignificant increase in the LP final products in the myocardium; with reoxygenation, the amount of these products significantly increases twofold in comparison with controls. Ionol reduced the amount of Schiff's bases during normal respiration and completely prevented them from accumulating during hypoxia and reoxygenation.

Finally, the third situation, in which the antioxidant protection of the

Table 58. Variations in the content of Schiff's bases in the myocardium during hypoxia and subsequent reoxygenation (fluorescence, relative units)

	Artificial respiration (aerobic conditions)	Respiratory standstill (hypoxia)	Restored respiration (reoxygenation)
Control	8.1 ± 2.0	13.4 ± 2.2	16.0 ± 3.2
Ionol	3.9 ± 1.9	3.3 ± 2.4	3.04 ± 0.99

heart may be traced specifically from reoxygenation damage, may be very definitely observed during transitory regional myocardial ischemia, a widespread model of coronary insufficiency attack. The curves in Fig. 83 characterize the dynamics of the pressure developed by the left ventricle and the intensity of structure functioning during a 30 min complete clamping of the ascending branch of the left coronary artery in rats and the effect

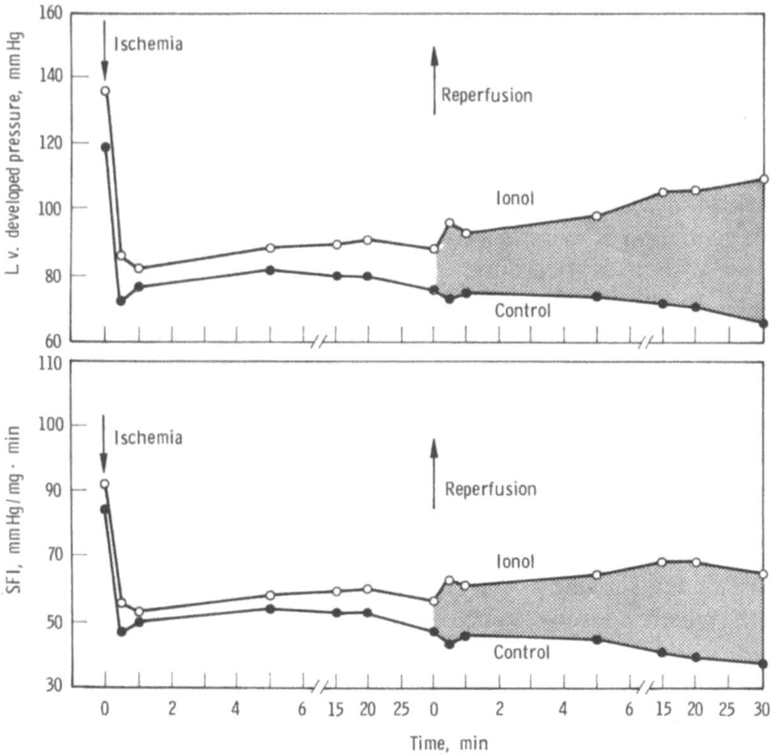

Fig. 83. Effect of ionol on depression of contractile function of heart left ventricle *(l. v.)* during transitional ischemia and subsequent reperfusion. *Shaded area* shows ionol effect; *SFI,* structure functional intensity

on this process of ionol. From Fig. 83, it is apparent that as in the studies of other investigators, ischemia itself caused considerable depression of the contractile function of the heart and subsequent stabilization of that function at a low level [278]. Elimination of the stop in the coronary artery and subsequent perfusion and reoxygenation of the myocardium did not restore the contractile function and even became the cause of its undergoing a further slow depression, which is most probably due to the fact that reoxygenation causes maximal LP activation and cell membrane degradation.

It is significant that the animals that were administered a relatively small dose of ionol (20 mg/kg) prior to the experiments with transitory ischemia responded to acute ischemia in approximately the same manner as control rats; however, ionol fully prevented reoxygenation depression of the contractile function and, moreover, considerably restored the contractile function of the heart during reoxygenation.

On the whole, the findings of our laboratory concerning the fact that antioxidants provide protection against reoxygenation more effectively than against anoxia agree with the results of biochemical studies simultaneously performed. The latter showed that during protracted anoxia of an isolated heart, the mycardium accumulates no malonic aldehyde, a peroxidation product, which upon subsequent reoxygenation rapidly accumulates in amounts proportional to the duration of the preceding anoxia. For example, when anoxia lasts 80 m, malonic aldehyde concentration increases fourfold [147, 148].

These facts unambiguously show that in principle, LP inhibitors can prevent reoxygenation disturbances in heart function. This concept seems to have definite significance for the cardiologic clinic since reoxygenation inevitably occurs after every attack of angina pectoris and, hence, a more or less pronounced reoxygenation damage is a usual event. During open-chest surgery, restored circulation in a myocardium damaged by excessively prolonged hypoxia may lead to the development of reoxygenation contracture of the human heart: The 'stone heart' [67] that forms as a result is the basis of acute cardiac insufficiency that occurs on the operation table.

The above evidence provides a certain basis for studying the possibility of using antioxidants clinically for the tratment of coronary disease and in cardiosurgery. At the same time, it indicates that peroxidation inhibitors can, in principle, protect the heart from three factors, namely, from excess catecholamines during serious stress, from moderate hypoxia, and from post-hypoxic or postischemic reoxygenation. Similarly, it would be desirable to examine possibilities for antioxidant protection of the heart during myocardial infarction, which is known to be of prime importance in contemporary cardiology.

With acute coronary occlusion, a most frequent direct cause of myocardial infarction, several factors are present and, when combined, should very probably cause LP activation. Indeed, the accumulation of restored carriers due to the switching off of the respiratory chain is combined with the presence of molecular oxygen, the tension of which even in the center of the ischemic focus does not drop below 5-10 mm Hg. During natural reoxygenation caused by the elimination of spasms or reactive hyperemia, this accumulation may considerably increase. We have, therefore, two basic conditions essential for the formation of free radicals and LP induction – excess electron donors, i.e., restored carriers, and the presence of oxygen, the electron acceptor. At the same time, the EPS that determines increased discharge and oxidation of catecholamines takes place. Under these conditions, the oxidation of adrenaline to adrenochrome results in adrenaline semiquinone, which may discharge the electron to oxygen and generate a superoxide radical, an important LP inductor.

It would be natural to expect that a combination of ischemia, reoxygenation and stress, something actually observed with any acute infarction, should result in abrupt LP activation in the myocardium. Indeed, it has been shown that as early as 24 h after myocardial infarction induced by clamping the descending branch of the left coronary artery in rats, a three- to four-fold increase in lipid thermochemiluminescence occurs in the ischemic zone of the myocardium. This increase may indirectly be indicative of LP activation and attains a maximum on day 3 after the clamping of the coronary artery and gradually disappears by day 30. The extent to which thermochemiluminescence is activated, the size of infarction, and the ECG changes characteristic of ischemia may be substantially decreased by prior administration to the animals of the antioxidants sodium selenite and α-tocopherol or the two substances combined [226]. This result was confirmed in a large number of studies performed in the same laboratory [248a, 249].

In assessing these important data, however, one should keep in mind that thermochemiluminescence of lipids extracted from the heart reflects the LP initiation taking place in vitro when oxygen is available. Hence, increased chemiluminescence of the lipids extracted from the ischemic zone is only indicative of the fact that ischemia created the prerequisites for the process to become more intensive in the presence of oxygen. This shift in itself

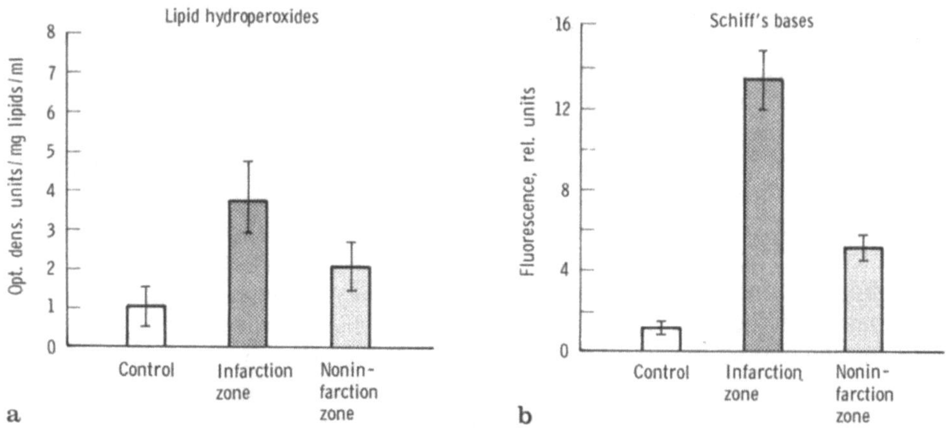

Fig. 84. Accumulation of lipid hydroperoxides and Schiff's bases in myocardium during experimental infarction

is not unambiguous proof of LP activation in vivo; it may take place, for instance, due to the increased content of unsaturated fatty acids in the lipids of the ischemic zone, the fatty acids being LP substrates, or it may be due to diminished activation of natural antioxidant system, and so on. It is obvious that to establish whether or not LP is activated during infarction in and beyond the ischemic zone, one must directly measure any LP products that might accumulate in the myocardial lipids. Hence, we determined the content of primary LP products – hydroperoxides and final LP products, the Schiff's bases – in and beyond the ischemic zone 24 h after having created an experimental myocardial infarction. At the same time, we evaluated the state of enzymic antioxidant systems in these zones [382].

The diagram in Fig. 84a shows that 24 h after occlusion of the coronary artery, the hydroperoxide content in and beyond the ischemic zone proved 3.5 and two times higher, respectively. The diagram in Fig. 84b shows that the content of the final LP products the Schiff's bases, had increased in and beyond the ischemic zone thirteen- and fivefold, respectively. This extraordinary activation of myocardial LP was accompanied during infarction by decreased activity of the principal enzymic antioxidant systems, which in normal conditions restrict LP intensity.

Figure 85 shows that the activity of superoxide dysmutase and catalase, the two antioxidant enzymes that restrict the formation of free oxygen forms, was reduced 18%–16% and the activity of glutathione peroxidase, which is responsible for utilization of lipid hydroperoxides, was reduced by more than twofold.

Thus, experimental myocardial infarction is really characterized by the considerable accumulation of early and final LP products and by the partial inactivation of certain enzymic systems that restrict LP under normal conditions. Hence, as was already indicated, one should keep in mind that this high LP activation in the membranes of myocytes containing polyene phospholipids, the principal LP substrate, cannot be an isolated event. It should inevitably combine with activation of phospholipases and the detergent action of fatty acids and lisophospholipids; in turn, this lipid triad should result in mass degradation of the membrane lipid bilayer. Experimental results [61] are in complete agreement with this. The authors showed that as early as 12 h after the beginning of ischemia the total phospholipid content in the heart had decreased by one-third, the amount of phosphatidyl ethanolamine, the predominant LP substrate being reduced by 50%, and the amount of phosphatidyl choline had decreased by only 25%. This loss of a considerable part of membrane lipids is accompanied by a tenfold increase in the sarcolemma permeability for calcium. This is understandably so, since it was established that the phospholipid content decreases most conspicuously in the sarcolemma, where degradation of the lipid bilayer apparently occurs the earliest [559].

Facts show that the activation of LP and degradation of the membrane lipid bilayer in the myocardium not only take place during experimental infarction but really do happen in humans. For example, clinical observations by Katzenovich [211] showed that the content of lipid hydroperoxides in the blood increased by more than fourfold after acute myocardial infarction.

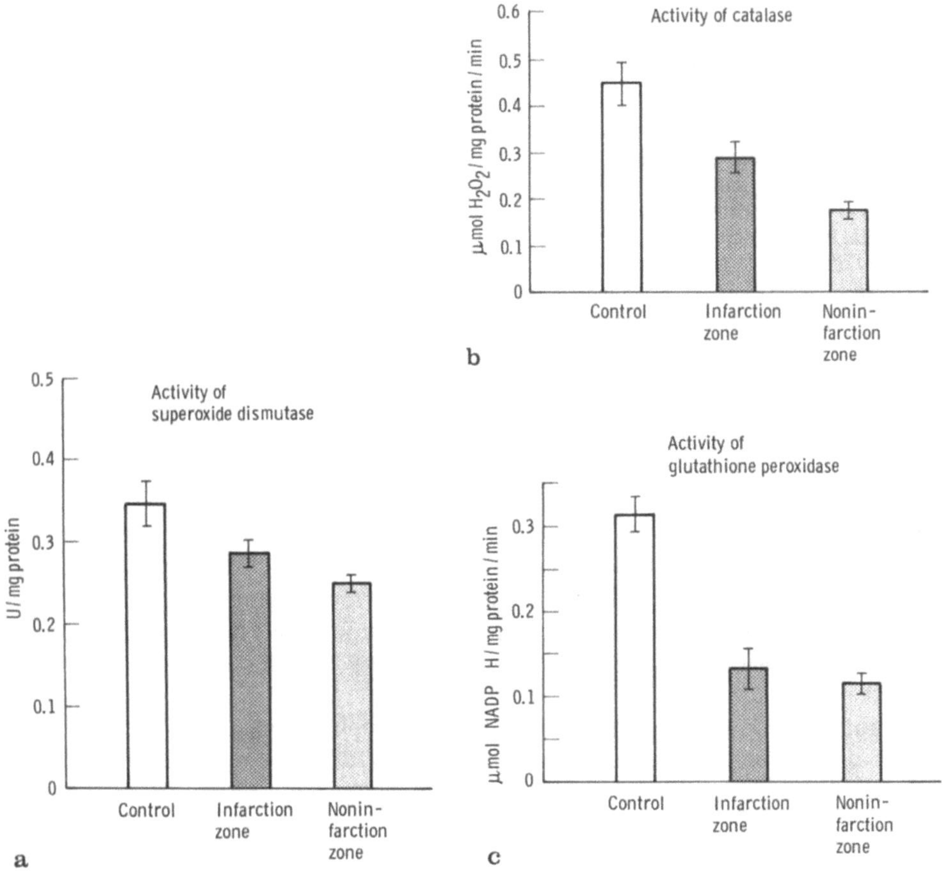

Fig. 85. Decreased activity of basic antioxidant enzymes of myocardium during experimental infarction

This agrees with the findings of a still earlier investigation [151], which revealed the changes in fatty acid content in the myocardia of people who died of infarction, these changes being indicative of a lower content of phosphatidyl ethanolamine, the principal LP substrate.

In connection with these findings on the role of LP in the degradation of cardiomyocytic membranes, it should be noted that the above experimental evidence indicates that during infarction LP is not only activated in the ischemic zone, it is also substantially increased in the nonischemic zone. This activation beyond the ischemic zone is apparently conditioned by stress, i.e., by the action of excess catecholamines, and also by natural reoxygenation caused by reactive hyperemia and the opening of collaterals. One cannot exclude the possibility that this LP activation combined with the reduced activity of antioxidant systems can play a role in enlarging the zone of myocardial necrosis, the basis of so-called continuously relapsing or prolonged infarctions quite familiar in cardiology. From this viewpoint, it would be

Table 59. Effect of preadministration of ionol on the necrotic area and fermentemia during experimental myocardial infarction

	Infarction (11)	Ionol + infarction (9)	Significance of differences
Infarction area under epicardium, % total left ventricular area	63.3 ± 2.0	59.5 ± 3.1	$p > 0.05$
Infarction area under endocardium, % total left ventricular area	50.9 ± 1.9	43.7 ± 3.2	$p > 0.5$
Activity of aspartate transaminase, units/l	130.4 ± 12.2	97.2 ± 7.1	$p < 0.05$

relevant to examine the results that we obtained when using ionol, a synthetic antioxidant, to protect the myocardium against experimental infarction [375].

The data in Table 59 show that prior administration of ionol insignificantly affected the infarction area 2 days after ligating the descending branch of the left coronary artery. Admittedly, ionol reduced the extent of fermentemia observed during infarction. However, this ionol effect was revealed earlier during fermentemia caused by EPS; hence, one may assume that fermentemia on day 2 after the experimental infarction was largely conditioned by stress damage to various tissues, its restriction being the result of the antistress effect of ionol.

Figure 86 shows the dynamics of the developing pressure in the left ventricle during relative rest and under maximal load by the resistance developed

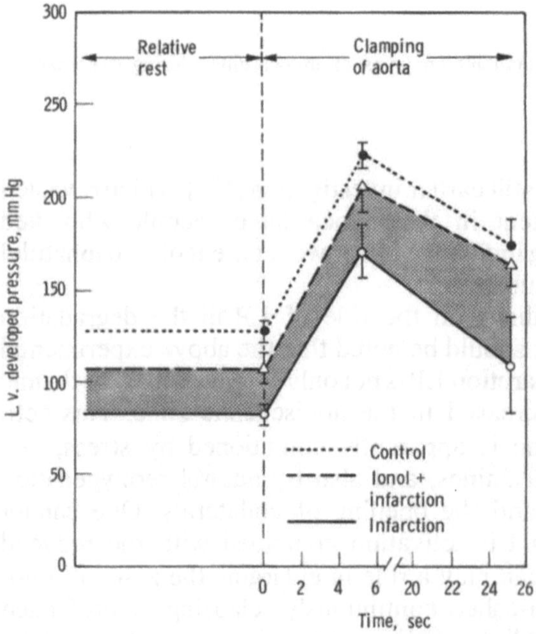

Fig. 86. Effect of prior injection of ionol on disturbance of contractile function of left ventricle (*L. v.*) during experimental infarction. *Shaded area* shows ionol effect

by clamping the aorta for 25 s. As is apparent from Fig. 86, in animals with experimental infarction the pressure and its development velocity were considerably reduced compared with controls. This infarction-caused depression proved to be at a maximum 24 s after clamping the aorta, when the unaffected myocardial regions develop fatigue. The principal result of the experiment was that prior administration of the antioxidant had largely prevented this defect from developing in the contractile function. The shaded area (Fig. 86) reflects this protective effect.

Since ionol reduces the area of the ischemic necrosis insignificantly, we assumed this effect to have been caused by the fact that the antioxidant protected the safe nonischemic myocardial regions from being damaged by the stress concomitant with infarction. In other words, it appeared probable that we were dealing with the protection of safe myocardial regions against stress damage.

To verify this assumption, we studied the contractile function of an a priori nonischemic division of the heart, namely the right auricle, in animals with experimental infarction of the left ventricle. From Table 59 it follows that 24 h after the infarction had occurred in the left ventricle, the contractile function of the right auricle was substantially disturbed: The maximal tension developed by an isolated auricle in isometric conditions and the intensity of its structural function determined as the ratio of tension to the auricular mass multiplied by contraction frequency, were reduced almost twofold. Administration of Inderal, a β-blocker, prior to inducing the infarction resulted in almost complete prevention of this defect in the contractile function. Consequently, auricular damage was caused by the effect of excess catecholamines, i.e., by the adrenergic stress concomitant with the infarction.

The data in Table 60 further show that ionol prevented disturbance of the contractile function of the auricle to the same extent as Inderal. Thus, an antioxidant prevented depression of the contractile function of the nonischemic region of the heart in the period of acute myocardial infarction; in this case, the antioxidant protection of the heart was essentially of an antistress nature. Hence, it appears advisable to remind the reader that the damage to the membrane lipid bilayer during stress and ischemia is based not on isolated LP activation but on the lipid triad that develops from LP

Table 60. Effect of prior administration of Inderal and ionol on contractile function of right auricle during infarction of left ventricle

	Developing tension, mg	Structure functioning intensity, g/mg · min
1. Control (10)	382 ± 12.5	3.8 ± 0.2
2. Infarction (15)	198 ± 4.3	1.8 ± 0.1
3. Inderal + infarction (11)	331 ± 15	2.9 ± 0.2
4. Ionol + infarction (10)	315.2 ± 15	2.9 ± 0.2
p_{1-2}	> 0.001	> 0.001
p_{1-3}	< 0.5	< 0.5
p_{1-4}	< 0.5	< 0.5

activation, phospholipase and lipase activation, and the detergent action of excess fatty acids and lisophosphatides.

In addition to antioxidants, we used, therefore, chloroquine, a phospholipase inhibitor, and nicotinamide, a lipase inhibitor, to protect the nonischemic regions of the myocardium during experimental infarction. Figure 87 shows that 24 h after creating an experimental infarction, the contractile function of a nonischemic myocardium, namely the right auricle, is characterized by a pronounced depression of Starling's curve, which is displaced downward and to the right. Administration to the animals of the membrane protectors mentioned above prior to infarction largely prevented this disturbance in the

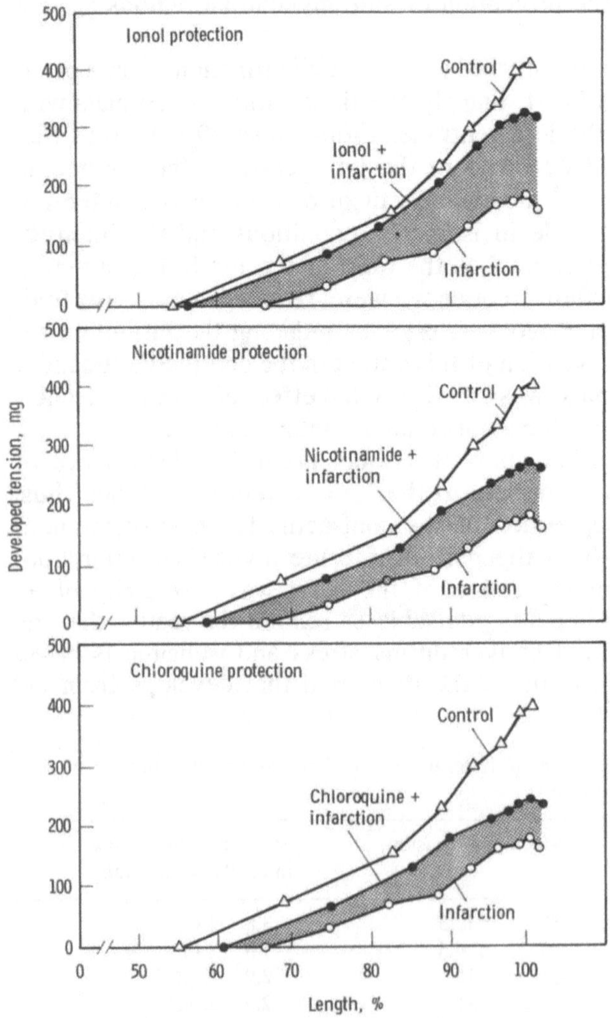

Fig. 87. Prevention by membranoprotectors of depression of Starling's curve in nonischemized myocardial region under infarction

contractile function. The shaded areas in Fig. 87 allow one to assess the protective effects of ionol, nicotinamide, and chloroquine.

In evaluating the fact that the inhibitors of various components of the lipid triad unidirectionally protect the undamaged divisions of the myocardium during experimental infarction, one should keep in mind that in case of a whole organism these components of the lipid triad are inseparably interconnected. As was already mentioned, LP activation causes labilization of the lysosomes, in which a considerable amount of cell phospholipase is concentrated. The membrane-bound phospholipases that are released from the lysosomes and also the activated ones can play an important part in degrading the membrane lipid bilayer and forming lysophospholipids and free fatty acids [209]. Like the high fatty acid concentrations observed in the blood during stress and myocardial infarction, lysophosphatides disturb the orderly arrangement of phospholipids in the membranes owing to their detergent effect and this may cause additional LP activation. The increase in membrane permeability for Ca^{2+} resulting from this chain reaction and the excess Ca^{2+} in the cell arising as a result in turn activate phospholipases [566]. This brings about the closure of the vicious circle that plays an important part in damaging cell membranes, in making those damages irreversible, and in destroying myocytes.

In this way, one can visualize the process in the ischemic zone during infarction as well as the foci of necrotic damage. Beyond the ischemic zone in infarction, in the principal mass of the myocardium under stress, the same cycle of events is less pronounced and may be arrested in one of its links during the inhibition of any component of the lipid triad. It is on this basis that one should understand the protective effects of LP, phospholipase and lipase inhibitors examined above.

As a result, we can underline the following concept, which is essential for the clinic: The function of the safe parts of the myocardium largely predetermines the fate of the myocardial infarction patient. These parts are damaged by stress, which is always concomitant with infarction; their protection by means of antioxidants and other membrane protectors appears promising. The prospects of effective utilization of antioxidants during IHD largely depend not only on the correct choice of LP inhibitors and their appropriate doses but also on well-grounded combined use with other factors. The above concept on the pathogenic chain of ischemic damage affords grounds to assume that the most promising are combinations of antioxidants: first, with factors that improve energy supply to the heart, i.e., with vasodilators, adenine nucleotide derivatives; second, with β-blockers; third, with lipase and phospholipase inhibitors; and fourth, with factors that block the slow electrogenic channel of calcium transport, i.e., with substances like verapamyl, diltiazeme, and infidepin. This investigation result for cardiac antioxidant protection is of certain interest in clinical cardiology but is not the principal one in this presentation. The consequence that depending on the antioxidant state of the organism, on the power of its antioxidant systems, the same stress or hypoxic action may either cause or not cause LP activation and heart injury appears more significant. This means that the organism's antioxidant systems represent in the direct sense a natural system for preventing stress and hypoxic damages.

Hence, it is highly probable that under definite conditions, the organism may change its antioxidant status and resistance to harmful factors, in the pathogenesis of which LP plays a role spontaneously without the administration of exogenous antioxidants. Earlier, we discussed an example in which the organism's antioxidant status changed during adaptation to the periodic action of hypoxia, which prevents LP activation during stress and enhances resistance to stress damage (see Chap. 3).

To assess the mechanism of this important phenomenon more specifically, we compared LP activation in the myocardium during stress and ischemic injury to the heart. We evaluated the effect of adaptation to hypoxia on LP during both injuries and clarified how prior adaptation to hypoxia affects the state of enzymic antioxidant systems in the myocardium during stress and ischemic damage [382]. From Table 61 it follows that as in the previously discussed experiments. EPS was accompanied by LP activation in the myocardium, namely by a 1.5–2-fold increase in the concentrations of hydroperoxides and Schiff's bases in the myocardium. This LP activation was not accompanied by changed superoxide dysmutase activity but by decreased activity of glutathione peroxidase-1, which utilizes hydrogen peroxide. The activity of glutathione peroxidase-2, which transforms organic peroxides, remained unchanged.

From Table 61 it further follows that infarction caused much higher LP activation than stress. Indeed, in the ischemic zone, the hydroperoxide content increased 3.7-fold and the content of LP final products – Schiff's bases – increased 13-fold. Beyond the ischemic zone, LP activation was lower; the hydroperoxide content increased twofold and that of Schiff's bases fivefold. At the same time, a pronounced reduction in the activity of the antioxidant enzymic systems studied was noted in and beyond the ischemic zone. For example, in the ischemic zone, the activities of superoxide dysmutase, catalase, glutathione peroxidase-1, and glutathione peroxidase-2 were lower by 18, 36, 57, and almost 3 times, respectively. Beyond the ischemic zone, this depressed activity of antioxidant enzymic systems was either equally or even more pronounced.

The previously examined scheme that reflects the role of antioxidant enzymes in LP control (Fig. 73) is important in assessing these facts. Figure 73 shows that superoxide dysmutase and catalase act on the process initiation level, while glutathione peroxidase-2 catalyzes the transformation of low-resistant lipid peroxides, capable of being sources of free radicals, to stable products; natural antioxidants (α-tocopherol) interact with radical intermediates of LP.

The above data show that in stress, the activity of superoxide dysmutase does not change, while that of its two related enzymes i.e., catalase and glutathione peroxidase-1, which degrade hydrogen peroxide, decrease. It may be assumed that changed enzyme activity should lead to increased H_2O_2 content and to the resultant increased formation of free-radical oxidation products of lipids in the myocardium, which actually occurs under stress.

Table 61. State of endogenous antioxidant systems of the myocardium during EPS, infarction, and adaptation to hypoxia

	I Control	II Stress (p_{I-II})	III Infarction (p_{I-III}): In ischemic zone	III Infarction (p_{I-III}): Beyond ischemic zone	Adaptation to hypoxia (p_{I-IV})	Adaptation to hypoxia plus stress (p_{IV-V})	Adaptation to hypoxia plus infarction (p_{IV-VI}): In ischemic zone	Adaptation to hypoxia plus infarction (p_{IV-VI}): Beyond ischemic zone
Hydroperoxides, optical density units per 1 mg lipids in 1 ml	1.03 ± 0.13	2.14 ± 0.16 (< 0.01)	3.71 ± 0.42 (< 0.005)	2.00 ± 0.18 (< 0.01)	0.96 ± 0.14	0.81 ± 0.19	1.68 ± 0.18 (< 0.01)	1.28 ± 0.30
Fluorescence of Schiff's bases, relative units	1.00 ± 0.22	1.74 ± 0.40 (< 0.05)	13.31 ± 1.39 (< 0.005)	5.11 ± 0.60 (< 0.005)	0.94 ± 0.17	0.95 ± 0.16	6.93 ± 0.81 (< 0.005)	4.99 ± 0.63 (< 0.005)
Superoxide dysmutase activity, conv. units/g protein	344 ± 25	386 ± 19	283 ± 18 (< 0.05)	247 ± 8 (< 0.05)	314 ± 18	303 ± 14	254 ± 22 (< 0.05)	272 ± 11
Glutathione peroxidase-1 activity, nmol NADPH/mg protein/min	315 ± 18	284 ± 24	134 ± 22 (< 0.01)	116 ± 8 (< 0.01)	213 ± 20 (< 0.05)	208 ± 31	150 ± 14 (< 0.01)	115 ± 7 (< 0.01)
Glutathione peroxidase-2 activity, nmol NADPH/mg protein/min	218 ± 11	214 ± 15	74 ± 13 (< 0.01)	45 ± 6 (< 0.005)	173 ± 8 (< 0.05)	189 ± 11	62 ± 13 (< 0.01)	44 ± 14 (< 0.005)
Catalase activity, nmole H_2O_2/mg protein/min	445 ± 43	363 ± 18 (< 0.05)	285 ± 28 (< 0.01)	178 ± 8 (< 0.005)	425 ± 30	360 ± 20 (< 0.05)	303 ± 35 (< 0.01)	178 ± 20 (< 0.01)
Tocopherol content, µg/mg lipids	2.67 ± 0.36	2.02 ± 0.21 (< 0.05)						

This activation of free-radical oxidation of lipids should lead to a more intense utilization of natural antioxidants in the bilayer of cell membranes and to their decreased content. Indeed, an analysis of the α-tocopherol content in stress-exposed animals showed its concentration to be reduced by about one-fourth compared with controls and equal to $2.67 \pm 0.36\ \mu g/kg$ lipids in controls and $2.02 \pm 0.21\ \mu g/mg$ lipids after stress. On the whole, the obtained results show that the decrease in catalase and Se-containing glutathione peroxidase-1 activities can play a role in activation LP during stress and that the prevention of stress injuries by exogenously administered antioxidants can very probably be due to the fact that they compensate for the shortage of α-tocopherol, a natural antioxidant.

During ischemic damage to the myocardium, the situation essentially differs. An LP activation immeasurably higher than under stress is accompanied by a general depression of the activities of all the three enzymic systems studied, not only in the ischemic zone but also beyond it – in the myocardium that was safe at the time the investigation was started. This has already been examined above.

In this presentation, the data in Table 60 are of importance; they characterize the effect of prior adaptation to the periodic action of hypoxia. As is apparent from Table 60, adaptation itself caused no significant change either in LP intensity or in superoxide dysmutase and catalase activities; however, it reduced the activity of the Se-containing glutathione peroxidase-1 by almost 33% and to a lesser degree (21%) the activity of glutathione peroxidase-2. At the same time, prior adaptation resulted in distinctly reduced LP activation, both during stress and myocardial infarction. For instance, after stress, adapted animals showed no increase in the content of hydroperoxides and Schiff's bases in the myocardium; hence, stress-induced LP activation was fully prevented. Accordingly, the stress depression of glutathione peroxidase activity observed in controls did not occur. During myocardial infarction, the preventive effect of adaptation to hypoxia was evident as LP activation in the ischemic zone in adapted animals being manifest to a considerably lower degree than in controls; the hydroperoxide content increased only to 1.68 against 3.71 units optical density/mg lipids in controls and the content of Schiff's bases increased to 6.93 against 13.31 relative units in controls. Adaptation had no substantial effect on the activities of the given antioxidant enzyme systems both in the ischemic zone and in the unaffected of the myocardium.

The principal outcome of these experiments was that they showed that prior adaptation to hypoxia, a factor preventing stress and ischemic injury, either prevents or decreases LP activation in the heart, both during stress and infarction; at the same time, however, it provides no protective effect for the principal antioxidant enzymic systems of the myocardium. This means that the preventive effect of adaptation materializes at the expense of some other mechanisms, not through increased power of the antioxidant systems. In studying this mechanism, one should keep in mind that greater membrane stability and the diminished probability of LP activation may depend on a decrease in the content of polyunsaturated fatty acids in the phospholipids, which are LP substrates, and may also be due to an increase in the cholesterol content and other shifts in lipid composition. Hence, the study of the effect of adaptation to the periodic action of hypoxia upon the lipid composition, specifically the phospholipid and fatty acid composition, of myocardial membranes acquires great significance in understanding the nature of the protective effect of such adaptation. A similar task appears to be very important in the study of adaptation to other environmental factors. For example, enhanced resistance to LP may apparently explain yet another very remarkable fact previously shown in the laboratory of Gorizontov, namely that adaptation to repeated electrical painful stress action, resulting in the so-called resistance phase, considerably increases animal tolerance to ionizing radiation [484, 599], the damaging effect of which is known to occur via activation of free-radical oxidation. Such facts give grounds to assume that gradual enhancement of the power of the organism's antioxidant systems is a mechanism of increased resistance in adaptation to hypoxia, physical loads, and repeated stress actions. This assumption implies an important role for the antioxidant systems in the organism's adaptation to a broad spectrum of environmental factors, the role of these systems in phenotypic adaptation.

Another more general aspect of the problem is that lipid peroxidation, controlled by the relationship of anti- and prooxidants in the organism, is only one of the factors that modify membrane lipid composition. Phospholipase activity and the spectrum of the fatty acids supplied to the organism in food are apparently no less important modifying factors. As has now been established, the resultant dynamics of membrane phospholipid composition substantially affects the activity of the organism's main enzymic systems localized in the membranes. For instance, Na^{+}-K^{+}-ATPase and sarcolemmal adenyl cyclase, SR Ca^{2+}-ATPase, and other enzymes that play a key role in the vital activity of the cell and the whole organism belong to the membrane-bound lipid-dependent enzymes [91, 143, 394].

This suggests that *changes in membrane-lipid composition occurring relatively quickly under the influence of external factors and subsequent changes in the activity of lipid-dependent enzymes can play the role of important links in the mechanism of adaptation to the environment. Such a rapidly realized adaptational link, independent of enzyme biosynthesis and increase in enzyme population, may apparently be achieved in the stage of urgent adaptation and thus precede long-term adaptation, which is based on the activation of protein biosynthesis and the formation of a structural trace.* Evidently, a verification of this assumption requires an experimental study of the dynamics of lipid metabolism in the cells of the systems of the organism responsible for adaptation to principal environmental factors.

The three systems of natural prevention of stress damage examined (GABA-ergic inhibitory system in the brain, prostaglandins, and antioxidant systems) naturally do not exhaust the store of means by which the organism manages to stay unharmed in serious and protracted stress situations. Yet, consideration for the role of these systems brings us closer to understanding the general mechanism of adaptation to serious stress, i.e., essentially extreme situations that either cannot or should not be avoided for a long time.

By imitating this natural mechanism which has become established in the course of evolution, it has now already proved possible chemically to prevent stress damage. Earlier, it was shown that prior adaptation of animals to hypoxia or cold may be used not only for preventing stress injuries but also for changing the behavior in stress situations in a direction favorable to the organism: Adapted animals were able to realize a vitally essential drinking reflex in spite of pain and they showed increased aggressiveness during intraspecific competition.

Taken together, these facts suggest that further study of adaptation to stress situations may help develop methods for preventing stress-induced diseases and methods for controlling behavior in extreme situations.

Conclusion

An ancient Sumerian legend tells that about 7,000 years ago Gilgamesh, hero and top priest of the town of Uruk, being outraged by the fact that all men are mortal went to the gods to protest and demand immortality. In referring today to that legend, we realize that strictly speaking, Gilgamesh's protest was unscientific, for death is an inevitable link in life. However, his protest did contain an important rational kernel, being in itself an expression of man's active attitude towards nature and an expression of his desire to control natural events, specifically those that constitute the foundation of human life. In our view, this desire, which is inherent in man, has played a tremendous role in the origin of science in general and biology in particular.

The organism's individual adaptation, to which this book is devoted, constitutes one of the most ancient and, at the same time, most important issues of biology. Some hundred odd years ago, the great German philosopher Georg Wilhelm Friedrich Hegel suggested one of the deepest definitions of adaptation when he wrote "The living prevents the cause from turning into action, i.e., eliminates it as a cause." In effect, this concept determines the ability to adapt to a given habitat, which is a distinctive feature of all living matter. Indeed, with the abundant diversity of adaptive reactions, all of them restrict or prevent the unfavorable action of the environment upon the organism. Actually, there are two types of reactions. First, reactions of simple avoidance or removal of an active factor; this type of reactions is highly diverse, from photochemotaxis to the creation of a whole system of shelters, as well as food stocks and the specialized production of food. Second, direct adaptive reactions, achieved as a result of the direct action of environmental factors and complex, not infrequently habitat-induced stress situations that effect living organisms. In this last class of reactions, which has been the major subject of this book, the elimination or restriction of environmental 'causal factors' is provided, as we saw, by profound changes in the structures and functions of the systems of the organism. Such adaptation enhances resistance to hypoxia, physical loads, cold, danger-induced stress situations, and so on.

Over the past century, studies of direct adaptive reactions to the effects of environmental factors have been distinguished by a remarkable inconsistency. On the one hand, fundamental discoveries have been made in studying the organism's reactions to biologic environmental factors, microbes and viruses. These discoveries first constituted the empirical basis for the successful prevention of the principal infectious diseases and then became its scientific foundation. On the other hand, the mechanism of adaptation to diverse physical factors of the environment, particularly extreme stress situations, has been much less studied. As a result, the prophylaxis of principal noninfectious diseases, specifically coronary disease, hypertension, diabetes, and blastomatous growth, the pathogenesis of which undoubtedly involves stress-induced injury, remains an issue open to modern medicine.

The concept developed in this book on the mechanism of phenotypic adaptation has become the basis for effective adaptational and chemical prophylaxis of stress damage, hereditary hypertension, and other experimentally induced diseases. This indicates that the most perspective approach to enhancing the organism's resistance and further improving the prophylaxis of noninfectious human diseases is in a deep study of the mechanism of adaptation, in understanding that mechanism, in reproducing the effects of adaptation by means of chemical factors, and in the chemical and adaptational prophylaxis of stress-induced injuries.

References

1. Adams DB, Baccelli G, Mancia G, Zanchetti A (1971) Relation of cardiovascular changes in fighting to emotion and exercise. J Physiol 212:321–335
2. Adams RJ, Cohen DW, Gupte S (1979) In vitro effect of palmityl carnitine on cardiac plasma membrane Na-K-ATPase and sarcoplasmic reticulum Ca^{2+}-ATPase and Ca transport. J Biol Chem 254:12404–12410
3. Agadzhanyan NA (1956) Extinction of conditioned motor electric defence conditioned reflexes in rarefied atmosphere. Zh Vyssh Nerv Deiat 6:260–268
4. Albert NR, Gordon MA (1968) Myofibrillar adenosine, triphosphatase activity in congestive heart failure. Am J Physiol 202:940–946
5. Anokhin PK (1948) Physiological prerequisites for understanding the pathogenesis of hypertension. Transactions of 4th Session of the USSR Academy of Medical Sciences, 34–48 (Russian) [Materialy 4-y Sessii Acad. Med. Nauk SSSR]
6. Anokhin PK (1975) Essays on the physiology of functional systems. Meditsina, Moscow (Russian)
7. Antony P, Weihe E (1977) Inhibition of myocardial ornithine decarboxylase activity by adenosine 5′-triphosphate. J Mol Cell Cardiol 9. Suppl. Biannual meeting of the International Study Group for Research in Cardiac Metabolism, p. 1 abstr. 1
8. Aoki K, Tazumi K, Kato S, Sato I, Hotta K, Yamashita K (1976) Ca^{++} binding ability and release of sarcoplasmic reticulum from arterial smooth muscle in spontaneously hypertensive rat. Jpn Heart J 13:352–353
9. Appelt AW, Wetty JD, Peterson MB (1976) Changes in sarcolemmal and sarcoplasmic reticulum ATPase activities with age in the cardiomyopathic Syrian hamsters. J Mol Cell Cardiol 8:901–907
10. Archakov AI (1975) Mikrosomalnoye okisleniye (Microsomal oxidation) Nauka, Moscow (Russian)
11. Arkhipenko YuV, Kagan VE, Kozlov YuP (1982) Modification of enzyme system of calcium ion transport in SR membranes. Molecular mechanisms of changes in ATPase Ca^{2+} activity. Biochemistry, v. 47, in press (Russian).
12. Ashenbrenner J, Zak R, Cutilletta AF, Rabinowitz M (1971) Effects of hypoxia on degradation of mitochondrial components in rat cardiac muscle. Am J Physiol 221:1418–1425
13. Ashmarin IP, Konarev VT, Sidorova VV, Simanovsky LN, Fedorova NL (1976) Variations in protein content of rat brain- and liver chromatin during functional cell reconstruction. Influence of adaptation and hypoxia. Dokl Akad Nauk SSSR 228:222–224 (Russian)
14. Azhgihin IS (1978) Prostaglandins: a new class of biologically active substances. In: Prostaglandiny (Prostaglandins) Meditsina, Moscow, pp 6–83 (Russian)
15. Bach AN (1897) On the role of peroxides in slow oxidation processes. Zh Ross Fizkhim O 29:373–398 (Russian)
16. Baranski S (1963) Incorporation of S^{35}-methionine into the CNS of white mice during altitude hypoxia. Bull Acad Pol Sci 11:255
17. Barany M, Gaetjeas E, Barany E, Karp E (1964) Comparative studies of rabbit cardiac and skeletal myosins. Arch Biochem Biophys 106:280–293
18. Baratz SS, Melnikova ZM (1962) Hypertension and coronary atherosclerosis among workers of Uralmashzavod. Sov Med 8:13–18 (Russian)

19. Barbashova ZI (1960) Acclimatization to hypoxia and physiological mechanisms. Nauka, Leningrad (Russian) [Akklimatizatziya k gipoksii i eje fisiologicheskije mekhanismy]
20. Barbashova ZI (1969) Some results in the study of the nature of body resistance and its variation mechanisms. Kosm Biol Med 4:6–12 (Russian)
21. Barkalaya AI (1968) Nature of response of pancreas insular apparatus under chronic tension. Probl Endokrinol (Mosk) 14:71–74 (Russian)
22. Barkroft J (1937) Basic features of the architecture of physiological functions. Biomedgiz, Leningrad (Russian translation) [Osnovnye cherty arkhitektury fisiologicheskikh funktziy]
23. Barondes SH, Cohen HD (1967) Delayed and sustained effect of acetoxycycloheximide on memory in mice. Proc Natl Acad Sci USA 58:157–164
23a. Barondes SH, Cohen HD (1968) Arousal and the conversion of 'short-term' to 'long-term' memory. Proc Natl Acad Sci USA 61:923–929
24. Bassett JR, Cairncross KD (1975) Endogenous levels of catecholamines in the rat myocardium following exposure to stress. Pharmacol Biochem Behav 4:35–38
25. Beikert A (1954) Tierexperimentelle Untersuchungen über die Leistungsfähigkeit arbeits-hypertrophischer Herzen. Klin Wochenschr 32:527
26. Belardinelli L, Rubia R, Berne RM (1979) Blockade of Ca^{2+}-dependent rat atrial slow action potentials by adenosine and lanthanum. Pflugers Arch 380:19–27
27. Bergmeyer HU (1970) Methoden der enzimatischen Analyse. Chemie, Weinheim
28. Beritov IS (1961) Nervous mechanisms of the higher vertebrate behaviour. Akad Nauk SSSR, Moscow [Nervnye mekhanismy povedeniya vysshikh zhivotnykh]
29. Beriya VP, Kagan VE, Arkhipenko YuV, Kozlov YuP (1975) Antioxidants as stabilizers of cytochromes in membranes of endoplasmic reticulum of rat liver *in vivo*. Biofizika 20:238–240 (Russian)
30. Berne RM (1980) The role of adenosine in the regulation of coronary blood flow. Circ Res 47:807–813
31. Berne RM, Rubio R, Duling BR (1977) Vasoactive substances affecting the coronary circulation. Excerpta Med Int Congr Ser 225:28–43
32. Bernet F, Denimal J (1974) Evolution de la response adrenosympathique a l'exercice au cours de l'entrainement chez le rat. Eur J. Appl Physiol 33:57–70
33. Bhan AK, Scheuer J (1972) Effects of physical training on cardiac actomyosin adenosin triphosphatase activity. Am J Physiol 223:1486–1490
34. Blagovestova NP, Loginova EV, Simonov EE, Fomenko MM (1968) Duration of bone marrow reaction to acclimatization to hypoxia. Problems of space biology. Nauka, Moscow (Russian) [Problemy kosmicheskoy biologii]
35. Block V (1970) Facts and hypothesis concerning memory consolidation processes. Brain Res 24:561–575
36. Bloor CM, Leon AS, Pasyk S (1968) The effects of exercise on organ and cellular development in rats. Lab Invest 19:675–680
37. Blyuger AF, Shuster YY (1963) Effect of strong stimulators on β-lipoproteid content in blood serum and some organs Bull Exp Biol 7:56–60 (Russian)
38. Bobrova AV (1962) Functional capacity of the kidney following resection and unilateral nephrectomy. Urologiia 4:14–17 (Russian)
39. Boden WE, Liang CS, Apstein CS, Hood WB (1978) Experimental myocardial infarction. XVI. The detection of inotropic contractile reserve with post-extrasystolic potentiation in acutely ischemic canina myocardium. Am J Cardiol 41:523–530
40. Boehmer H, Shortman K (1973) The separation of different cell classes from lymphoid organs. IX. A simple and rapid method for removal of damaged cells from lymphoid cell suspensions. J Immunol Methods 2:293–301
41. Bogdanova ED, Kagan VE, Kuliev IY, Meerson FZ, Prilipko LL (1981) Activation of lipid peroxidation in brain and the appearance of antibodies to the brain antigens in stress. Immunologiia 2:65–66 (Russian)
42. Bogomolov AF (1975) Variations in the contents and synthesis of nucleic acids and proteins in the lungs in adaptation to hypoxia. Bull Exp Biol 9:33–35 (Russian)
43. Bogoyavlensky VF, Salikhov IG, Amirov NK, Antagulova RS, Bikbulatova KS, Kopylov AM (1977) The ischemic heart disease in managers of large industrial complexes. In: Smetnev AS

(ed) Abstracts of reports of 2nd all-Russia congress of cardiologists. Saratov. No publs. (Russian) [Tezisy dokladov vtorogo vserossiyskogo kongressa kardiologov]
44. Bolland JL, Koch HP (1975) The course of autooxidation reaction in polyisoprenes and allied compounds. Part 9. The primary thermal oxidation product at ethyl linoleate. J Chem Soc 7:445–450
45. Boroyan RG (1977) Changes in prostaglandin release in the heart under varying experimental conditions. Circulation 10:9–12 (Russian)
46. Bors W, Michel C, Seran M, Lengfelder E (1978) The involvement of oxygen radicals during the autooxidation of adrenalin. Biochim Biophys Acta 540:162–172
47. Botviniev OK, Veltishev YV (1969) A method for separate determination of hydrocortisone and corticosterone in small plasma volumes. Lab Prakt 6:341–344 (Russian)
48. Brodie DA (1963) Literature review of restraint-induced ulcers. Neuropsychiatry J 4:388–390
49. Brumberg VA (1969) On changes in the bulks of motor and sensitive neurons of the spinal cord and the surrounding glyal cells in various states of motor activity. Dokl Akad Nauk SSSR 184:1231–1234 (Russian)
50. Bures J, Buresova O (1963) Cortical spreading depression as a memory disturbing factor. J Comp Physiol Psychol 56:268
51. Burlakova EB, Alesenko AV, Molochkina EM (1975) Bioantioxidants in radiation injury and malignant growth. Nauka, Moscow (Russian) [Bioantioxidanty v luchevom porazhenii i zlokachestvennom roste]
52. Burton K (1968) Determination of DNA concentrations with diphenylamine. Methods Enzymol 12B:163–166
53. Butcher RW, Baird CE (1968) Effects of prostaglandin on cyclic AMP levels in fat and other tissues. J Biol Chem 243:1713–1718
54. Caldarera CM, Orlandini G, Casti A, Moruzzi G (1974) Polyamine and nucleic acid metabolism in myocardial hypertrophy of the overloaded heart. J Mol Cell Cardiol 6:95–104
55. Cairncross KD, Bassett JR (1975) Changes in myocardial function as a consequence of prolonged emotional stress. Prog Brain Res 42:313–318
55a. Chance B (1965) The energy-linked reaction of calcium with mitochondria. J Biol Chem 240:2729–2748
56. Chance B, Sies H, Boveris A (1979) Hydroperoxide metabolism in mammalian organs. Physiol Rev 59:527–605
57. Chapman D, Urbina J, Keough KM (1974) Biomembrane phase transitions. Studies of lipid-water systems using differential scanning calometry. J Biol Chem 249:2512–2521
58. Chazov EI (1975) Emotional stress and cardiovascular diseases. Vestn Akad Med Nauk SSSR 8:3–8 (Russian)
59. Chernyshova GV, Stoida LV (1969) Phosphorous compound content in the right- and left ventricle tissues in dogs in cutting vagus nerves. Bull Exp Biol Med 8:51–54 (Russian)
60. Chesnokov VN, Mertvetzov NP, Salganic RI (1974) Selective sensitivity of induced isozyme in rat liver hexokinase to the action of proteolytic enzymes. Biochem. 39; 294–298 (Russian)
61. Chien KR, Richard G, Pfaw BA, Farber JL (1979) Ischemic myocardial cell injury. Am J Pathol 97:215–219
62. Chin AK, Seaman, R, Kapileshwarker M (1973) Plasma catecholamine response to exercise and cold adaptation. J Appl Physiol 34:409–412
63. Chohan IS, Singh I, Balakrishnan K, Talwar GP (1975) Immune response in human subjects at high altitude. Int J Biometeorol 19:137–143
64. Clô C, Orlandini GD, Casti A, Guarnieri C (1976) Polyamines as growth stimulating factors in eukariotic cells. Ital J Biochem 25:94–114
65. Cohen MM (1976) 15-Methyl prosaglandin E_2 protects gastric mucosal barrier. In: Samuelsson B, Paoletti L(eds) Advances in prostaglandin and thromboxan research, vol. 2. Raven Press, New York, pp 937–938
66. Cohen M, Sztokalo J, Hinsch E (1973) The antihypertensive action of arachidonic acid in the spontaneous hypertensive rat and its antagonism by anti-inflammatory agents. Life Sci 13:317–325
67. Cooley DA, Reul GJ, Wukasch DC (1972) Ischemic contracture of the heart: 'stone heart'. Am J Cardiol 29:575–577

68. Covell JW, Pool PE, Braunwald E (1967) Effect of acutely induced ischemic heart failure on myocardial high energy phosphate stores. Proc Soc Exp Biol Med 124:126–131
69. Craddock VM, Henderson AR (1978) De novo and repair replication of DNA in liver of carcinogen treated animals. Cancer Res 38:2135–2143
70. Croncholm B, McLander L (1964) Memory disturbances after electroconvulsive therapy. 5. Conditions one month after a series of treatments. Acta Psychiatr Neurol Scand 40:211
71. Csallany AS, Ayal KL (1976) Quantitative determination of organic solvent soluble lipofuscin pigments in tissues. Lipids 11:412–417
72. Cutiletta AF, Dowell RT, Rudnik M (1975) Regression of myocardial hypertrophy. I. Experimental model changes in heart weight nucleic acid collagen. J Mol Cell Cardiol 7:767–781
73. Dahl LK, Tuthill R (1974) Further evidence of the toxicity of NaCl. Increased blood pressure and mortality in the spontaneously hypertensive rat. J Exp Med 139:617–628
74. Danscher G, Fjerdingstad EJ (1975) Diethyl dithiocarbamate (antabuse): decrease of brain heavy metal staining pattern and improved consolidation of shuttle box avoidance in coldfish. Brain Res 83:143–155
75. Davis RA, Kernk J, Showalter R (1978) Alteration of hepatic Na-K-ATPase and bile flow by estrogen: effect in liver surface membrane lipid metabolism and function. Proc Natl Acad Sci USA 75:4130–4134
76. Dawson DM (1970) Creatine kinase from brain: kinetic aspects. J Neurochem 17:65–74
77. Dembo AG, Pinchuk VM, Levina LI (1973) Heart dilatation and myocardial hypertrophy in athletes. In: Heart dilatation and myocardial hypertrophy in athletes. VNIIFK, Moscow, pp 42–66 (Russian) [Dilyatatziya serdtza i gipertrofiya miokarda u sportsmenov]
78. Dergachev VV, Kruglikov RI, Meerson FZ (1967) Effect of cofactors of nucleic acid and protein synthesis and precursors on forming and retaining the defence conditioned reflex. Bull Exp Biol Med 64:10–13 (Russian)
79. Desiderato O, Testa M (1976) Shock-stress, gastric lesions and habituation in the chronic gastric fistula rat. Physiol Behav 16:67–73
80. Desiderato O, MacKinnon JR, Hissom HJ (1974) Development of gastric ulcers in rats following stress termination. Comp Physiol Psychol 87:208–214
81. Devyatkina TA (1978) Effect of ionol on development of experimental atherosclerosis. Dokl Akad Nauk SSSR 242:449–452 (Russian)
82. Dhalla NS, Singh JN, Bajusz E, Jasmin G (1976) Comparison of heart sarcolemmal enzyme activities in normal and cardiomyopathic (UM-X7.1) hasters. Clin Sci Mol Med 51:233–242
83. Dhalla NS, Tomlinson CW, Singh JN, Lee SL, McNamara DB, Harrow JA, Yates JC (1976) Role of sarcolemmal changes in cardiac pathophysiology. In: The sarcolemma: recent advances in studies on cardiac structure and metabolism. Urban & Schwarzenberg, Munich Berlin Vienna, pp 377–394
84. Di Giusto EL, Carncross K, King M (1971) Hormonal influences on fear-motivated responses. Psychol Bull 75:432–444
85. Diliberto EJ, Allen PL (1981) Mechanism of dopamine-β-hydroxylation. Semidehydroascorbate as the enzymic oxidation product of ascorbate. J Biol Chem 256:3385–3393
86. Dirks JF, Kinsman RA, Horton DJ, Fross KH, Jones ENF (1977) Panic-fear in asthma: rehospitalization after long-term treatment. Psychosom Med 39:52–53
87. Dobson JG (1980) Adenosine inhibition of catecholamine-elicited contractile and glycogenolytic responses in rat heart. Fed Proc 39:1173
88. Dorofeichuk VG, Shchukova NP, Klyueva ZI, Surkova AA (1968) Indices of complementary activity of blood serum in virtually healthy children. In: Trans Gorky Inst Pediatr 6:96–98 (Russian)
89. Douglass HO, Le Veen HH (1970) Stress ulcers. Arch Surg 100:178–181
90. Dubinin NP (1973) The philosophy of dialectic materialism and problems of genetics. Vopr filos 4:96–107 (Russian)
91. Duck-Chong CG (1976) A reassessment of the phospholipid dependence of membrane-bound enzymes, with special reference to glucose-6-phosphatase and Na^+, K^+-dependent adenosine triphosphatase. Enzyme 21:174–192
92. Duncan CP (1948) Habit reversal induced by electroshock in the rat. J Comp Physiol Psychol 41:11–16

93. Duncan CP (1949) The retroactive effect of electroshock on learning. J Comp Physiol Psychol 42:32–44
94. Dunn AJ (1976) Biochemical correlates of training experiences: a discussion of the evidence. In: eds Rosensweigand H, Bennett EL Neural mechanisms of learning and memory. MIT Press, Cambridge Mass pp 311–320
95. Durnova GN, Kaplansky AS, Roshchina NA (1966) Phagocytary activity of blood leucocytes in reticuloendothelial cells and production of antibodies in mice under declined atmospheric pressure. Bull Exp Biol Med 11:80–83 (Russian)
96. Dybdal NO, Gribble D, Madigan JE, Stabenfeldt GH (1980) Alterations in plasma corticosteroids, insulin and selected metabolites in horses used in endurance rides. Equine Vet J 12:131–140
97. Dzhalagonia SL (1972) Reproduction of neurotic state in monkeys with intact brain and damaged frontal lobes. Zh Vyssh Nerv Deiat 22:708–717
98. Eccles JC (1953) The neurophysiological basis of mind. Oxford Clarendon Press
99. Eichenbaum H, Quenon BA, Heacock A, Agranoff BW (1976) Differential behavioral and biochemical effects of regional injection of cycloheximide into mouse brain. Brain Res 101:171–176
100. Eisen AZ, Weissman S, Karon M (1962) Pseudouridine metabolism. I. Isolation and the effect of diet on urinary excretion. J Lab Clin Med 59:620–629
101. Ellis EF, Oelz O, Roberts LJ, Payne AN, Sweetman BJ, Nies AS, Oates JA (1976) Coronary arterial smooth muscle contraction by a substance released from platelets: evidence that it is thromboxane A_2. Science 193:1135–1137
102. Emanuel NM, Denisov EG, Mayzus ZK (1965) Chain reactions of carbohydrate oxydation in the fluid phase. Nauka, Moscow (Russian) [Tzepnye reaktzii okisleniya uglevodorodov v zhidkoy faze]
103. Emanuel NM, Obukhova LK, Smirnova LD, Bunto TB (1976) Effect of prolonged life expectancy in experiment under the action of chloralhydrate 2-ethyl-6-methyl-3-oxypyridine. Dokl Akad Nauk SSSR 226:961–963 (Russian)
104. Euler US von (1956) Noradrenaline (chemistry, physiology, pharmacology and clinical aspects). Thomas, Springfield
105. Euler US von (1959) The development and applications of the trihydroxyindole method for cathecholamines. Pharmacol Rev 11: 262–268
106. Falconer IR, Hetzel BS (1964) Effect of emotional stress and TSH on thyroid vein hormone level in sheep with exteriorized thyroids. Endocrinology 75:42–48
107. Fallen ET, Elliot W, Gorlin R (1967) Apparatus for study of ventricular function and metabolism in the isolated perfused rat heart. J Appl Physiol 22:836–839
108. Farutina LM (1964) Changes in rat kidneys after removal of varying amounts of tissue. In: Regeneration and cell reproduction in animals. Nauka, Moscow pp 95–110 (Russian) [Regeneratziya i kletochnoe razmnozhenie u zhivotnykh]
109. Fedorov IV (1975) Biochemical changes in skeletal muscles in hypokinesia and possible ways of restoring them. Kosm Biol Med 1:3–8 (Russian)
110. Filippovich SI (1963) Compensatory processes in the digestive system after resection of the stomach and small intestine (an experimental study). Meditsina, Moscow (Russian) [Kompensatornye protzessy v pishjevaritelnoy sisteme posle rezektzii zheludka i tonkogo kishechnika: Eksperimentalnye issledovanya]
111. Fitts RN, Booth FW, Winder WW, Holloszy JO (1975) Skeletal muscle respiratory capacity, endurance and glycogen utilization. Am J Physiol 228:1029–1033
112. Fizel A, Fizelova A (1971) Cardiohypertrophy and heart failure: dynamics of changes in high energy phosphate compounds, glycogene and lactic acid. J Mol Cell Cardiol 2:187–192
113. Fleckenstein A (1969) Inhibitory effect of isoptin on the metabolic processes in myocardium. Mechanism of action and possible uses. Symposium on assessment of isoptin treatment effect. A. L. Myasnikov Institute of Cardiology, USSR Academy of Medical Sciences, Moscow, pp 12–19 (Russian) [Simpozium, posvyashjennyi otzenke lechebnogo deystviya izoptina]
114. Fleckenstein A, Nakayama K, Fleckenstein-Grün G, Byon JK (1976) Interaction of hydrogen ions, calcium antagonistic drugs with excitation-contraction coupling of vascular smooth muscle. In: Betz B (ed) Ionic action on vascular smooth muscle. Springer, Berlin Heidelberg New York, pp 117–137

115. Fletcher DG, Harkins HN (1954) Acute peptic ulcers as a complication of major surgery, stress or trauma. Surgery 36:212–226
116. Flood JP, Bennett EL, Ormea AE, Rosenzweig MR, Jarvik ME (1978) Memory: modification of anisomycin-induced amnesia by stimulants and depressants. Science 199:324–326
117. Folch J, Lee M, Stanley GHS (1957) A simple method for the isolation and purification of total lipids from animal tissues. J Biol Chem 226:497–509
118. Forrester T, Williams GA (1977) Release of adenosine triphosphate from isolated adult heart cells in response to hypoxia. J Physiol (Lond) 248:371–390
119. Frankenhaeuser M, Patkai P (1965) Interindividual differences in catecholamine excretion during stress. Scand J Physiol 6:117
120. Franson R, Patriarea P, Elsbach P (1974) Phospholipid metabolism by phagocytic cells. Phospholipases A_2 associated with rabbit polymorphonuclear leukocyte granules. J Lipid Res 15:380–388
121. Frenkl RL, Csakvåry G, Csalay L (1969) A study of the stress reaction elicited by muscular exertion in trained and untrained man and rats. Acta Physiol Acad Sci Hung 36:365–370
122. Fridovich I (1975) Oxygen: boon and bane. Am Sci 63:54
123. Fridovich I (1979) Oxygen radicals, hydroperoxide and oxygen toxicity. In: Free radicals in biology, vol 1. Mir, Moscow, pp 272–298 (Russian translation) [Svobodnye radikaly v biologii]
124. Furkalo AK, Kuchuk AA, Nikishin LF (1980) Criterion for assessing tests with loads in patients with ischemic heart disease with intact and atherosclerotic coronary arteries. Kardiologiia 3:20–24 (Russian)
125. Furuta Y (1977) Studies on sodium and calcium contents of cardiovascular tissue in experimental hypertension. Jpn Circ J 41:19–28
126. Galbo H, Christensen HJ, Hoist JJ (1977) Catecholamines and pancreatic hormones during autonomic blockade in exercising man. Acta Physiol Scand 101:428–437
127. Galkin VV, Berdyshev GD (1968) On the question of quantifying nucleic acids by biochemical methods in the tissues of different animals. Biokhimiia 33:66–76 (Russian)
128. Gandel VG, Azhgikhin IS (1978) Obtainment of Prostaglandins. In: Prostaglandins. Meditsina, Moscow, pp 269–331 (Russian) [Prostaglandiny]
129. Gandelsman AB, Gracheva RP, Prokopovich NB (1960) Motor hypoxemia. In: Problems of sport physiology. Fizkultura i sport, Moscow, pp 240–249 (Russian) [Problemy fisiologii sporta]
130. Ganelina IE, Iljinsky BV (1960) The nervous system and atherosclerosis. Arkh Patol 11:5–13 (Russian)
131. Gazenko OG, Demin NN, Malkin VB, Pevzner LZ (1968) On adaptation changes in the cerebellum and motor structures of the spinal cord during various hypoxic conditions. Dokl Akad Nauk SSSR 179:997–1000 (Russian)
132. Georgiev GP (1970) Regulation of RNA synthesis in animal cells. Usp Sovrem Biol 3:331–353 (Russisch)
133. Gerasimov AM, Panchenko LF, Koen YM, Antonenkov VD (1974) Correlation of glutathione peroxidase and catalase activities during phenol barbital-induced liver hyperplasia. Doklady AN SSSR, 216:1175–1177 (Russian)
134. Gladkova AI, Rykova NI (1974) On the question of the hormone mechanisms conditioning sex distinctions during experimental myocardial infarction. Krovoobrashcheniie 20:64–66 (Russian)
135. Goldberg AL, Dice JF (1974) Intracellular protein degradation in mammalian and bacterial cells. Ann Rev Biochem 43:835–869
136. Goldberg VA (1964) The role of professional factors and nutrition in the epidemiology of (the) coronary atherosclerosis. In: Spreading of hypertension and coronary atherosclerosis and life conditions. Meditsina, Leningrad, pp 53–63 (Russian) [Rasprostranenie gipertonicheskoy bolezni i koronarnogo ateroskleroza i usloviya zhizni]
137. Goldman H, Rosoff C (1968) Pathogenesis of acute gastric stress ulcers. Am J Pathol 52:227–243
138. Goldman SS, Albers RW (1973) Sodium-potassium-activated adenosine triphosphatase. IX. The role of phospholipids. J Biol Chem 293:379

139. Gollnick PD, King DW (1969) Effect of exercise and training on mitochondria of a rat skeletal muscle. Am J Physiol 216:1502
140. Gordon R, Spector S, Sjoerdsma A, Udenfriend S (1966) Increased synthesis of norepinephrine and epinephrine in the intact rat during exercise and exposure to cold. J Pharmacol Exp Ther 155:440–447
141. Gorizontov PD (1976) Lymphoid tissue and non-specific resistance of the organism. Arkh Patolog 38:3–13 (Russian)
142. Gorlin R (1967) Angina pectoris with normal coronary arteries. JAMA 201:27–28
143. Gribanov GA (1975) Structure and biological significance of phospholipids. Usp Sovrem Biol 80:382–398 (Russian)
144. Grover RP (1963) Pulmonary vasoconstriction in stress at high altitude. J Appl Physiol 18:567–574
145. Groves BM, Marcus FI, Ewy GA, Phibbs BP (1977) Coronary arterial spasm in variant angina pectoris produced by hyperventilation and valsalva-documented by coronary arteriography. In: Horman JC (ed) Coronary artery medicine and surgery. Appleton-Century-Crofts, New York, pp 311–320
146. Guarnieri G, Ferrari R, Visioli O, Caldarera CM, Nayler WG (1978) Effect of α-tocopherol on hypoxic-perfused and reoxygenated rabbit heart muscle. J Mol Cell Cardiol 10:893–906
147. Guarnieri G, Flamigni F, Rossoni-Caldarera CM (1979) Glutatione-peroxidase activity and release of glutatione from oxygen-deficient perfused rat heart. Biochem Biophys Res Commun 89:678–684
148. Guarnieri G, Flamigni F, Caldarera CM (1980) Role of oxygen in the cellular damage induced by reoxygenation of hypoxic heart. J Mol Cell Cardiol 12:797–808
149. Gubachev YM, Iovlev BV, Karvasarski BD, Rasumov SA, Stabrovski EM (1976) Emotional stress in man under normal and pathological conditions. Meditsina, Leningrad [Emotzionalnyi stress v usloviyakh normy i patologii cheloveka]
150. Gudbjarnason S (1971) Acute alterations in energetics of ischemic heart muscle. Symposium: Metabolism of the hypoxic and ischemic heart. Geneva
151. Gudbjarnason S, Hallgrimsson J (1978) Cardiac lipids and myocardial ischemia. Proceedings of the VIII world congress of cardiology, Tokyo
152. Gudbjarnason S, Doell B, Oskursdottir G, Hallgrimsson J (1978) Modification of cardiac phospholipids and catecholamine stress tolerance. In: De Duve C, Hayaishi O (eds) Tocopherol, oxygen, and biomembranes. Elsevier/North Holland Biomedical Press, Amsterdam, pp 297–310
153. Guillemin R, Varge T, Rossier J, Minck S, Ling N, River C, Vale W, Bloom F (1977) β-endorphin and adrenocorticotropin are secreted concomitantly by pituitary gland. Science 197:1367–1369
154. Gukasov VM, Sergeev PV, Seyfulla PD, Vladimirov YA (1974) On the antioxidant effect of steroid hormones on peroxidation of the lipids of mitochondrial membranes in vivo and in vitro. Bull Exp Biol Med 11:54 (Russian)
155. Gullis RJ, Rowe CE (1975) The stimulation by transmitter substances and putative transmitter substances of the net activity of phospholipase A_2 of synaptic membranes of cortex of guinea-pig brain. Biochem J 148:197–208
156. Haft JI, Fain K (1973) Intravascular platelet aggregation in the heart induced by stress. Circulation 47:353–358
157. Hagnell O (1965) The premorbid personality of persons who develop cancer in a total population investigated in 1947 and 1957. Ann NY Acad Sci 125:846–855
158. Holden JS, Priestly JG (1937) Respiration. Biomedgiz, Moscow Leningrad (Russian translation) [Dykhanie]
159. Hamamdzic M, Malik KU (1977) Prostaglandins in adrenergic transmission of isolated perfused rat pancreas. Am J Physiol 232:E201–E209
160. Hamberg M, Samuelsson B (1966) Prostaglandins in human seminal plasma. Prostaglandins and related factor 46. J Biol Chem 241:257–263
161. Hamberger CA (1963) Difference between isolated neuronal and vascular glia with respect to respiratory activity. Acta Physiol Scand 58:1–58
162. Hamberger CA, Hyden H (1949) Production of nucleoproteins in the vestibular ganglion. Acta Otolaryngol [Suppl] (Stockh) 75:53–81

163. Hamberger CA, Hyden H (1963) Increase of enzymatic changes in neurons and glia during increased function and hypoxia. J Cell Biol 16:521–525
164. Hammond GL, Wieben E, Markert CL (1979) Molecular signals for initiating protein synthesis in organ hypertrophy. Proc Natl Acad Sci USA 76:2455–2459
165. Hardvicke PMD, Green NM (1974) The effect of delipidation on the adenosine triphosphatase of sarcoplasmic reticulum. Electron microscopy and physical properties. Eur J Biochem 42:183–193
166. Hatefi Y, Hanstein WG (1970) Lipid oxidation in biological membranes. I. Lipid oxidation in submitochondrial particles and microsomes induces by chaotropic agents. Arch Biochem Biophys 136:73–86
167. Hearse DJ (1980) The pathophysiology of irreversible damage in myocardial ischemia. J Mol Cell Cardiol 12:54
168. Hearse DJ, Humphrey SM, Chain EB (1973) Abrupt reoxygenation of the anoxic potassium-arrested perfused heart: a study of myocardial enzyme release. J Mol Cell Cardiol 5:395–407
169. Hearse DJ, Humphrey SM, Nayler WG, Slade A, Border D (1975) Ultrastructural damage associated with reoxygenation of the anoxic myocardium. J Mol Cell Cardiol 7:315–324
170. Hechter O, Yoshinaga K, Halkerston ID (1967) Estrogen-like anabolic effects of cyclic 3′–5′ adenosine monophosphate and other nucleotides in isolated rat uterus. Arch Biochem Biophys 122:449–465
171. Hedquist P (1977) Basic mechanisms of prostaglandin action on autonomic neurotransmission. Ann Rev Pharmacol Toxicol 17:259–279
172. Heiss HW, Barmeyer J, Wink K, Huber G, Hagemann G, Beiter G, Keul J, Reindell H (1975) Durchblutung und Substratumsatz des gesunden menschlichen Herzens in Abhängigkeit vom Trainingszustand. Verh Dtsch Kreislaufforsch 41:247–252
173. Hidaka H, Hara F, Harada N, Hashizume J, Yano M (1974) Selective inhibition of dopamine--hydroxylase in the peripheral tissues by 5-dimethyldithiocarbamylpicolinic acid; its effect on stress-induced ulcer, ethanol-induced sleep and blood pressure. J Pharmacol Exp Ther 191:384
174. Highman B, Altland PD (1967) Effect of dimethyl sulfoxide, exercise and training on rat serum enzymes and tissues. Am J Physiol 213:779–782
175. Hinesley RK, Norton JA, Aprison MH (1968) Serotonin, norepinephrine and 3,4-dihydroxyphenylethylamine in rat brain parts following electroconvulsive shock. J Psychiat Res 6:143–152
176. Hirata F, Axelrod J (1980) Phospholipid methylation and biological signal transmission. Science 209:1082–1090
177. Hirata F, Strittmatter WJ, Axelrod J (1979) β-adrenergic receptor agonists increase phospholipid methylation, membrane fluidity, and β-adrenergic receptor–adenylate. Proc Natl Acad Sci USA 76:368–372
178. Hjalmarson AC, Isaksson O, Källfelt B (1972) In vitro mechanical activity and heart protein synthesis. European symposium on heart overloading. Paris, pp 313–314
179. Holloszy JD (1967) Biochemical adaptation in muscle. J Biol Chem 242:2278–2282
180. Horgan VG, Philpot GSL (1961) Peroxide-like substance in alarmed mice. Nature 192:662–663
181. Horowicz P, Larrabee MG (1958) Glucose consumption and lactate production in a mammalian sympathetic ganglion at rest and in activity. J Neurochem 2:102–118
182. Houser VP, Cash RJ, Van Hart DA (1975) Effects of manipulating cholinergic tone upon the activity-stress ulcer. Pharmacol Biochem Behav 3:825–831
183. Hsueh W, Isakson PC, Needleman P (1977) Hormone selective lipase activation in the isolated rabbit heart. Prostaglandins 13:1073–1091
184. Hubner K (1967) Kompensatorische Hypertrophie, Wachstum und Regeneration der Rattenniere. Ergeb Allg Pathol 48:1–80
185. Hultgren HM, Miller H (1967) Human heart weight at high altitude. Circulation 35:207–218
186. Hurtado A (1942) Chronic mountain sickness. JAMA 120:1278
187. Hurtado A (1964) Acclimatization to high altitudes. In: The physiological effects of high altitude. Pergamon, Oxford London New York Paris, pp 1–17
188. Hyden H (1958) Biochemical changes in glial cells and nerve cells at varying activity. In:

Brücke K (ed) Biochemistry of the central nervous system. Fourth international congress of biochemistry, Vienna, vol. 3, p 64
189. Hyden H (1962) Cytophysiological aspects of the nucleic acids and proteins of nervous tissue. In: Elliott KAC, Page JH, Quastel JH, Thomas CC (eds) Neurochemistry, Springfield, pp 331–375
190. Hyden H (1963) Neuron. In: Functional morphology of the cell. Inostrannaya Literatura, Moscow, pp 185–260 (Russian translation) [Funktzionalnaya morfologiya kletki]
191. Hyden H, Egyhazi E (1964) Changes in RNA content and base composition in cortical neurons of rats in a learning experiment involving transfer of handness. Proc Natl Acad Sci USA, 52:1030–1035
192. Hyden H, Lange PW (1966) A genetic stimulation with production of adenic-uracyl rich RNA in neurons and glia in learning. The question of transfer of RNA from glia to neurons. Naturwissenschaften 53:64–70
193. Hyden H, Lange PW (1972) Protein changes in different brain areas as a function of intermittent training. Proc Natl Acad Sci USA 69:1980–1984
194. Jacob R, Ebrecht G, Halubarsh C, Rupp H, Kissling C (to be published) Mechanics and energetics in cardiac hypertrophy as related to isoenzyme. Basic Res Cardiol
194a. Jacob F, Monod I (1961) Genetic regulatory mechanisms in the synthesis of proteins. J Mol Biol 3:318–356
195. Jain SK, Shohet SB (1981) Calcium potentiates the peroxidation of erythrocyte membrane lipids. Biochim Biophys Acta 642:46–54
196. Jerne NK, Nordin AA (1963) Plaque formation in air by single antibody-producing cells. Science 140:405
197. Jones KW, Kinross J, Moitland N, Noval M (1979) Normal human tissues contain RNA and antigens related to infectious adeno-virus type 2. Nature 277:274–279
198. Kagan VE, Kotelevtsev SV, Kozlov YuP (1974) The role of enzymic peroxidation of phospholipids in the stripping mechanism of endoplasmic reticulum membranes in vivo. Dokl Akad Nauk SSSR 217:213–216 (Russian)
199. Kagan VE, Novikova KN, Belousova LV, Bratkovskaya LB, Chernova TL, Galushenko IV (1976) Lipid peroxidation in photoceptor membranes. In: Free radical lipid oxidation in norm and pathology. Moscow, pp 57–58 (Russian) [Svobodnoradikalnoe okislenie lipidov v norme i patologii]
200. Kagan VE, Azizova OA, Arkhipenko YuV, Klaan NK, Kozlov YuP, Vladimirov YA (1977) Relationship between structural and functional reconstructions in SR membranes under lipid peroxidation. Biofizika 22:625–639 (Russian)
201. Kagan VE, Shvedova AA, Novikova KN (1978) On phospholipase participation in 'reparation' of photoceptor membranes subjected to peroxidation. Biofizika 23:279–284 (Russian)
202. Kagan VE, Lankin VE, Shvedova AA, Novikova KN, Dobrina PK, Bratkovskaya LB, Kuliev IY (1979) Enzymic and non-enzymic systems of photoceptor membrane protection against active oxygen species and lipid peroxides. Bull Exp Biol Med 8:164–166 (Russian)
203. Kagan VE, Arkhipenko YuV, Pisarev VA, Belousova LV, Kozlov YuP (1981) Injury of SR membranes in peroxidation and its role in the development of muscle pathologies. In: Membrane biophysics. Nauka, Moscow, pp 88–95 [Biofizika membran]
204. Kako K, Bing RJ (1958) Contractility of actomyosin bands prepared from normal and failing human hearts. J Clin Invest 37:465–470
205. Kalyuzhny IG, Narbekov ON, Bedekova RB, Baymuratova RK (1976) Functional state of thyroid gland during short-term adaptation of healthy persons to high-mountain conditions of Tien-Shan and the Pamirs. In: Trans Kirghiz Med Inst 110:94–100 (Russian) [Sbornik trudov kirgizskogo meditzinskogo instituta]
206. Kapelko VI (1964) Negative chronotropic vagus effect during compensatory heart hyperfunction. Dokl Akad Nauk SSSR 155:715–718 (Russian)
207. Karlsson J, Saltin B (1970) Lactate, ATP, and CP in working muscles during exhaustive exercise in man. J Appl Physiol 29:598–602
208. Katz A, Maxwell JB (1964) Actin from heart muscle: sulfhydryl groups. Circ Res 14:345–350
209. Katz A, Messineo FC (1981) Lipid-membrane interactions and the pathogenesis of ischemic damage in the myocardium. Circ Res 48:1–16

210. Katz AM, Nash-Adler P, Miceli Y (1979) Inhibition of Ca efflux from the sarcoplasmic reticulum by free fatty acids. Circulation Suppl 2 60:11–12
211. Katzenovich ER (1980) Antiradical activity of lipids and the content of polyunsaturated fatty acids in the blood serum in myocardial infarction. Med Zh Uzb 11:32–36 (Russian)
212. Kawarada Y, Lambek J, Matsumoto T (1975) Pathophysiology of stress ulcer and its prevention. II. Prostaglandin E, and microcirculatory responses in stress ulcer. Am J Surg 129:217–222
213. Keim KL, Sigg EB (1976) Physiological and biochemical concomitants of restraint stress in rats. Pharmacol Biochem Behav 4:289–297
214. Kerr A, Diasio RB, Bommer WJ (1965) Effect of altitude (hypoxia) on coronary artery size in the white rat. Am Heart J 69:841–842
215. Kesner RP, D'Andrea JA (1971) Electroconvulsive shock disrupts both information shortage and retrieval. Physiol Behav 7:73–76
216. Kety S (1970) Biogenic amines in the central nervous system: their possible role in arousal, emotion and learning. In: Quarton GC (ed) The neurosciences second study program. Rockefeller Univ Press, New York, pp 324–336
217. Khaidarlin SK (1967) Nucleic acid content in small weighed portions and individual cells of the spinal cord during different functional states. Biokhimiia 32:677–682 (Russian)
218. Khomulo PS (1963) On the significance of lengthy functional tension of the nervous system in the development of atherosclerosis in dogs. Dokl Akad Nauk SSSR 152:727–729 (Russian)
219. Khomulo PS (1964) Atherosclerosis resulting in young dogs in lengthy functional tension of the central nervous system. Dokl Akad Nauk SSSR 156:976–977 (Russian)
220. Khomulo PS (1974) Neurogenic atherosclerosis and its development mechanisms. Kardiologiia 5:140–147 (Russian)
221. Kloner RA, Braunwald E (1980) Observation on experimental myocardial ischemia. Cardiovasc Res 14:371–395
222. Klygul TA, Krivopalov VA (1966) Installation with automatic device for recording rat behaviour for experimental assessment of small tranquiliser doses. Farmakol Toksikol 29:2 (Russian)
223. Kobrick JL (1976) Effects of prior hypoxia exposure on visual target detection during later more severe hypoxia. Percept Mot Skills 42:751–761
224. Kogan AK (1961) A new simple method of graduated stenosis of renal and other arteries in small laboratory animals in chronic experiment. Bull Exp Biol Med 1:112–113 (Russian)
225. Kogan AK (1977) On lipid and other mediators of carcinogenesis. In: Problems of malignant growth. Transactions of medical faculty. Patrice Lumumba Peoples' Friendship University, Moscow, pp 46–53 (Russian) [Sbornik nauchnykh trudov meditzinskogo fakulteta Ordena Druzhby Narodov Universiteta Druzhby narodov im. Patrisa Lumumby]
226. Kogan AK, Kudrin AN, Nikolaev SM (1976) Free-radical lipid peroxidation and pathogenesis of coronary occlusive myocardial infarction. In: Free-radical oxidation in norm and pathology. Moscow, pp 68–71 (Russian) [Svobodnoradikalnoe okislenie lipidov v norme i patologii]
227. Kolbeck R, Paradise NF (1975) Localisation of rapidly exchangeable calcium in mammalian heart muscle. Am J Physiol 288:345–351
228. Kolberg A (1959) Relations of renal tubular and glomerular function as influenced by 75 per cent reduction of nephron number. A pathophysiological study. Scand J Clin Lab Investig Suppl 41 11:1–152
229. Kolloas J, Buskirk ER, Akers RF, Prokop EK, Baker PT, Picon-Reategui S (1968) Work capacity of long-time residents and newcomers to altitude. J Appl Physiol 24:792–799
230. Kolokolchikova EG, Korolkov VI (1976) Variations in the mass of the right ventricle and in its RNA content during 'inverse' development of hypertrophy. Bull Exp Biol Med 12:1421–1422 (Russian)
231. Koob GF, Annau Z (1974) Behavioural and neurochemical alterations induced by hypoxia in rats. Am J Physiol 227:73–78
232. Korchagin VP, Bratkovskaya LB, Shvedova AA, Arkhipenko YuV, Kagan VE, Shumakov SA (1980) Oligomerization of integral membranous proteins in lipid peroxidation. Biokhimia 45:1767–1772 (Russian)
233. Korolkov VI, Satilov AA (1971) Altitude resistance of rats to hypoxic hypoxia during different

post-training periods. In: vital issues of space biology and medicine. Moscow, p 147 (Russian) [Aktualnye voprosy kosmicheskoy biologii i meditziny]
234. Korolkov VI, Gaevskaya MS, Serova LV, Nosova EA, Lunev IY, Britvan II, Lopenko MA (1973) Physiological reactions in white rats during readaptation after training to hypoxic hypoxia. Kos Biol Med 7:20–23 (Russian)
235. Korotaev MM, Bobrovnitsky MP, Motorin YA (1974) Variations in general and tissue gas metabolism in humans during bicycle ergometry following mountain acclimatization and pressure chamber training. In: Hypercapnia, hyperoxia and hypoxia. Medical Institute, Kuibyshev, pp 128–129 (Russian) [Giperkapniya, giperoksiya, gipoksiya: Tezisy dokladov Vsesoujsnoy konferentzii]
236. Kovalenko EA (1975) Pathophysiologic aspects of the problem of lengthy hypokinesia. Patol Fiziol Exp Ter 3:11–24 (Russian)
237. Kozlov YuP (1975) Free-radical oxidation of lipids in biomembranes in norm and pathology. Transactions of Moscow Society of Naturalists, vol. 52. Moscow, pp 5–14 (Russian)
238. Krasnovskaya IA (1974) Changes in the hypothalamic and hypophiseal systems of rats under lengthy hypoxia. Probl Endokrinol 20:53–57 (Russian)
239. Krayenbühl HP (1977) Effect of hypertrophy on the contractile function of the heart in man. Basic Res Cardiol 72:184–189
240. Krebs EG, Stull JT, England PJ, Huang TS, Brostrom CO, Vandenheede J (1973) The regulation of muscle metabolism and function by protein phosphorylation. In: Abstract of the 15th annual Riami winter symposium, p 245
241. Kreps EM (1959) Oxyhemometry. Meditsina. Moscow (Russian) [Oksigemometriya]
242. Kreps EM, Verzhbinskaya NA, Chenykaeva EY, Chirkovskaya EV, Gavurina ZK (1956) On animal adaptation to chronic hypoxia. Fiziol Zh SSSR 42:148–158
243. Kristt DA, Freimark SJ (1973) Histopathology and pathogenesis of behaviorally induced gastric lesions in rats. Am J Pathol 73:411–424
244. Kruchinina NA, Chernigovskaya SV (1976) On the question of the incidence of the ischemic heart disease in persons engaged in intensive brain work. In: Bluzhas YN (ed) Cardiology (Kardiologiia). Institute of Cardiology, p 46 (Russian)
245. Kruglikov RI (1975) Neurochemical mechanisms of the forming and fixing of time relationships. Results of science and technology. Physiology of man and animals. VINITI Akad Nauk SSSR, Moscow 16:59–118 (Russian) [Itogi nauki i tekhniki. Fiziologiya cheloveka i zhivotnykh]
246. Krupina TN, Fedorov BN, Benevolskaya TV, Baikova OI (1971) Variations in cardiac activity in prolonged restriction of motor activity. Kosm Biol Med 2:76–81 (Russian)
247. Kübler W, Katz AM (1977) Mechanism of early 'pump' failure of ischemic heart. Possible role of adenosine triphosphate depletion and inorganic phosphate accumulation. Am J Cardiol 40:467–471
248. Kudrin AN, Kogan AK, Strukov AI, Katursky LV, Nikolaev SM (1975) Antioxidants and myocardial infarction. In: Second international congress of pathological physiology, Praha, p 216
248a. Kudrin AN, Nuzhny VP, Nikolaev SM (1977) Effects of sodium selenite, α-tocopherol and their combinations on certain indices of the functional state of the adrenergic and cholinergic systems of cardiac regulation in experimental myocardial infarction. Farmakol Toksikol 6:706–710 (Russian)
249. Kudrin AN, Kogan AK, Korolev VV, Nikolaev SM, Beskrovnov NN (1978) Free-radical lipid oxidation in the origin of myocardial infarction and the remedial and preventive role of sodium AO-selenite and its combinations with vitamin E. Kardiologiia 2:115–118 (Russian)
250. Kurth K, Mikschy U (1978) Human antibodies reactive with purified envelope antigens of primate type C tumor viruses. Proc Natl Acad Sci USA 75:5692–5696
251. Kushko VM, Panchenko LF (1959) Incorporation of ^{35}S methionine in the proteins of the central nervous system and other organs of different-age animals under hypoxia. Patol Fiziol Exp Ter 3:22–32 (Russian)
252. Lammerant J, Becsei I (1973) Left venticular contractility and developed tension in the intact dog submitted to an intracoronary infusion of adenosine. J Physiol (Lond) 229:41–49
253. Lang GF (1948) Hypertension and the central nervous system. In: Transactions of fourth session of USSR Academy of Medical Sciences. Moscow, pp 6–18 (Russian)

254. Lang GF (1950) Hypertension. Medgiz, Leningrad (Russian) [Gipertonicheskaya boleznj]
255. Langer GA (1978) The structure and function of the myocardial cell surface. Am J Physiol 235:H461–H468
256. Langer GA (1980) The role of calcium in the control of myocardial contractility: an update. J Mol Cell Cardiol 12:231–239
257. Laychock SG, Shen SC, Carmens EL, Rubin RR (1978) The effect of corticotropin on phospholipid metabolism in isolated and renocortical cells. Biochim Biophys Acta 528:355–369
258. Lebedev AV, Levitsky DO, Loginov VA (1980) Oxygen: an inductor of calcium ion transfer via bilayer lipid membranes. Dokl Akad Med Nauk SSSR 252:1494–1497 (Russian)
259. Le Blanc J (1969) Stress and interstress adaptation. Fed Proc 28:996–1000
260. Lee JB (1969) Hypertension, natriuresis, and the renal prostaglandins. Ann Intern Med 70:1033–1038
261. Lee SZ, Balasubramerion V, Dhalla NS (1976) Excitation-contraction coupling in heart. XIX. Effect of hypoxia on calcium transport by subcellular particles in the isolated perfused rat heart. Can J Physiol Pharmacol 54:49–58
262. Lefer AM, Ogletree ML (1978) Beneficial action of glucocorticoids and prostaglandins in myocardial ischaemia. In: Hjarimson A, Wilhelmsen L (eds) Acute and long-term medical management of myocardial ischaemia. Lindgren & Soner, pp 229–237
263. Lefkowitz RJ, Haber E (1971) A fraction of the ventricular myocardium that has the specificity of the cardial beta-adrenergic receptors. Proc Natl Acad Sci USA 68:342–349
264. Legato MJ, Langer GA (1969) The subcellular localization of calcium ion in mammalian myocardium. J Cell Biol 41:401–423
265. Lenard J, Singer SJ (1966) Protein conformation in cell membrane preparations as studied by optical rotatory dispersion and circular dichroism. Proc Natl Acad Sci USA 56:1828
266. Lepran I, Nemecz G, Koltai M, Szekeres L (1981) Effect of a linoleic acid-rich diet on the acute phase of coronary occlusion in conscious rats: influence of indomethacin and aspirin. J Cardiovasc Pharmacol 3:847–853
267. Le Shan L (1959) Psychological states as factor in the development of malignant disease: a critical review. J Natl Cancer Inst 22:1–18
268. Le Shan L, Worthington RE (1955) Some psychologic correlates of neoplastic disease: a preliminary report. J Clin Exp Psychopathol 16:281–288
269. Le Shan L, Worthington RE (1956) Loss of cathexes as a common psychodynamic characteristic of cancer patients: an attempt at statistical validation of a clinical hypothesis. Psychol Rep 2:183–193
270. Levey GS (1971) Restoration of glucagon responsiveness of solubilized myocardial adenyl cyclase by phosphatidyl serine. Biochem Biophys Res Commun 43:108–113
271. Levey GS (1971) Restoration of norepinephrine responsiveness of solubilized myocardial adenylate cyclase by phosphatidylinositol. J Biol Chem 246:405–410
272. Libby P, Maroko PR, Bloor C, Sobel BE, Braunwald E (1973) Reduction of experimental myocardial infarct size by corticosteroid administration. J Clin Invest 52:599–607
273. Lifshits AV (1949) Effect of hypoxia on higher nervous activity. Fiziol Zh SSSR 35:3–15
274. Limas CJ (1974) Stimulation by angiotensin of myocardial prostaglandin synthesis. Biochim Biophys Acta 337:417–420
275. Limas CJ, Cohn JN (1973) Isolation and properties of myocardial prostaglandin synthesis. Cardiovasc Res 7:623–628
276. Limas CJ, Fugan D, Freis ED (1974) Effect of acute cardiac overload on intramyocardial cyclic 3′,5′-AMP: relation to prostaglandin synthesis (38289). Proc Soc Biol Med 147:103–109
277. Lindenmayer GE, Sordahl LA, Schwartz A (1969) Reevaluation of oxidative phosphorylation in cardiac mitochondria from normal animals and animals in heart failure. Circ Res 23:439–450
278. Litvitsky PF, Kogan AK, Kudrin AN, Lukyanova LO (1981) Pathogenic role of activation of lipid peroxidation and protective effect of sodium selenite in ischemia and myocardial reperfusion. Bull Exp Biol Med 3:271–274 (Russian)
279. Liungqvist A, Unge G (1973) The proliferative activity of the myocardial tissue in various forms of experimental cardiac hypertrophy. Acta Pathol Microbiol Scand 81:233–240

280. Löwe H, Schultz U, Rosner T (1980) Disturbance of the functional intercapillary distance during a myocardial hypoxia induced by high doses of isoproterenol. J Mol Cell Cardiol 12:97
281. Loginova EV (1968) Re-exposure at 'altitude': a method for disclosing concealed traces of adaptation to hypoxia. In: Gazenko OG (ed) Problems of space biology, vol 8. Nauka, Moscow pp 188–198 (Russian) ed. by O. G. Gazenko [Problemy kosmicheskoy biologii]
282. Loubier R, Pfister A (1966) Metabolic changes in cerebellum in experimental hypoxia. Rev Med Aeronaut 5:23
283. Lucy JA (1972) Functional and structural aspects of biological membranes: a suggested structural role of vitamin E in the control of membrane permeability and stability. Ann NY Acad Sci 203:4–11
284. Luderitz B, Naumann d'Alnoncourt C, Steinbeck G (1976) Effects of free fatty acids on electrophysiological properties of ventricular myocardium. Klin Wochenschr 54:309–313
285. Lullman H, Peters T (1977) Plasmolemmal Ca in excitation-contraction coupling in the heart. Clin Exp Pharmacol Physiol 4:49–57
286. Lukomsky PV, Meerson FZ, Soloviov VV, Shenderov SM, Zharov EI, Markovskaya GI, Mdinaradze YS, Arshakuni RO, Taraeva NG (1967) Disturbances in cardiac contractile function during myocardial infarction and use for therapeutic purposes of synthesis cofactors and nucleic acid precursors. Kardiologiia 1:3–11 (Russian)
287. Luzikov VA, Rakimov MM, Saks VA, Berezin LV (1967) Protective effect of substrates during inactivation of NAD · H_2-oxydase and succinate oxydase with cobra venom and trypsin. Biokhimiia 32:1234–1241 (Russian)
288. Makarov GA (1974) On the mechanism of disturbances of structural processes in tissues during lengthy hypoxia. Patol Fiziol Exp Ter 4:41–45 (Russian)
289. Makman RS, Sutherland EW (1965) Adenosin 3′,5′-phosphate in excherichia coli. J Biol Chem 240:1309–1314
290. Malkin VB, Gippenreimer EB (1977) Acute and chronic hypoxia. In: Gazenko OG (ed) Problems of Space Biology, vol 35. Nauka, Moscow, p 35 (Russian)[Problemy kosmicheskoy biologii]
291. Mallorga P, Tallman JF, Hennberry RC, Hirata F, Strittmatter WT, Axelrod J (1980) Mepacrine blocks beta-adrenergic agonist-induced desensitization in astrocytoma cells. Proc Natl Acad Sci USA 74:1341
292. Malyshev VV, Lifantjev VI, Meerson FZ (1982) Functional state of heart mitochondria in dynamic of emotional-painful stress. Kardiologiia 6:118–120 (Russian)
293. Manankova NM, Salganik RI (1977) Intensified cholesterol hydroxylation in rat liver resulting from induction of microsomal enzymes. Vopr Med Khim 4:458–463 (Russian)
294. Mandache E, Penaru A, Pambuccian C, Chergut A, Cretulescu G, Stoica A, Slavescu V (1978) The role of the adrenal medulla in the psychosomatic activity and reactivity to stresses. Rev Roum Med Ser Endocrinol 16:171–178
295. Manina AA (1976) Ultrastructural foundations of brain activity. Meditsina, Leningrad (Russian)
296. Manning GW, McEackern CG, Hall GE (1939) Reflex coronary artery spasm following sudden occlusion of other coronary branches. Arch Intern Med 64:661–674
297. Marchbanks VH (1960) Flying stress and urinary 17-hydroxycorticosteroid levels during twenty-hour missions. Aerosp Med 31:639
298. Margreth A, Novelle F (1964) Observations on the chemical determination and distribution of ribonucleic acid in several striated muscles. Exp Cell Res 35:38–51
299. Markovskaya GI, Meerson FZ (1967) Effect of the cofactors of nucleic acid synthesis and precursors on heart contractile function in experimental myocardial infarction. Patol Fiziol Exp Ter 6:30–33 (Russian)
300. Marticorena EA, Severino J, Hultgren HN (1967) I. Cardiopulmonary pathology at high altitude. II. How the people become old at high altitude. III. Treatment of high altitude pulmonary edema by bed rest alone. WHO. In: Abstracts of international biological program: Population Biology of Altitude. Washington p 10
301. Martin LC, Werterbercer GE, Bullard RW (1971) Thyroidal changes in the rat during acclimatization to simulated high altitude. Am J Physiol 22:1057–1063

302. Martynenko FP, Rozhko OT, Novikova NP (1974) Incorporation of leucin ^{14}C in the somatotrophic hormone of rats depending on the thyroid hormone level in the body. Prob Endokrinol 20:107–110 (Russian)
303. Maseri A, Mimmo R, Chierchia S, Marchesi C, Pesola A, L'Abbate A (1975) Coronary artery spasm as a cause of acute myocardial ischemia in man. Chest 68:625–633,
304. Matlina ES, Rakhmanova TB (1967) A method for determining epinephrine, norepinephrine, dopamine, and DOPA contents in tissues. In: Banshikov VM, Menshikov VV (eds) Transactions on novel equipment and techniques, vol 5. 1st Moscow Medical Institute, Moscow, pp 136–142 (Russian) [Metody issledovaniya nekotorykh sistem gumoralnoy regulatzii]
305. Matsuda M, Hoshino M (1976) Conversion in vivo of succinic semialdehyde γ-hydroxybutyric acid in the mouse. Jikeikai Med J 23:189–193
306. Matsui H, Schwartz A (1966) Purification and properties of a highly active oubain-sensitive Na^+, K^+-dependent adenosine triphosphatase from cardiac tissue. Biochem Biophys Acta 128:386–390
307. Matsumoto S, Krasnow N (1968) Nucleic acid metabolism of cardiac diaphragmatic and quadriceps muscle. Am J Physiol 214:620–626
308. Matthies H (1974) Dem Gedächtnis auf der Spur. Wiss Forsch 24:412–415
309. Maynert EW, Levi R (1964) Stress-induced release of brain norepinephrine and its inhibition by drugs. J Pharmacol Exp Ther 143:90–95
310. McIlwain H (1970) Metabolic adaptation in brain. Nature 226:803
311. McKenzie SG, Bar HP (1973) On the mechanism of adenyl cyclase inhibition by adenosine. Can J Physiol Pharmacol 51:190–196
312. Medugorac J, Kammereit A, Jacob R (1975) Einfluss eines chronischen Schwimmentrainings auf Struktur und Enzymaktivität von Myosin beim Rattenmyokard. Z Physiol Chem 356:1161–1171
313. Meerson FZ (1959) Compensatory adaptations. In: Bakulev AN (ed) Big Medical Encyclopedia of the USSR, vol 73, pp 762–772 (Russian) [Bolshaya Meditzinskaya Entziklopediya]
314. Meerson FZ (1960) Compensatory hyperfunction and insufficiency of the heart. Meditsina, Moscow (Russian) [Kompensatornaya giperfunktziya i nedostatochnostj serdtza]
315. Meerson FZ (1963) On the interrelation of physiological function and the genetic apparatus. Meditsina, Moscow (Russian) [O vsaimosvyazi fiziologicheskoy funktzii i geneticheskogo apparata kletki]
316. Meerson FZ (1965) The myocardium during cardiac hyperfunction, hypertrophy, and insufficiency. Meditsina, Moscow (Russian) [Miokard pri giperfunktzii, gipertrofii i nedostatochnosti serdtza]
317. Meerson FZ (1967) Structural provision of body function. Nauka, Moscow (Russian) [Plasticheskoe obespechenie funktzii organizma]
318. Meerson FZ (1968) Cardiac hyperfuntion, hypertrophy, and insufficiency. Meditsina, Moscow (Russian) [Giperfunktziya, gipertrofiya, nedostatochnostj serdtza]
319. Meerson FZ (1971) Mechanism of hypertrophy of the heart and experimental prevention of acute cardiac insufficiency. Br Heart J Suppl 33:100–108
320. Meerson FZ (1973) General mechanism of adaptation and prophylaxis. Meditsina, Moscow (Russian) [Obshjiy mekhanizm adaptatzii i profilaktiki]
321. Meerson FZ (1975) Adaptation of the heart to big loads and cardiac insufficiency. Nauka, Moscow (Russian) [Adaptatziya serdtza k bolshoy nagruzke i serdechnaya nedostatochnostj]
322. Meerson FZ (1978) Adaptation, deadaptation, and cardiac insufficiency. Meditsina, Moscow (Russian) [Adaptatziya, deadaptatziya i nedostatochnostj serdtza]
323. Meerson FZ (1980) Pathogenesis and prophylaxis of heart damages caused by stress. J Mol Cell Cardiol 12:105
324. Meerson FZ, Alekhina GM (1967) Effect of intensity of contractile function of cardiac myocytes on DNA synthesis in the cells of the myocardial interstitial tissue. Dokl Akad Nauk SSSR 173:1222–1226 (Russian)
325. Meerson FZ, Bogomolov AF (1973) Adaptational changes in the lungs and in the oxygen utilisation system during adaptation to altitude hypoxia. In: Amosova VV (ed) Volga conference of physiologists, and also of biochemists, pharmacologists and morphologists. Ab-

stracts of Reports. Cheboksary, p 199 (Russian) [Povolzhskaya konferentziya fiziologov s uchastiem biokhimikov, farmakologov, morfologov: Tezisy dokladov]
326. Meerson FZ, Dmitrieva TM (1962) Influence of heart hyperfunction on realisation of the negative chronotropic vagus effect. Dokl Akad Nauk SSSR 145:1184–1187 (Russian)
327. Meerson FZ, Kobozev GV (1955) On the method of inducing experimental aortal stenosis. Bull Exp Biol Med 5:50–53 (Russian)
328. Meerson FZ, Kruglikov RI (1966) On the role of the relationship between the genetic apparatus and the neuron physiological function in the mechanisms of conditioned reflexes and memory. Zh Vyssh Nerv Deiat 6:274–291 (Russian)
329. Meerson FZ, Larionov NP (1975) Depression of the contractile function and reduced effectivity of oxygen utilisation during compensatory heart hypertrophy. Kardiologiia 4:107–114 (Russian)
330. Meerson FZ, Pshennikova MG (1979) Effects of adaptation to high altitude hypoxia on the contractile function and adrenoreactivity of the heart. Basic Res Cardiol 74:142–154
331. Meerson FZ, Rozanova LS (1966) Effect of actinomycin 2703 and a set of stimulators of nucleic acid synthesis on development of fatigue. Dokl Akad Nauk SSSR 166:496–499 (Russian)
332. Meerson FZ, Saltykova VA (1977) Effect of adaptation to altitude hypoxia on resistance of resistive vessels. Kardiologiia 17:83–87 (Russian)
333. Meerson FZ, Solomatina ES (1975) Correlation of adaptation to altitude hypoxia and memory. In: Korobkov AV (ed) Physiological and clinical problems of adaptation to hyperthermia, hypoxia and hypodynamia. Transactions of symposium. Patrice Lumumba University Press, Moscow, pp 178–179 (Russian) [Fiziologicheskie i klinicheskie problemy adaptatzii k gipertonii, gipoksii i gipodinamii]
334. Meerson FZ, Solomatina ES (1977) Adaptation to altitude hypoxia, a control factor of memory and behaviour. In: Kozlov YuP (ed) Compensatory and adaptational processes in the central nervous system. Abstracts of USSR symposium. Irkutsk University Press, Irkutsk, 139, pp 85–88 (Russian) [Kompensatornye i adaptivnye protzessy v tzentralnoy nervnoy sisteme: Tezisy Vsesoujznogo simpoziuma]
335. Meerson FZ, Trikhpoeva AM (1980) Prevention of heart muscle contractile function arising after emotional-painful stress with gamma-hydroxybutyric acid and antioxidants. Bull Exp Biol Med 11:531–533 (Russian)
336. Meerson FZ, Ugolev AA (1980) Impairment of membrane calcium transport: a common pathogenetic link in various forms of cardiac insufficiency. Kardiologiia 1:68–75 (Russian)
337. Meerson FZ, Ustinova EE (1981) Effect of adaptation to cold and hypoxia on aggressive responses in mice (abstr.) Bull Exp Biol Med 10:512 (Russian)
338. Meerson FZ, Ustinova EE (1982) Prevention of stress-induced heart damage and its hypoxic contracture by natural antioxidant α-tocopherol. Kardiologiia 7:89–93 (Russian)
339. Meerson FZ, Ustinova EE (to be published) Effect of long-term and short-term stress action on the heart resistance to hypoxia. Bull Exp Biol Med 9 (Russian)
340. Meerson FZ, Vasiljev VK (1981) Damage and reparation of the heart muscle DNA in emotional-painful stress. Bull Exp Biol Med 9:297–299 (Russian)
341. Meerson FZ, Vasiljev VK (1982) Prevention of disturbance of the heart muscle DNA structure caused by emotional-painful stress by blockade of beta-adrenoceptors and lipid peroxidation. Vopr Med Khim 2:115–118 (Russian)
342. Meerson FZ, Yavich MP (1980) Creatine phosphokinase activity and its isoenzyme state in the myocardium in heart adaptation to prolonged load. Vopr Med Khim 26:599–604 (Russian)
343. Meerson FZ, Zaletaeva TA, Laguchev SS, Pshennikova MG (1964) On the correlation between the masses of energy-forming and functional structures during adaptation of differentiated cells to lengthy increase in level of functioning. Dokl Akad Nauk SSSR 157:668–671 (Russian)
344. Meerson FZ, Kruglikov RI, Kolomeitseva IA (1965) On the role of nucleic acid and protein synthesis in the mechanism of conditioned reflexes and memory. Bull Exp Biol Med 60:3–7 (Russian)
345. Meerson FZ, Mayzelis MY, Malkin VB, Leikina EM, Kruglikov RI, Popko NA (1965) Activa-

tion of protein and DNA synthesis in the brain: a factor of adaptation to altitude hypoxia. Dokl Akad Nauk SSSR 187:697–700 (Russian)

346. Meerson FZ, Simonyan NA, Bazardzhyan AG, Lekina EM (1965) Effect of nucleic acid inhibitor and stimulator on development of compensatory hypertrophy of a sole kidney. Dokl Akad Nauk SSSR 165:1409–1412 (Russian)
347. Meerson FZ, Dergachev VV, Kruglikov RI (1966) Effect of cofactors of nucleic acid synthesis and precursors on the forming of conditioned defence reflex of avoidance. In: Chernukh AM (ed) Problems of physiology and pathology of the nervous system. Scientific conference. Institute of General Pathology and Pathophysiology, Moscow, pp 159–160 (Russian) [Voprosy fiziologii i patologii nervnoy sistemy: Materialy nauchnoy konferentzii]
348. Meerson FZ, Kapelko VI, Radzievsky SA (1968) Correlation of contractile function and left- and right-ventricle metabolism. Kardiologiia 2:146–158 (Russian)
349. Meerson FZ, Mayzelis MY, Malkin VB (1969) On the role of nucleic acid and protein synthesis in adaptation to altitude hypoxia. Izv Akad Nauk SSSR 6:819–831 (Russian)
350. Meerson FZ, Kruglikov RI, Meerson AZ, Mayzelis MY, Leikina EM (1970) Activation of RNA and protein synthesis in the brain and enhancement of memory resistance to extreme irritants under the influence of training for altitude hypoxia. Kosm Biol Med 4:56–59 (Russian)
351. Meerson FZ, Isabaeva VA, Ivanshina AZ, Kruglikov RI, Glumov RM (1971) Conditioned reflexes in massive and prolonged trainings of animals of two different strains in the process of adaptation to altitude hypoxia. Zh Vyssh Nerv Deiat 21:470–477 (Russian)
352. Meerson FZ, Isabaeva VA, Ivanshina AZ (1972) Speedier development and longer retainment of conditioned reflexes in hypoxia-adapted animals. Dokl Akad Nauk SSSR 202:982–984 (Russian)
353. Meerson FZ, Malkin VB, Annanurova LA, Loginova EV, Mayzelis MY (1972) Effect of RNA synthesis inhibitor on development of adaptation to altitude hypoxia. Izv Akad Nauk SSSR 3:405–412 (Russian)
354. Meerson FZ, Pomoinetsky VD, Yampolskaya BA (1972) The role of mitochondria biogenesis in heart adaptation to altitude hypoxia. Dokl Akad Nauk SSSR 203:973–976 (Russian)
355. Meerson FZ, Pomoinetsky VD, Yampolskaya BA (1972) Dynamics of DNA, RNA and protein synthesis in myocardial mitochondria and nuclei during adaptation to altitude hypoxia. Vopr Med Khim 5:533–542 (Russian)
356. Meerson FZ, Golubeva LYu (1973) Effect of prior adaptation to basic environmental factors on ATP concentration and phosphorylation potential in the myocardium during acute heart overload. Dokl Akad Nauk SSSR 210:989–993 (Russian)
357. Meerson FZ, Krantz D, Sadyraliev TS (1973) Dynamics of protein synthesis in the neurons and glias of rat brain during adaptation to altitude hypoxia. Tsitologiia 15:324–329 (Russian)
358. Meerson FZ, Solomatina ES, Vikhlyaev YI, Klygul TA (1973) Effect of adaptation to high altitude hypoxia on animal behavior in conflicting situations. Zh Vyssh Nerv Deiat 23:751–756 (Russian)
359. Meerson FZ, Yavich MP, Breger AM, Lerman MI (1974) The mechanism of heart adaptation to prolonged load and dynamics of RNA synthesis in the myocardium. Basic Res Cardiol 69:484–499
360. Meerson FZ, Pshennikova MG, Kapelko VI, Radzievsky SA, Novikova NA (1975) Nerve regulation of the heart in adaptation to altitude hypoxia. Kardiologiia 12:50–58 (Russian)
361. Meerson FZ, Kapelko VI, Pfaifer KF (1976) Contraction and relaxation of the heart muscle in adaptation to physical loads. Fiziol Zh SSSR 62:793–795 (Russian)
362. Meerson FZ, Kruglikov RI, Solomatina ES (1976) Effect of adaptation to altitude hypoxia on fixation of temporal connections. Zh Vyssh Nerv Deiat 26:1026–1031 (Russian)
363. Meerson FZ, Barbarash NA, Shorin YP (1977) On the mechanism of preventive effect of adaptation to altitude hypoxia on development of hypertension. Kardiologiia 12:71–79 (Russian)
364. Meerson FZ, Radzievsky SA, Giber LM, Kogan AK, Rozhitskaya II, Volovinskaya OI, Pshennikova MG (1977) Heart damage in emotional stress and its prevention by preliminary adaptation to altitude hypoxia. Dokl Akad Nauk SSSR 237:977–980 (Russian)
365. Meerson FZ, Barbarash NA, Dvurechenskaya GY (1978) Inhibitory effect of adaptation to

hypoxia on the development of hereditary hypertension in rats. Dokl Akad Nauk SSSR 241:1472 (Russian)

366. Meerson FZ, Lifshits RI, Pavlova VI, Yakushev VS, Kamilov FK (1978) Activity of GABA biosynthesis in the brain as a protective mechanism in emotional-painful stress. Dokl Akad Nauk SSSR 239:743–746 (Russian)
367. Meerson FZ, Golubeva LYu, Kagan VE, Shimkovich MV, Ugolev AA (1979) Activation of lipid peroxidation as the decisive link in the pathogenesis of stress damage to the heart, and prevention of stress and hypoxic damage by the antioxidant ionol. In: Energy transport, protein synthesis, and hormonal control of heart metabolism. Fourth USA–USSR Joint symposium on myocardium, Tashkent, USSR, September 14-16, 1979. NIH Publication N 80-2017, August 1980, USA, pp 277–292 (Russian)
368. Meerson FZ, Gorina MS, Zeland AM, Giber LM (1979) Effect of emotional-painful stress on the contractile function and adrenoreactivity of heart muscle. Bull Exp Biol Med 11:528–530 (Russian)
369. Meerson FZ, Kagan VE, Golubeva LYu, Ugolev AA, Shimkovich MV, Giber LM, Rozhitskaya II (1979) Prevention of stress-induced and hypoxic heart damages with the antioxidant ionol. Kardiologiia 8:108–111 (Russian)
370. Meerson FZ, Kagan VE, Prilipko LL, Rozhitskaya II, Giber LM, Kozlov YuP (1979) Activation of lipid peroxidation under emotional-painful stress. Bull Exp Biol Med 10:404–406 (Russian)
371. Meerson FZ, Krause EG, Pshennikova MG, Golubeva LYu, Karchevsky P, Wollenberger A (1979) Effect of adaptation to altitude hypoxia on adrenoreactivity of the heart and the state of adenyl cyclase and phosphodiesterase systems of the myocardium. Fiziol Zh SSSR 65:727–732 (Russian)
372. Meerson FZ, Malyshev VV, Popova NS, Markovskaya GI (1979) Decrease of stress reaction and stomach ulcerous lesions under the influence of GOBA. Bull Exp Biol Med 12:659–661 (Russian)
373. Meerson FZ, Pavlova VI, Yakushev VS, Kamilov FK (1979) Application of sodium gamma-hydroxybutyrate for prevention of damage induced by emotional-painful stress. Patol Physiol Exp Ter 3:26–31 (Russian)
374. Meerson FZ, Barbarash NA, Dvurechenskaya GY. Prokina NS, Saltykova VA (1980) Natriuretic and antihypertensive effects of acute hypoxia in animals with spontaneous hereditary hypertension. Bull Exp Biol Med 26:599–604 (Russian)
375. Meerson FZ, Belkina LM, Ugolev AA, Golubeva LYu, Abdikaliev NA, Solomatina ES (1980) Application of antioxidants for prevention of experimental myocardial infarction and reoxygenational heart disturbances. Kardiologiia 10:80–86 (Russian)
376. Meerson FZ, Evseev VA, Davydova TV, Giber LM (1980) Effect of adaptation to high-altitude hypoxia on the indices of non-specific resistance, production of hemagglutines and development of adjuvant arthritis in rats. Bull Exp Biol Med 89:272–274 (Russian)
377. Meerson FZ, Malyshev VV, Kagan VE, Treshuk LI, Rozhitskaya II (1980) Activation of lipid peroxidation and focal contractural lesions in heart muscle in emotional-painful stress. Arkh Patol 42:9–12 (Russian)
378. Meerson FZ, Pavlova IV, Korobeinikova EN (1980) Prevention of stress damages of the organism with antioxidants and beta-blocker inderal. Vopr Med Khim 26:827–832 (Russian)
379. Meerson FZ, Shimkovich MV, Khorunshij VA (1980) Effect of emotional painful stress on heart muscle reactivity to changes in calcium concentration. Bull Exp Biol Med 9:281–283 (Russian)
380. Meerson FZ, Abdikaliev NA, Golubeva LYu (1981) Prevention of hypoxic heart damage by antioxidant from the class of oxypyridines. Bull Exp Biol Med 9:281–283 (Russian)
381. Meerson FZ, Arkhipenko YuV, Rozhitskaya II, Kagan VE (1981) Injury of the Ca^{2+}-transport system of the heart sarcoplasmic reticulum in emotional-painful stress. Bull Exp Biol Med 4:405–406 (Russian)
382. Meerson FZ, Kagan VE, Arkhipenko YuV, Belkina LM, Rozhitskaya II (1981) Prevention of lipid peroxidation activation and myocardial antioxidant system damage in stress and experimental infarction. Kardiologiia 12:55–60 (Russian)
383. Meerson FZ, Pavlova VI, Yakushev VS, Sinitsyn SP (1981) Activation of GABAergic in-

hibitory system in cerebral hemispheres under the action of various stress factors. Bull Exp Biol Med 1:23–24 (Russian)
384. Meerson FZ, Kagan VE, Kozlov YuP, Belkina LM, Arkhipenko YuV (1982) Role of lipid peroxidation in the pathogenesis of ischemic injury and antioxidant protection of the heart. Kardiologiia 2:81–92 (Russian)
385. Meerson FZ, Medvedev LN, Golubeva LYu, Ustinova EE (1982) Effect of emotional-painful stress on Na-K-ATPase activity in heart muscle. Bull Exp Biol Med 8:61–62 (Russian)
386. Meerson FZ, Krasikov SI, Boev VM, Kagan VE (to be published) Effect of antioxidants on untrained body resistance to maximal physical loads. Bull Exp Biol Med 8 (Russian)
387. Meerson FZ, Malyshev VV, Petrova VA, Lifantjev VI (1982) Prevention of hypophyseal-adrenal system activation and heart damages in stress with the antioxidant ionol. Kardiologiia 9:85–88 (Russian)
388. Meerson FZ, Pavlova VI, Sukhikh GT, Pshennikova MG, Efimenko GP, Solomatin VB, Frolov BA, Vorontsova EYa (1982) Post-stress activation of nucleic acid and protein synthesis and its role in the organism's adaptation responses. Patol Fiziol Exp Ter 5:3–14 (Russian)
389. Meerson FZ, Samosudova NV, Glagoleva EV, Shimkovich MV, Belkina LM (to be published) Disturbance of heart contractile function and myocardium ultrastructure after emotional painful stress. Arkh Anat Gistol Embriol (Russian)
390. Meerson FZ, Saulya AI, Guski G, Vasiljev VK, Belkina LM, Markovskaya GI, Shimkovich MV, Saltykova VA (to be published) Heart contractile function and myocyte ultrastructure under prolonged hypokinesia in growing animals. Patol Fiziol Exp Ter (Russian)
390a. Meerson FZ, Vorontsova EYa, Pshennikova MG (to be published) Effect of sustained stress on myocardium resistance to hypoxia. Bull Exp Biol Med (Russian)
391. Mekler LB (1978) Mechanism of tumor induction in the light of the general theory of oncogenesis. Usp Sovrem Biol 85:134–151 (Russian)
392. Mertvetsov NP, Saprykin VA, Chesnokov VN, Salganic RI (1974) Induction in rat liver under cortisol action of tyrosine aminotransferase isoenzymes highly sensitive to protease action. Biokhimiia 39:3–8 (Russian)
393. Michell RH, Jones CM (1974) Enhanced phosphatidyl inositol labelling in rat parotid fragments exposed to α-adrenergic stimulation. Biochem J 138:47–52
394. Mikelsaar K, Severina II, Skulachev VP (1974) Phospholipids and oxidative phosphorylation. Usp Sovrem Biol 78:348–370 (Russian)
395. Mikulaj L, Kvetnansky R, Murgas K (1974) Changes in adrenal response during intermittent and repeated stress. Rev Czech Med 20:162–169
396. Miller DG, Malov S (1977) Quantitative determination of stress-induced myocardial damage in rats. Pharmacol Biochem Behav 7:139–143
397. Miminoshvili DI, Magakyan GO, Kokaya GY (1956) Experience in obtaining experimental models of hypertension and coronary insufficiency in monkeys. In: Theoretical and applied problems of medicine and biology in experiments on monkeys. Meditsina, Moscow, pp 85–97 (Russian) [Teoreticheskie i prakticheskie voprosy meditziny i biologii v eksperimente na obezjyanakh]
398. Mirrakhimov MM (1964) Essays on the effect of the mountain climate of Central Asia on the organism. Akad Nauk Kirghiz SSR, Frunze (Russian) [Ocherki vliyaniya gornogo klimata Sredney Azii na organizm]
399. Mirrakhimov MM (1968) The cardiovascular system under alpine conditions. Meditsina, Leningrad (Russian) [Serdechno-sosudistaya sistema v usloviyakh vysokogorjya]
400. Mirrakhimov MM (1977) Treatment of internal diseases by mountain climate. Meditsina, Leningrad (Russian) [Lechenie vnutrennikh bolezney gornym klimatom]
401. Mirrakhimov MM, Aydaraliev AA, Dzhunushev ML (1972) Retention periods of increased resistance of the body under various conditions of acclimatization to mountains. Kosm Biol Med 6:14–18 (Russian)
402. Monge G, Lozano R, Marchena C, Whittemburg J, Torres C (1969) Kidney function in the high-altitude native. Fed Proc 28:1199–1203
403. Morad M, Goldman J (1973) Exitation-contraction coupling in heart muscle membrane control of development of tension. Prog Biophys Mol Biol 27:259–313
404. Moret P, Covarrubias E, Coudert J, Duchosal F (1972) Cardiocirculatory adaptation to chronic

hypoxia. Comparative study of coronary flow, myocardial oxygen consumption and efficiency between sea level and high altitude residents. Acta Cardiol (Brux) 27:283–305
405. Morkin E, Ashoford TP (1968) Myocardial DNA synthesis in experimental cardiac hypertrophy. Am J Physiol 215:1409–1413
406. Motorin YA (1971) Changes in certain indications of external respiration during reacclimatization after 14- and 30-day stay in medium alpine regions. Vital problems of space biology and medicine. Moscow, pp 186–187 (Russian)
407. Musin BS (1966) Significance of cholinergic mechanisms in the activity of the skeleton-muscular and respiratory systems and the hearts of rats during various conditions of development. In: Chernukh AM (ed) 24th conference of young scientists. Publs: Institute of normal and pathological physiology, USSR academy of Medical Sciences. Moscow, pp 49–50 (Russian) [24-ya konferentziya molodykh uchenykh)
408. Myles WS, Ducker AJ (1971) The excretion of catecholamines in rats during acute and chronic exposure to altitude. Can J Physiol Pharmacol 49:721–726
409. Naets JP, Wittek M (1969) Erythropoietic activity of marrow and disappearance rate of erythropoietin in the rat. Am J Physiol 217:297–301
410. Naill WA, Hattenhauer M (1975) Impairment of myocardial O_2 supply due to hyperventilation. Circulation 52:854–858
411. Nayler WG, Poole-Wilson P, William A (1979) Hypoxia and calcium. J Mol Cell Cardiol 11:683–706
412. Nayler WG, Poole-Wilson P (1981) Calcium antagonists: definition and mode of action. Basic Res Cardiol 76:1–15
413. Neely JR, Liebermeister H, Battersby EJ, Morgan HE (1967) Effect of pressure development of oxygen consumption by isolated rat heart. Am J Physiol 212:804–814
414. Nicolson AN, Wright CM (1975) Effect of mild hypoxia on delayed differentiation in the monkey (Macaca mulata). Exp Neurol 47:535–543
415. O'Brien PJ (1969) Intracellular mechanisms for the decomposition of a lipid peroxide. Decomposition of a lipid peroxide by metal ions, heme compounds and nucleophiles. Can J Biochem 47:485–492
416. Ogunro EA, Hearse DJ, Shillingford JP (1977) Creatine kinase isoenzymes: their separation and quantitation. Cardiovasc Res 11:94–102
417. Okamoto K (1969) Spontaneous hypertension in rats. Int Rev Exp Pathol 7:227–232
418. Oliva PB, Breckenridge JC (1977) Arteriographic evidence of coronary arterial spasm in acute myocardial infarction. Circulation 56:366–375
419. Olsson R, Gentry MK, Townsend RS (1973) Adenosine metabolism: properties of dog heart microsomal 5′-nucleotidase. In: Eloor CM, Olsson RA (eds) Current topics in coronary research. Plenum, New York, pp 27–39
420. O'Neill JA, Pruitt BA, Moncrief JA, Switzer WE (1967) Studies related to the pathogenesis of Curling's ulcer. J Trauma 7:275–287
421. Onishi T, Onishi H (1962) Contractile protein in mitochondria. M Biochem 51:380–383
422. Oppelt W (1963) Kleines Handbuch technischer Regelvorgänge. Technik, Berlin
423. Orloff J, Handler JS, Bergstrom S (1965) Effect of prostaglandin (PGE) on the permeability response of road bladder to vasopressin, theophyline and adenosine 3′,5′-monophosphate. Nature 209:397–403
424. Osipov VP (1926) A specific teaching on mental diseases. Gosizdat, Moscow Leningrad (Russian) [Chastnoe uchenie o dushevnykh boleznyakh]
425. Ostrovsky AY (1976) Changes in function of cardiovascular system and in lipid metabolism in seamen during lengthy voyages. In: 7th international symposium on marine medicine. Meditsina, Moscow, p 171 (Russian) [7-y mezhdunarodnyi simpozium po morskoy meditzine]
426. Ott T, Lossner B, Matthies H (1972) Die Wirkung von Nucleotid-Monophosphaten auf die Akquisition und Extinktion bedingter Reaktionen. Psychopharmacologia (Berlin) 23:261–271
427. Ottowicz J, Nazar K, Wolowska A (1967) Effect of physical effort on the content of corticosteron and ascorbic acid in the adrenals in rats. Wychow Fiziol Sport 11:119–126
428. Ou LC, Tenney SM (1970) Properties of mitochondria from heart of cattle acclimatized to high altitude. Respir Physiol 8:151–154

429. Pare WP, Livingston A (1970) Brain norepinephrine and stomach ulcers in rats exposed to chronic conflict. Physiol Behav 5:215–220
430. Pavlov AD (1970) On adaptation of macromolecule synthesis to 'altitude' in bone marrow cells. In: Predtechensky SA (ed) Geographic medium and health. Kabardino-Balkarsky State University, Nalchik, pp 84–86 (Russian) [Geograficheskaya sreda i zdorovje naseleniya]
431. Pavlov IP (1951) Twenty years of objective studies of higher nervous activity in animals. Nauka, Moscow (Russian) [Dvadtzatiletniy opyt objektivnogo izucheniya vysshey nervnoy deyatelnosti zhivotnykh]
432. Pavlov IP (1952) Lectures on the work of the large cerebral hemispheres. USSR Academy of Medical Sciences, Moscow (Russian) [Lektzii o rabote bolshikh polushariy golovnogo mozga]
433. Penpargkul S, Repke DI, Katz A (1977) Effect of physical training on calcium transport by rat cardiac sarcoplasmic reticulum. Circ Res 42:134–137
434. Perriard J-C, Perriard ER, Eppenberger HM (1978) Detection and relative quantitation of mRNA for creatine kinase isoenzymes in RNA from myogenic cell cultures and embryonic chicken tissues. I Biol Chem 253:6529–6536
435. Perrone MH, Greer TL, Hinkle PM (1980) Relationships between thyroid hormone and glucocorticoid effects in GHZ pituitary cells. Endocrinology 106:600–605
436. Perry SV (1971) Development and specialization in muscle and biochemistry of dystrophies. J Neurol Sci 12/3:298–306
437. Petren T, Sjostrand T, Sylven B (1973) Der Einfluss des Trainings auf die Häufigkeit der Capillaren in Herz – und Skeletmusculatur. Arbeitsphysiologie 9:376–386
438. Petrosyan YS, Ioseliani DT, Kpahidze NN (1979) Spasm of coronary arteries in patients with acute myocardial infarction. Kardiologiia 1:47–51 (Russian)
439. Pitt B, Elliot EC, Gegs DE (1967) Adrenergic receptor activity in the coronary arteries of the anaesthetized dog. Circ Res 21:75–84
440. Pohle W, Matthies H (1974) Incorporation of RNA precursors into neuronal and glial cells of rat brain during a learning experiment. Brain Res 65:231–237
441. Pollard I, Basett JR, Cairncross KD (1979) Plasma thyroid hormone and glucocorticosteroid concentrations in the male rat following prolonged exposure to stress. Aust J Biol Sci 32:237–242
442. Pool-Willson PA (1978) Continuous measurement of $^{47}Ca^{2+}$ uptake during and after hypoxia in rabbit myocardium. J Mol Cell Cardiol 10:79
443. Popov VG, Kharkevich DA, Syrkin AL, Basistov AF, Lazebnik LB, Lazutin VK (1968) Experience in applying sodium oxybutyrate in patients with acute coronary insufficiency and disturbed heart rhythm. Ter Arkh 40:44–48 (Russian)
444. Popova EI (1969) On the analysis of the phenomenon of effector transfer. In: Asratyan EA (ed) 22nd conference on problems of nervous activity. Abstracts of reports. Ryazan Medical Institute, Ryazan, p 194 (Russian) [22-e soveshjanie po problemam vysshey nervnoy deyatelnosti. Tezisy dokladov]
445. Posner BI, Mierzwinski L, Fallen EL (1973) Studies on amino acid levels and transport in the mechanically stressed rat heart. J Mol Cell Cardiol 5:221–233
446. Potts WJ (1971) The effect of different environments prior to electroconvulsive shock on the gradient of retrograde amnesia. Physiol Behav 7:161–164
447. Poupa O, Rakusan K (1966) The terminal microcirculatory bed in the heart of athletic and non-athletic animals. In: Physical activity in health and disease. Universitets Forlaget, Oslo, pp 18–29
448. Poupa O, Krofta K, Prochazka J (1966) Acclimation to simulated high altitude and acute cardiac necrosis. Fed Proc 25:1243
449. Preedy VR, Pain VM, Garlick PJ (1980) The effect of glucagen on muscle protein synthesis in the perfused rat hemicorpus. Biochem Soc Trans 8:367–368
450. Prilipko LL, Orlov ON, Ivanova SM, Kagan VE, Meerson FZ, Ushakov AS, Gazenko OG (1982) Activation of lipid peroxidation in stress in humans estimated by pentan content in exhaled air. Dokl Akad Nauk SSSR 265:1010–1013 (Russian)
451. Prinzmetal M, Kemnamer R, Merliss R, Wado T, Bor H (1959) Angina pectoris. I. The variant form of angina pectoris. Am J Med 27:375–379

452. Prokhazka I, Khavkina IV, Barbashova ZI (1973) Effect of prolonged hypokinesia on rat heart muscle. Fiziol Zh SSSR 59:1237–1241 (Russian)
453. Prokhorova MI, Tupikova ZN (1959) Methods of determining radioactive carbon (^{14}C) in components of carbohydrate and lipid metabolism. Meditsina, Leningrad (Russian) [Metody opredeleniya radioaktivnogo ugleroda (C^{14}) v komponentakh uglevodnogo i lipidnogo obmena]
454. Pshennikova MG (1973) The protein synthesis in neurons and glia cells of stellate ganglia in rats during adaptation to intermittent and continuous hypoxia. Fiziol Zh SSSR 59:421–427 (Russian)
455. Pshennikova MG, Meerson FZ, Taraeva NG (1966) Effect of folic and orotic acids and of actinomycin on the contractile function of the myocardium during experimental cardiac hyperfunction. Kardiologiia 4:54–58 (Russian)
456. Quartermain D, McEwen BS, Azmitia EC (1972) Recovery of memory following amnesia in rat and mouse. J Comp Physiol Psychol 79:360
457. Rabinowitz M (1973) Protein synthesis and turnover in normal and hypertrophied heart. Am J Cardiol 31:202–212
458. Rabinowitz M, Nair KG, Zak R (1970) Cellular and subcellular basis of cardiac hypertrophy. Med Clin North Am 54:211–220
459. Rabkin IK, Abugov AM, Matevosov AL (1972) Selective coronarography in diagnosis of chronic ischemic heart disease. Vestn Rentgenol Radiol 6:31–41 (Russian)
460. Rakhimov MM, Almatov KT (1978) Effect of calcium ions on the interaction of phospholipase D with phospholipases of mitochondrial membranes. Biokhimiia 43:1390–1403 (Russian)
461. Randall DC, Kaye MP, Randall WC, Brady JV, Martin KH (1978) Response of primate heart to emotional stress before and after cardiac denervation. Am J Physiol 230:988–995
462. Randt C, Quartermain D, Goldstein M, Anagnoste B (1971) Norepinephrine biosynthesis inhibition: effects on memory in mice. Science 172:498–499
463. Rapp JP, Dahl LK (1971) Adrenal steroidogenesis in rat bred for susceptibility and resistance to the hypertensive effect of salt. Endocrinology 88:52–55
464. Rathbun W, Betlach V (1969) Estimation of enzymically produced orthophosphate in the presence of cysteine and adenosine triphosphate. Anal Biochem 28:436–445
465. Razumov SA, Galperin IA, Krasnovskaya IA, Pipchuk VM, Stabrovsky EI, Volkova LA, Dukhno ZI (1974) Neuroendocrine manifestations of emotional stress modelled in cocks. In: Transactions of all-Russia conference on neuroendocrinology. Institute of Evolutionary Physiology and Biochemistry, Leningrad, pp 142–143 (Russian) [Materialy Vsesoujsnoy konferentzii po neyroendokrinologii]
466. Reading HW (1972) Effect of some adrenocorticotropin analogs on protein synthesis in brain. Biochem J 127:7
467. Rees HD, Brogan LL, Endingh DJ, Dunn AJ, Shinkman KG, Damstra-Enting T, Wilson JE, Classman E (1974) Effect of sensory stimulation on the uptake and incorporation of radioactive lysine into protein of mouse brain and liver. Brain Res 68:143–156
468. Repin YM, Startsev VG (1975) Mechanism of selective lesion of the cardiovascular system under psychoemotional stress. Vestn Akad Med Nauk SSSR 8:71–76 (Russian)
469. Rethy A, Tomasi V, Trevisani A, Barnabic O (1972) The role of phosphatidyl serine in the hormonal control of adenylate cyclase of rat liver plasma membranes. Biochim Biophys Acta 200:58–59
470. Reuter H, Scholz H (1974) The regulation of the calcium conductance of cardiac muscle by adrenaline. J Physiol (Lond) 264:49–62
471. Ritter S, Pelzer NL (1978) Magnitude of stress-induced brain norepinephrine depletion varies with age. Brain Res 152:170–175
472. Roberts R, Henry PD, Sobel BE (1975) An improved basis for enzymatic extimation of infarct size. Circulation 52:743–754
473. Rodbell M, Londos C (1976) Regulation of hepatic adenylate cyclase by glucagon, GTP, divalent cations and adenosine. Metabolism 25: 1347–1351
474. Rodbell M, Birnbaumer L, Pohl SL (1971) Hormones receptors and adenyl cyclase activity in mammalian cells. In: Rall TW, Rodbell M, Condliffe PG (eds) The role of adenyl cyclase and cyclic 3′,5′-AMP in biological systems. National Institute of Health, Bethesda, pp 59–76

475. Rona G, Chappel CI, Balazs T, Gaudry R (1959) An infarct-like myocardial lesion and other toxic manifestations produced by isoproterenol in the rat. Arch Pathol 67:443–455
476. Rona G, Chappel CJ, Kahn DS (1963) The significance of factors modifying the development of isoproterenol-induced myocardial necrosis. Am Heart J 66:389–395
477. Ross G (1976) Adrenergic responses of the coronary vessels. Circ Res 39:461–465
478. Rotta A, Morales S, Battilana G (1965) Valores de colesterol sanguineo y presion arterial en sujetos sanos, de nivel del mar y de la altura. Rev Peru Cardiol 12:9–15
479. Rowe GG, Castillo CA, Crumpton CW (1962) Effect of hyperventilation on systemic and coronary hemodynamics. Am Heart J 63:67–73
480. Rowell LB, Taylor HL, Wang J, Carlson WS (1964) Saturation of arterial blood with oxygen during maximal exercise. J Appl Physiol 19:284–286
481. Rozental VM (1969) Variations in ACTH content in the hypophysis and its secretion during first day after stress. Probl Endokrinol (Mosk) 15:49–54 (Russian)
482. Rubin H (1975) Central role for magnesium in coordinate control of metabolism and growth in animal cells. Proc Natl Acad Sci USA 72:3551–3555
483. Rubio R, Berne RM, Dobson JG (1973) Sites of adenosine production in cardiac and skeletal muscle. Am J Physiol 225:938–953
484. Rudakov IA, Leonov BV, Lomova MA (1963) Relation of rat radiosensitivity and antioxidant activity of bone marrow and content of non-etherified fatty acids in the blood in stress state. Radiobiologiia 3:518–522 (Russian)
485. Rudman D, Scott JW, del Rio AE, Houser DH, Sheen S (1974) Effects of melanotropic peptides on protein synthesis in mouse brain. Am J Physiol 222:687–692
486. Ruoff HJ (1977) Rat gastric mucosal cAMP and cGMP after adrenergic stimulation and blockade. Eur J Pharmacol 44:349–354
487. Rupp H (1981) The adaptive changes in the isoenzyme pattern of myosin from hypertrophied rat myocardium as a result of pressure overload and physical training. Basic Res Cardiol 76:79–88
488. Rusin VY (1973) Activity and reactivity of hypophysial-adrenal system in chronic stress. In: Stress and its pathogenetic mechanisms. Meditsina, Kishinev, pp 116–167 (Russian) [Stress i ego patogeneticheskie mekhanizmy]
489. Russek HI (1960) Emotional stress and coronary heart disease in American physicians. Am J Med Sci 240:711–721
490. Russo JV, Margolis S (1972) Hemodynamic effects of free fatty acid augmentation following myocardial infarction. Circulation [Suppl 2] 46:215
491. Ryabinina ZA (1964) On some cytologic features of a regenerating kidney. In: Regeneration and cell reproduction in animals. Nauka, Moscow, pp 56–77 (Russian) [Regeneratziya i kletochnoe razmnozhenie u zhivotnykh]
492. Salganik RI, Manankova NM, Semenova LA (1977) Hypocholesterolemic effect of induction of liver microsomal enzymes in rats. Vopr Med Khim 4:468–473 (Russian)
493. Salzmann SH, Hirsch EZ, Hellerstein H, Bruell JH (1970) Adaptation to muscular exercise myocardial epinephrine ^{3}H uptake. J Gen Appl Physiol 29:92–95
494. Samosudova NV, Lyudkovskaya RG (1979) Inter-relationship of muscle transversal-striated structure and distribution of calcium therein. II. Ultrastructural localization of calcium in frog muscle fibres during rest and calcium contracture. Tsitologiia 21:391–396 (Russian)
495. Samosudova NV, Skrebnitskaya LK, Klimova AA (1981) Localisation of calcium and its redistribution in muscles with dual calcium regulation. Biofizika 26:319–323 (Russian)
496. Sanov SS, Satpaeva RA (1976) An epidemiologic study of the ischemic heart disease among certain groups of people residing in Kalinin District, Alma-Ata. Zdrav Kaz 2:5–6 (Russian)
497. Sara SJ (1974) Delayed development of amnesic behaviour after hypoxia. Physiol Behav 13:693–696
498. Sarkisov DS, Vtyurina BV (1969) Electron microscopy analysis of increase in heart endurance. Meditsina, Moscow (Russian) [Elektronno-mikroskopicheskiy analiz povysheniya vynoslivosti serdtza]
499. Sato M, Nagao T, Yamaguchi I (1971) Pharmacological study on a new 1,5-benzothiazepine derivate (CRD-401-Diltiasem). Arzneim Forsch 21:1338–1343
500. Schrader J, Rubio R, Berne RM (1974) Inhibition of slow-action potentials of guinea pig atrial muscle by adenosine: a possible effect of Ca^{2+} influx. J Mol Cell Cardiol 7:427–433

501. Schrader J, Baumann G, Gerlach E (1977) Adenosine as inhibitor of myocardial effects of catecholamines. Pflügers Arch 372:29–35
502. Schreiber SS, Oratz M, Evans C, Silver E, Rothschild MA (1968) Effect of acute overload on cardiac muscle mRNA. Am J Physiol 215:1250–1259
503. Schwartz A (1971) Studies on mitochondria from normal, hypertrophied and failing myocardium. In: Alpert J (ed) Cardiac hypertrophy. Academic Press, New York, pp 511–537
504. Segawa T, Hayashi H, Nakano M, Murakami H, Kasahara C (1977) Prostaglandin and norepinephrine metabolism effect of indometacin on prostaglandin synthesis and norepinephrine turnover rate. Jpn J Pharmacol 24:478–480
505. Seifter E (1976) Of stress, vitamin A, and tumors. Science 193:74–75
506. Selye H (1952) The story of the adaptation syndrome. Acta Inc Med Publ Montreal
507. Selye H (1961) The pluricausal cardiopathies. Thomas, Springfield
508. Semenov NN (1934) Chain reactions. Goskhimtekhizdat, Leningrad (Russian) [Tzepnye reaktzii]
509. Sergeev PV, Denisov YP (1978) On the role of the lipid component of biologic membranes in the action of steroid hormones. In: Klimov AN (ed) Lipid structure, biosynthesis and transformation in the organisms of animals and humans, vol 78 (Russian) [Struktura, biosintez i prevrashjeniya lipidov v organizme zhivotnykh i cheloveka]
510. Severson JA, Fell RD, Griffith DR (1978) Adrenocortical function in response to myocardial necrosis in exercise-trained rats. J Appl Physiol 44:104–108
511. Shabadash AL, Zelikina TI, Agracheva ND (1963) The cytology and cytochemistry of ribonucleoproteins of mitochondria and of tigroid in the cells of the central nervous system in latent period of radiation injury. Arkh Anat Gistol Embriol 2:3–6 (Russian)
512. Shad SD, Gryer RE, Haymond MW, Santiage JV (1976) Norepinephrine and epinephrine release and adrenergic medication of smoking-associated. N Engl J Med 295:573–588
513. Shadruchek M, Vaćek M (1971) On the question of stress and the ischemic heart disease. Zdravokhranenie 14:45–59 (Russian)
514. Shatunova NF, Sytinsky IA (1962) Effect of semicarboside on GABA content in brain tissue. In: The nervous system, vol 3. Leningrad University Press, Leningrad, pp 12–17 (Russian) [Nervnaya sistema]
515. Sheitanov MP (1973) Use of food as an efficiency-enhancing factor (Abstr). Medical Dissertation, University of Sofia
516. Shkolnik MI (1959) On the rate of protein renovation in cerebral hemisphere cortex under hypoxia. In: Pruss GM (ed) Transactions of Vitebsk medical institute, vol 10. Vitebsk Medical Institute, Vitebsk, p 150 (Russian) [Zbornik nauchnykh rabot Vitebskogo meditzinskogo instituta]
517. Shumskaya IA, Korochkin LI (1975) A study of the intensity of RNA synthesis in rat hippocamp during training. Zh Vyssh Nerv Deiat 25:778–783 (Russian)
518. Shvedova AA, Sidorov AS, Novikov KN, Galushenko IV, Kagan VE (1979) Lipid peroxidation and electric activity of the retina. Vision Res 19:49–55
519. Shvedova AA, Kagan VE, Kuliev IM, Dobrina SK, Prilipko LL, Meerson FZ, Kozlov YuP (1982) Lipid peroxidation and retina damage in stress. Bull Exp Biol Med 4:24–26 (Russian)
520. Simanovsky AN, Krasnovskaya TA, Prozorovskaya MP, Tavrovskaya TV (1973) Fiziol Zh SSSR 59:828–836 (Russian)
521. Simek J, Erbenova L, Deml F, Dvorackova J (1968) Liver regeneration after partial hepatectomy in rats exposed before the operation to stress. Experientia 24:1166–1167
522. Simonov PV (1970) The physiology of emotions and conditioned reflex activity. In: The physiology of higher nervous activity. A handbook on physiology. Nauka, Moscow, pp 97–110 (Russian) [Fiziologiya vysshey nervnoy deyatelnosti]
523. Sinichkin AA, Prokofiev VN (1979) A study of bromtimole blue – prealbumin in rat blood serum in hyperbarooxygenation. Bull Exp Biol Med 5:417–419 (Russian)
524. Sirotinin NN (1964) Effect of adaptation to hypoxia and acclimatisation to high mountain climate on animal resistance to some extremal actions. Patol Fiziol Exp Ter 5:12–15 (Russian)
525. Sklafer M, Gelband H, Sung RS, Palmer RF, Bassett AL (1978) Time-dependent alterations of myocardial microsomal yield and calcium accumulation in experimentally-induced right ventricular hypertrophy and failure. J Mol Cell Cardiol 10:395–407

526. Smith WL, Lands W (1971) Stimulation and blockade of prostaglandin biosynthesis. J Biol Chem 246:6700–6704
527. Smookler HM, Buckley JP (1969) Relationships between brain catecholamine synthesis, pituitary, adrenal function and the production of hypertension during prolonged exposure to environmental stress. Int J Neuropharmacol 8:33–41
528. Somova L (1977) The role of adrenergic mechanisms and prostaglandins in hypertensive states (Abstr). Medical Dissertation, Medical Academy, Sofia
529. Sonnenblick EH (1962) Force-velocity relations in mammalian heart muscle. Am J Physiol 202:931–939
530. Sperelakis N, Schneider GA (1976) A metabolic control mechanism for calcium ion influx that may protect the ventricular myocardial cell. Am J Physiol 37:1079–1085
531. Spiekerman PG, Gebhard MM, Nordbeck H (1975) Myocardial high energy phosphates and enzyme release during anaerobiosis in the isolated perfused dog heart. Abstracts of international symposium. International study group for research in cardiac metabolism. Bruxelles, p 41
532. Spirichev VB (1974) Processes of peroxidation and the role of alimentary factors in its control. Vopr Pitan 3:9–19 (Russian)
533. Squire LR, Barondes SH (1973) Memory impairment during prolonged training in mice given inhibitors of cerebral protein synthesis. Brain Res 56:215–255
534. Starke K, Peskar BA, Schumacher KA, Taube HD (1977) Bradykinin and postganglionic sympathetic transmission. Naunyn Schmiedebergs Arch Pharmacol 299:23–32
535. Steinberg D, Khoo JC (1977) Hormone-sensitive lipase of adipose tissue. Fed Proc 36:1986–1990
536. Steinberg D, Vaughan M, Strand O, Bergstrom S (1964) Effects of the prostaglandins on hormone-induced mobilisation of free fatty acids. J Clin Invest 43:1533–1540
537. Stolyarova NA (1968) Effects of muscle activity and certain biologically active substances on the corticosteroid content in the blood and adrenals. Fiziol Zh SSSR 54:838–842 (Russian)
538. Sullivan JM, Alpers JB (1971) In vitro regulation of rat heart 5′-nucleotidase by adenine nucleotides and magnesium. J Biol Chem 246:3057–3063
539. Sytinsky IA, Shan-Katsin (1966) Activity of glutamate decarboxylase and gamma-aminobutyric acid level in rat brain during general X-ray radiation. The nervous system, vol 7. Leningrad University Press, Leningrad, pp 47–51 (Russian) [Nervnaya sistema]
540. Szasz C, Gruber W (1978) Creatine kinase in serum. 4. Differences in substrate affinity among the isoenzymes. Clin Chem 24:245–249
541. Takeguchi C, Sin CJ (1972) A rapid spectrophotometric assay for prostaglandin synthetase: application to the study of nonsteroidal anti-inflammatory agents. Prostaglandins 2:169–184
542. Tappan DV, Reynafarie BD, Potter VR, Hurtado A (1957) Tissue pigment manifestation of adaptation to high altitudes. Am J Physiol 190:93–103
543. Tarusov BN (1957) Initial processes of radiation injury, Gosatomizdat, Moscow (Russian) [Pervichnye protzessy luchevogo porazheniya]
544. Taylor SL, Lamden MP, Tappela AL (1976) Sensitive fluorometric method for tissue tocopherol analysis. Lipids 11/7:530–538
545. Tedeschi E, Tedeschi DH, Mucha A, Cook Z, Mattis PA, Fellows EJ (1959) Effect of various centrally acting drugs on fighting behavior of mice. J Pharmacol Exp Ther 125:28–34
546. Tennant R, Wiggers CJ (1935) The effect of coronary occlusion on myocardial contractions. Am J Physiol 112:351–361
547. Thierry A-M, Javoy F, Glowinski J, Kety SS (1968) Effects of stress on the metabolism of norepinephrine, dopamine, and serotonin in the central nervous system of the rat. I. Modification of norepinephrine turnover. J Pharmacol Exp Ther 163:163–171
548. Tobian L, Redeaf PD (1957) Effect of hypertension on arterial wall electrolytes during desoxycorticosterone administration. Am J Physiol 189:451–460
549. Tomplinson CW, Lee SL, Dhalla NS (1976) Abnormalities in heart membranes and myofibrils during bacterial infective cardiomyopathy in the rabbit. Circ Res 39:82–92
550. Tsanev RG, Markov GG (1960) On the question of spectrophotometric quantitation of nucleic acid. Biokhimiia 25:151–159 (Russian)

551. Turkmenov MG, Imankulov DI (1975) Human efficiency in mountainous areas. Kosm Biol Med 9:70–78 (Russian)
552. Ugolev AA, Golubeva LYu, Meerson FZ (1980) Prevention of heart stress damages with preliminary adaptation to altitude hypoxia. Kardiologiia 20:91–94 (Russian)
553. Ukhtomsky AA (1966) Study on the dominant. Nauka, Moscow (Russian) [Uchenie o dominante]
554. Uphouse LL, McInnes IW, Schlezinger K (1974) Role of RNA and proteins in memory storage: a review. Behav Genet 4:29–85
555. Vacher JM, Miller AT (1968) Altitude-acclimatization: its effects on hypoxia induced performance decrements. Psychopharmacologia (Berlin) 12:250–257
556. Valdman AV (1975) Psychopharmacologic aspects of emotional stress. Vestn Akad Med Nauk SSSR 8:26–33 (Russian)
557. Vanderhock JY, Lands W (1973) Acetylenic inhibitors of sheep vesicular gland oxygenase. Biochim Biophys Acta 296:374–381
558. Van Dorp DA (1976) Essential fatty acids and prostaglandins. Acta Biol Med Germ 35:1041–1049
559. Vasdev SC, Kako KJ, Biro JP (1979) Phospholipid composition of cardiac mitochondria and lysosomes in experimental myocardial ischemia in the dog. J Mol Cell Cardiol 11:1195–1200
560. Vasiliev PV, Uglova NN (1967) Effect of adaptation to declined partial oxygen pressure on resistance to overloads. In: Problems of space biology, vol 6. Nauka, Moscow, pp 215–234 (Russian) [Problemy kosmicheskoy biologii]
561. Vasiliev PV, Malkin VB, Babchinsky FV, Loginova EV, Roshina NA, Yukhnovsky GD (1968) Comparative estimation of the effectivity of various adaptation modes to hypoxia. In: Problems of space biology, vol 8. Nauka, Moscow, pp 122–129 (Russian) [Problemy kosmicheskoy biologii]
562. Vatner SF, Higgins CB, Braunwald EH (1974) Effect of norepinephrine on coronary circulation and left ventricular dynamics in the conscious dog. Circ Res 34:812–824
563. Viar WN, Lombardo TA (1952) Abdominal aortic aneurysm with rupture into the inferior vena cava. Circulation 5:287–289
564. Viru AA (1974) Adrenocortical activity during repeated stress actions. In: Ester KM (ed) Voprosy endokrinologii (Problems of Endocrinology), USSR Ministry of Public Health, Tartu, p 139 (Russian)
565. Vladimirov GE (1954) Functional biochemistry of the brain. Nauka, Moscow (Russian) [Funktzionalnaya biokhimiya mozga]
566. Vladimirov YA, Archakov AI (1972) Lipid peroxidation in biologic membranes. Nauka, Moscow (Russian) [Perekisnoe okislenie lipidov v biologicheskikh membranakh]
567. Volkov NI (1967) Biochemical foundations of athlete stamina. Teor Prakt Fiz Kult 4:19–23 (Russian)
568. Voronin LG, Doronin GP (1965) A study of the effect of rarefied atmosphere on the function of the central nervous system by the method of chain motor conditioned reflexes. Zh Vyssh Nerv Deiat 15:831–837 (Russian)
569. Voskresensky ON, Devyatkina TA, Borisenko AN, Brechko VV, Goloborodko TI (1976) A search for antiatherosclerotic agents among antioxidants and their effects on free-radical mechanisms of atherogenesis. In: Valdman YA et al (eds) Pharmacology to Public Health. Abstracts of reports of 4th USSR congress of pharmacologists. First Leningrad Medical Institute, Leningrad, p 99 (Russian) [Farmakologiya – zdravookhraneniyu: Tezisy dokladov IV Vserossiyskogo sjezda farmakologov]
570. Vyalykh MF (1966) Adaptation of myocardial enzymes to various functional states of the heart. Problems of biochemical adaptation. Meditsina, Moscow, pp 86–93 (Russian) [Problemy biokhimicheskoy adaptatzii]
571. Vyalykh MF, Meerson FZ (1960) Metabolic processes in the myocardium during experimental heart disease. Vopr Med Khim 6:19–24 (Russian)
572. Waldenström AP, Hjalmarson AC, Thornell L (1978) A possible role of noradrenaline in the development of myocardial infarction. Am Heart J 95:43–51
573. Walpurger C, Anger H (1970) Die enzymatische Organisation des Energiestoffwechsels im Rattenherzen nach Schwimm- und Lauftraining. Z Kreislaufforsch 59:438–449

574. Warbanow W (1970) Morphologische und functionelle untersuchungen der Hypothermiebedingten Hypertropie des embryonalen Hühnerherzens. Acta Biol Med Germ 25:281–293
575. Watkinson WP, Foley DH, Rubio R, Berne RM (1979) Myocardial adenosine formation with increased cardiac performance in the dog. Am J Physiol 236:H13–H21
576. Webb YL, Hollander PB (1956) Metabolic aspects of the relationship between the contractility and membrane potentials of the rat atrium. Circ Res 4:618–626
577. Weid D de, Sarantakis D, Weinstein B (1973) Behavioral evolution of peptides related to scotophobin. Neuropharmacology 12:1109–1115
578. Weiss JM (1970) Somatic effects of predictable and unpredictable shock. Psychosom Med 32:394–408
579. Weiss JM (1971) Effects of coping behavior in different warning signal conditions on stress pathology in rats. J Comp Physiol Psychol 77:1–13
580. Weiss JM (1971) Effects of punishing the coping response (conflict) on stress pathology in rats. J Comp Physiol Psychol 77:4–21
581. Weitzien HU (1979) Cytolytic and membrane-perturbing properties of lysophosphatidylcholine. Biochim Biophys Acta 559:259–287
582. Wennmalm A (1976) Endogenous prostaglandins as regulators of autonomic neurotransmission in the rabbit heart. Acta Biol Med Germ 35:1127–1133
583. Wennmalm A (1978) Prostaglandin-mediated inhibition of noradrenaline release. Prostaglandins 15:113–121
584. Westfall TC (1977) Local regulation of adrenergic neurotransmission. Physiol Rev 57:659–728
585. Wickman-Coffelt JR, Zelis R, Mason DT (1974) Regulation of myosin ATPase activity in three red muscle types: cardiac right ventricle, cardiac left ventricle and skeletal muscle gastrocnemius. In: Fleckenstein A (ed) International study group for research in cardiac metabolism. 7th annual meeting. University Park Press, Quebec, Abstr no 3
586. Wickman-Coffelt J, Walsh R, Fenner C, Kamiyama T, Salel A, Mason DT (1976) Effects of severe hemodynamic pressure overload on the properties of canine left ventricular myosin: mechanism by which myosin ATPase activity is lowered during chronic increased hemodynamic stress. J Mol Cell Cardiol 8:263–271
587. Williams JN, Thorp SL (1970) Influence of degradative procedures, salts, respiratory inhibitors, and gramicidin on binding of cytochrome-C by liver mitochondria. Arch Biochim 141/2:622–633
588. Wood WG, Lindenmayer GE, Schwartz A (1971) Myocardial synthesis of ribonucleic acid. 1. Stimulation by isoproterenol. J Mol Cell Cardiol 3:127–138
589. Wyatt H, Parmley W, Rabinowitz B, Tyberg J, Chuck L, Kligerman M (1976) Enhanced cardiac response to catecholamines in physically trained cats. Circulation [Suppl 2] 54:
590. Yamashita K, Field JB (1973) The role of phospholipids in thyroid-stimulating hormone (TSA) stimulation of adenylate cyclase in thyroid plasma membranes. Biochim Biophys Acta 304:686–692
591. Yasue H, Omote S, Takizawa A, Nagao M, Miwa K, Tanaka S (1979) Circadian variation of exercise capacity in patients with Prinzmetal's variant angina: role of exercise-induced coronary arterial spasm. Circulation 59:938–948
592. Yoshimoto A, Ito H, Tomita L (1970) Cofactor requirements of the enzyme synthesizing prostaglandin in bovine seminal vesicles. J Biochem (Tokyo) 68:487–499
593. Yudaev NA, Pokrovsky BV, Protasova TN (1976) Hormone action mechanisms. The biochemistry of hormones and hormone control. Nauka, Moscow, pp 326–373 (Russian) [Biokhimiya gormonov i gormonalnoy regulyatzii]
594. Yuldashev KY, Babadzhanov AS (1962) Dynamics of the heart rhythm and conductability in persons with acute myocardial infarction under sodium hydroxybutyrate analgesia. Med Zh Uzb 10:52–56 (Russian)
595. Zak R (1962) Activity and RNA content of muscle. Fed Proc 21:319
596. Zak R (1974) Development and proliferative capacity of cardiac muscle cells. Cardiac hypertrophy and cardiomyopathy. Am Heart Assoc Monogr 43:17–26
597. Zak R, Dowell T, Martin A, Rabinowitz (1972) Protein synthesis in work-overloaded myocar-

dium. In: Hatt PY (ed) European symposium on heart overloading. L'Institut nationale de la sante et de la recherche medicale, Paris, pp 283–295
598. Zak R, Martin A, Rabinowitz M (1977) Comparison of turnover of several myofibrillar proteins and critical evolution of double isotope method. J Biol Chem 252:3430–3435
599. Zimin YI (1969) Change of haemopoesis in rats in stress reaction. Bull Exp Biol Med 7:19–22 (Russian)
600. Zimin YI, Khaitov RM (1975) Migration of T-lymphocytes to bone marrow during early stress reaction. Bull Exp Biol Med 12:68–70 (Russian)
601. Zvorykin VN (1953) Changes in higher nervous activity under rarefied air. Fiziol Zh SSSR 39:677–684 (Russian)
602. Zubin J, Barrera SE (1941) Effect of electric convulsive therapy on memory. Proc Soc Exp Biol Med 48:596–597

Subject Index

MIX
Papier aus verantwortungsvollen Quellen
Paper from responsible sources
FSC® C105338

If you have any concerns about our products,
you can contact us on
ProductSafety@springernature.com

In case Publisher is established outside the EU,
the EU authorized representative is:
Springer Nature Customer Service Center GmbH
Europaplatz 3, 69115 Heidelberg, Germany

Printed by Libri Plureos GmbH
in Hamburg, Germany